Deep Learning Concepts in Operations Research

The model-based approach for carrying out classification and identification of tasks has led to the pervading progress of the machine learning paradigm in diversified fields of technology. **Deep Learning Concepts in Operations Research** looks at the concepts that are the foundation of this model-based approach. Apart from the classification process, the machine learning (ML) model has become effective enough to predict future trends of any sort of phenomena. Such fields as object classification, speech recognition, and face detection have sought extensive application of artificial intelligence (AI) and ML as well. Among a variety of topics, the book examines:

- An overview of applications and computing devices
- Deep learning impacts in the field of AI
- Deep learning as state-of-the-art approach to AI
- Exploring deep learning architecture for cutting-edge AI solutions

Operations research is the branch of mathematics for performing many operational tasks in other allied domains, and the book explains how the implementation of automated strategies in optimization and parameter selection can be carried out by AI and ML. Operations research has many beneficial aspects for decision making. Discussing how a proper decision depends on several factors, the book examines how AI and ML can be used to model equations and define constraints to solve problems and discover proper and valid solutions more easily. It also looks at how automation plays a significant role in minimizing human labor and thereby minimizes overall time and cost.

Dr. Biswadip Basu Mallik is an associate professor of Mathematics in the Department of Basic Science & Humanities at Institute of Engineering & Management, University of Engineering & Management, Kolkata, India.

Dr. Gunjan Mukherjee is an associate professor in the Department of Computational Science, Brainware University, Barasat, India.

Rahul Kar holds a Master's degree in Pure Mathematics from Burdwan University and is currently working as a SACT-II Mathematics faculty member of Kalyani Mahavidyalaya, Kalyani, Nadia, West Bengal, India.

Aryan Chaudhary is the chief scientific advisor at BioTech Sphere Research, India.

Advances in Computational Collective Intelligence

Edited by

Dr. Subhendu Kumar Pani
Principal, Krupajal Group of Institutions, India

Published

Technologies for Sustainable Global Higher Education
By Maria José Sousa, Andreia de Bem Machado, and Gertrudes Aparecida Dandolini
ISBN: 978-1-032-262895

Forthcoming

Artificial Intelligence and Machine Learning for Risk Management of Natural Hazards and Disasters
By Cees van Westen, Romulus Costache, Dimitrios A. Karras, R. S. Ajin, and Sekhar L. Kuriakose
ISBN: 978-1-032-232768

Computational Intelligence in Industry 4.0 and 5.0 Applications: Challenges and Future Prospects
Joseph Bamidele Awotunde, Kamalakanta Muduli, and Biswajit Brahma
ISBN: 978-1-032-539225

Deep Learning for Smart Healthcare: Trends, Challenges and Applications
K. Murugeswari, B.Sundaravadivazhagan, S. Poonkuntran, and Thendral Puyalnithi
ISBN: 978-1-032-455815

Edge Computational Intelligence for AI-Enabled IoT Systems
By Shrikaant Kulkarni, Jaiprakash Narain Dwivedi, Dinda Pramanta, and Yuichiro Tanaka
ISBN: 978-1-032-207667

Explainable AI and Cybersecurity
By Mohammad Tabrez Quasim, Abdullah Alharthi, Ali Alqazzaz, Mohammed Mujib Alshahrani, Ali Falh Alshahrani, and Mohammad Ayoub Khan
ISBN: 978-1-032-422213

Machine Learning in Applied Sciences
By M. A. Jabbar, Shankru Guggari, Kingsley Okoye, and Houneida Sakly
ISBN: 978-1-032-251721

Social Media and Crowdsourcing
By Sujoy Chatterjee, Thipendra P Singh, Sunghoon Lim, and Anirban Mukhopadhyay
ISBN: 978-1-032-386874

AI and IoT Technology and Applications for Smart Healthcare
By Alex Khang
ISBN: 978-1-032-684901

Innovations and Applications of Technology in Language Education
By Hung Phu Bui, Raghvendra Kumar, and Nilayam Kamila
ISBN: 978-1-032-560731

Data-Driven Modelling and Predictive Analytics in Business and Finance
By Alex Khang, Rashmi Gujrati, Hayri Uygun, RK Tailor and Sanjaya Singh Gaur
ISBN: 978-1-032-60191-5

Deep Learning Concepts in Operations Research

Edited by
Biswadip Basu Mallik
Gunjan Mukherjee
Rahul Kar
Aryan Chaudhary

CRC Press
Taylor & Francis Group
Boca Raton London New York

CRC Press is an imprint of the
Taylor & Francis Group, an **informa** business

First edition published 2025
by CRC Press
2385 NW Executive Center Drive, Suite 320, Boca Raton FL 33431

and by CRC Press
4 Park Square, Milton Park, Abingdon, Oxon, OX14 4RN

CRC Press is an imprint of Taylor & Francis Group, LLC

© 2025 Taylor & Francis Group, LLC

British Library Cataloguing-in-Publication Data
A catalogue record for this book is available from the British Library

ISBN: 9781032553795 (hbk)
ISBN: 9781032559971 (pbk)
ISBN: 9781003433309 (ebk)

DOI: 10.1201/9781003433309

Typeset in Times
by Newgen Publishing UK

Contents

Preface

The immense progress of Science and Technology has been prominently obvious for bringing human civilization to the forefront of the modern era. The extensive research work in the field of computing has brought about many changes in human lives. The technological progress timeline has been redefined with the advent of artificial intelligence. The innovative introduction of AI and its allied branches are gradually evolving to turn many impossible scenarios into possibilities. AI and its sub-branch, machine learning has simulated human perception, concepts and behavior to a great extent and with high precision.

The construction of the model and its application in many classification and identification tasks has materialized due to continuous research and progress in the domain of artificial intelligence and machine learning. Apart from the classification process, the machine learning model has become effective enough to predict future trends of any sort of phenomena. Fields such as object classification, speech recognition, face detection, and the like, have sought extensive application of artificial intelligence. The application of AI and ML has extended to almost all disciplines. The domains of agriculture, health sectors, insurances are just a few of these.

Optimization has taken a turn in the field of artificial intelligence and machine learning. Operations research is a branch of mathematics and can be used for performing so many operational tasks in many domains. The implementation of automated strategies in optimization and parameter selection can be carried out by means of artificial intelligence and machine learning.

Operations research has many beneficial aspects in decision making. A proper decision depends on a number of factors. Mathematical equations and the definition of constraints can easily solve problems and identify a proper and valid solution. The involvement of human labour in such processes can be mitigated by means of automation. The application of AI and ML has not only provided fruitful solutions, but also the concepts can successfully be used in tackling many problems typically with good result accuracy.

Artificial Intelligence (AI) and Machine Learning (ML) are the emerging topics in the computer world, and for good reason - they help companies streamline operations and unearth data so that they can make better business decisions. Business growth is directly or indirectly managed by proper trend analysis and the predictive approach to growth to the ultimate level of development. The very concept is boosting almost every industry by allowing employees to work more effectively, and they're rapidly becoming a critical piece of technology for businesses to remain competitive with others. The gradual advancement of technology and its influences on overall business growth has led the new path towards stability of the modern economy.

Other core technologies such as face recognition on smart phones, personalized online shopping experiences, virtual assistants in homes, and even sickness diagnosis are all feasible due to extensive research work in these fields of technology. The demand for these technologies, as well as experts who are familiar with them, are on the rise. The average number of AI projects in place at a business is likely to more than treble over the next two years, according to a report by research firm, Gartner. Organizations face challenges because of this exponential expansion. They cite a lack of expertise, difficulties understanding AI use cases, and worries about data scope or quality as their top issues with these technologies. AI and machine learning, which were once the stuff of science fiction, are now becoming ubiquitous in enterprises. While these technologies are closely linked, there are significant distinctions between them. Here's a closer look at AI and machine learning, as well as some of the most popular vocations and talents, and how you can get started. Once again, the start-up policies of technology are being backed up by consistent persuasion from research along with the fruitful benefits of such notions. The novel application of the machine learning cum deep learning concept to smooth out the automation process for many mathematical models has become a real game changer in the new paradigm of technology.

Computing time reduction is a difficult problem to overcome. The complexity of any problem in any domain is based on the interrelationships between its involved entities and the effects due to its interactions with others even lying outside the defined peripheries, for example, social distance detection, voice and speech recognition, smooth traffic flow prediction and controlling congestions, prediction of the stock prices. All such emancipation of problems can be mitigated by research contributions collected in this book, providing enough insights into optimization and its mathematical connections through a research oriented approach towards a better goal, with the novel objective of presenting a conducive societal ambience to human beings.

Contributors

Gurjapna Anand
The NorthCap University
Haryana, India

Minal Aggarwal
Ramarao Adik Institute of Technology
D.Y. Patil (deemed to be University)
Nerul, Navi, Mumbai, India

Shivam Baghel
Vellore Institute of Technology
Vellore, Tamil Nadu, India

Yash Bhardwaj
GBPIET
Pauri Garhwal Uttarakhand, India

Arpitam Chatterjee
Department of Printing Engineering
Jadavpur University
West Bengal, India

Aswathy K. Cherian
SRM Institute of Science and Technology
    Kattankulathur, Chennai, India

R. Narmada Devi
Vel Tech Rangarajan Dr. Sagunthala R&D
Institute of Science and Technology
Avadi, Chennai, Tamil Nadu, India

Subhasmita Ghosh
Surendranath Evening College
Sealdah, India

Arpan Ghoshal
Seacom Engineering College
West Bengal, India

Nimish Goel
Vellore Institute of Technology
Vellore, Tamil Nadu, India

Reshma Gulwani
Ramrao Adik Institute of Technology
D.Y. Patil (Deemed to be University)
Nerul, Navi Mumbai, India

Dipti Jadhav
Ramarao Adik Institute of Technology
D.Y. Patil (deemed to be University)
Nerul, Navi, Mumbai, India

Subrata Jana
Jadavpur University
West Bengal, India

Jogendra Kumar
GBPIET
Puri Garhwal Uttarakhand, India

Sathish Kumar Kumaravel
Vel Tech Rangarajan Dr. Sagunthala R&D
Institute of Science and Technology
Avadi, Chennai, Tamil Nadu, India.

S. Lavanya
Karpagam College of Engineering
Coimbatore, India

Binay Maji
Maulana Abul Kalam Azad University of
    Technology
West Bengal, India

Biswadip Basu Mallik
Department of Basic Science and
    Humanities
Institute of Engineering & Management,
    University of Engineering & Management
Kolkata, India

Mayank Mehra
GBPIET
Pauri Garhwal Uttarakhand, India

Sucharita Mitra
Netaji Nagar Day College
West Bengal, India

Santanu Modak
Bengal College of Engineering and
    Technology
Durgapur, India

Kala Raja Mohan
Vel Tech Rangarajan Dr. Sagunthala R&D
Institute of Science and Technology
Avadi, Chennai, Tamil Nadu, India

Soumik Kumar Mohanta
Silicon Institute of Technology
Bhubaneswar, Odisha, India

Anita Mohanty
Silicon Institute of Technology
Bhubaneswar, Odisha, India

Ambarish G. Mohapatra
Silicon Institute of Technology
Bhubaneswar, Odisha, India

Parthiban Krishna Moorthy
Vellore Institute of Technology
Vellore, Tamil Nadu, India

Gunjan Mukherjee
Department of Computational Sciences
Brainware University
Kolkata, West Bengal, India

Anjali Munde
Southampton Malaysia Business School
University of Southampton, Malaysia

Regan Murugesan
Vel Tech Rangarajan Dr. Sagunthala R&D
Institute of Science and Technology
Avadi, Chennai, Tamil Nadu, India

Nagadevi Bala Nagaram
Vel Tech Rangarajan Dr. Sagunthala R&D
Institute of Science and Technology
Avadi, Chennai, Tamil Nadu, India

Bhaskar Nandi
Seacom Engineering College
West Bengal, India

Sasmita Nayak
Government College of Engineering
Kalahandi, Odisha, India

Gokarna Patil
Ramrao Adik Institute of Technology
D.Y. Patil (Deemed to be University)
Nerul, Navi Mumbai, India

Suman Patra
Netaji Nagar Day College
West Bengal, India

E. Poovammal
SRM Institute of Science and Technology
Kattankulathur, Chennai, India

Harshita Rana
GBPIET
Pauri Garhwal Uttarakhand, India

Siddharth Sahasrabuddhe
Ramrao Adik Institute of Technology
D.Y. Patil (Deemed to be University)
Nerul, Navi Mumbai, India

Sanat Kumar Sahu
Govt. V.Y.T. PG Autonomous College
Durg (C.G.) Chhattisgarh, India

Sushil Kumar Sahu
Christ College
Jagdalpur (C.G.), India

Anirban Sarkar
Department of Management and Marketing
West Bengal State University
Barasat, West Bengal, India

Pooja Swaroop Saxena
DIT University
Dehradun India

Divyanshu Semwal
GBPIET
Pauri Garhwal, Uttarakhand, India

Wasswa Shafik
School of Digital Science
Universiti Brunei Darussalam
Gadong, Jalan Tungku Link, Brunei
    Darussalam

Pooja Sharma
IIMT University
Meerut, India

Chinmay Shirsath
Ramrao Adik Institute of Technology
D.Y. Patil deemed to be University
Nerul, Navi Mumbai, India

Arpita Shome
Abacus Institute of Engineering & Management
Natungram, Dist. Hooghly, Mogra, West
    Bengal, India

Prabhatkumar Singh
Ramrao Adik Institute of Technology
D.Y. Patil deemed to be University
Nerul, Navi Mumbai, India

Simar Preet Singh
Bennett University
Greater Noida, Uttar Pradesh, India

A. Suganya
Karpagam College of Engineering
Coimbatore, India

Shubhada Tambe
Ramrao Adik Institute of Technology
D.Y. Patil deemed to be University
Nerul, Navi Mumbai, India

Purva Tekade
Ramrao Adik Institute of Technology
D.Y. Patil deemed to be University
Nerul, Navi Mumbai, India

Bipan Tudu
Dept of Instrumentation & Electronics
    Engineering
Jadavpur University, WB, India

M. Vaidhehi
SRM Institute of Science and Technology
Kattankulathur, Chennai, India

N. Varatharajan
Karpagam College of Engineering
Coimbatore, India

Priyanka Vashisht
Amity University
Noida, Uttar Pradesh, India

Sangeet Vashishtha
IIMT University
Meerut, India

R. Vikkram
Karpagam College of Engineering
Coimbatore, India

1 Deep Learning
Overview, Applications and Computing Devices

N. Varatharajan, S. Lavanya, A. Suganya, and R. Vikkram

1.1 INTRODUCTION

Machine learning has recently gained popularity in research and is used for a diversity of functions, together with content analysis, identifying spam, video inciting, picture dilution, and mixed media surface recovery. One of several machine learning calculations that is widely used in these applications is noteworthy learning (DL) [1]. One more name given to deep learning is representational learning. The best progress, such as in execution computing, which is making information obtaining capability intangible [2-5], may be the cause of the constant circulation of unneeded information within the ranges of important learning and dispersed learning. Deep learning is derived from traditional neural networks, but is much more efficient than its predecessors. Also, deep learning uses both transformations and graphs to create multi-layered learning models. Recently developed DL techniques include sound and discourse handling, visual information preparation, common dialect handling (NLP), and the like. It has achieved very good performance in many applications, including [4,6]. In general, the concept of the learning procedure depends upon the accuracy of any contribution information. Appropriate data depiction has been shown to provide better performance compared to poor data representation. For this reason, one of the important research interests in machine learning in recent years has been engineering studies which have seen a lot of research. This tactic emphasizes the creation of traits from natural data. Furthermore, it is very particular and demands a lot of work. For instance, distinct highlight types, such as the Situated Angles Histogram (Hoard) [5], the Scale Invariant Include Change (Filter) [7], and the Pack of Words [8], are depicted and differentiated in terms of dialect computers. When a new method is introduced and found to be effective, it becomes a research project that has been followed for decades. In contrast, feature extraction is used automatically throughout the DL algorithm [9]. This encourages researchers to eliminate discrimination with as little human effort and knowledge as possible. The primary layer of the multi-layer data representation engineering of these strategies evacuates low-level information, whereas the final layer expels high-level highlights [10]. It is worthy of note that this sort of plan, which empowers this method to happen in pivotal zones of the human brain, was at first based on incorrect pieces of knowledge (AI). The human brain has the capacity to extricate specialist information under different conditions. The data is particularly acknowledged as a representation of the input, but the result of this activity gets to be a special focus [11]. This strategy mirrors the capabilities of the human brain. In this way, it represents the key centers of DL. Due to its success DL has emerged as one of the foremost critical basic hypotheses in reaction to ML. This section provides an overview of deep learning from various perspectives, including fundamentals, architecture, competition, applications, computing tools, and transformation matrices which have published many deep learning analysis papers over the past few years. Nevertheless, deep learning [12], deep object search training [13],

DOI: 10.1201/9781003433309-1

interactive learning [14], and assessments present significant concerns. However, they serve the purpose of providing comprehensive information on DL concepts, including definitions, thorough investigations, computational disobedience, and DL organizations.

Prior to executing programs, DL factors such as utilization, concurrence, as well as substance need to be understood. To do this, one has got to have a thorough understanding of profound learning, especially its endless gaps and suggestions. As a result, to provide readers with a more grounded basis on which to understand profound learning, we advise studying a text that gives an in-depth outline of the subject. The objective is to talk about the most important components of DL, such as open-source issues, arrangements, and tools. In addition, our examination will serve as a springboard work with DL data. Here is an illustration of what we have provided: Practically all of the essential elements of profound learning are thoroughly examined in depth from various perspectives. This section helps students and trainees to fully understand the topic in hand. We discuss the current problems (limitations) with deep learning, such as the lack of training data, conflicting data, and information analysis: explicit parametric estimation, recall destruction, modeling compression, over fitting, during-problem, gradients solution dump, and inadequate specification. We investigate deep learning methods on computing systems (central processing units, graphical processing units, and field programmable gate arrays) by contrasting the characteristics of each device.

1.2 DEEP LEARNING: OVERVIEW

The background of deep learning is described in this section. We briefly introduce DL before describing how it differs from machine learning (ML). We'll give example that require DL.

The basis for a DL proposal are the last topic we cover. Mapping information found in an individual's brain leads to DL, a part of ML (Figure 1.1). The DL uses a great deal of metadata to represent input given to tags instead of human-written code to carry out its operations. A number of layers of simulated neural networks, or ANNs, have been utilized to build Deep Learning (DL) [15–18]. Each layer provides a different interpretation of the input data. Classical machine learning techniques require the sequential steps of prioritization, elimination, decision hiring, learning,

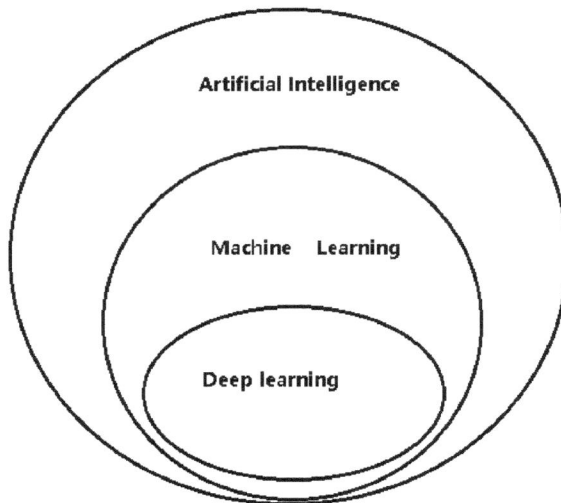

FIGURE 1.1 Deep Learning Family [**16**].

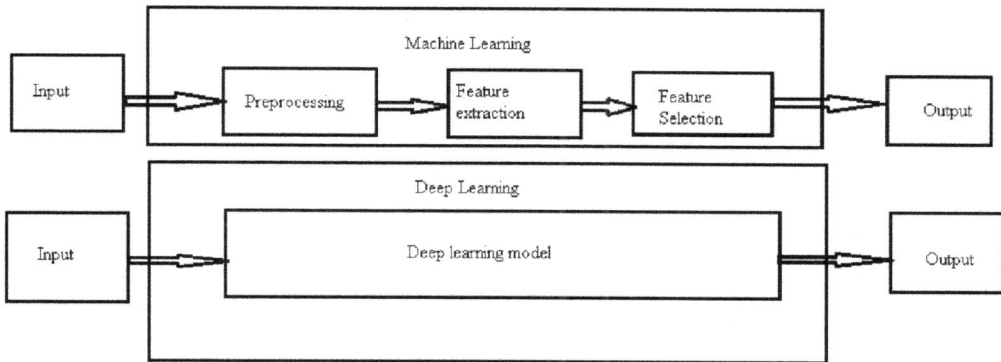

FIGURE 1.2 DL versus Traditional ML [18].

and implementation to solve classification issues. Attribute mixes have huge impacts on how well machine learning techniques work as well.

Inaccurate class distinctions may result from neutral selection. Deep learning can learn strategies for many tasks, in contrast to conventional machine learning techniques [18]. With DL, learning and categorization can be achieved in one pass (Figure 1.2). DL has gained popularity as a machine learning algorithm recently [15] as a result of the enormous growth and evolution of big data. It has not yet aided development in many ML jobs' [19] new performance in many areas, including picture resolution, object detection, and image recognition. Performance on tasks like picture categorization has recently been surpassed by deep learning. This technology has had an impact on virtually every branch of science. Due to the use of DL, the majority of enterprises and sectors have already been impacted [20].

Global businesses and technology companies are focusing on competition to advance deep learning. Even now, people's level of expertise and performance capacity fall short of DLs in a number of categories, including predicting rating, choosing a cutoff to assess loan demand, and calculating truck duration. The 2019 Nobel Prize winners – also referred to as the Turing Prize – were three deep learning pioneers [20]. Even though many goals have been achieved; the DL field is still developing. DL has the power to improve people's lives by providing more diagnosis, such as predicting natural disasters, discovering new drugs, and testing for cancer. Deep learning is capable of recognising ailments based on dermatologist suggestions using 129,450 pictures of 2032 disorders. Additionally, GoogleAI [21] outperformed these professionals in the US Pathology Board's estimated accuracy test for cancer diagnosis, scoring 70% versus 61%. For the detection of COVID-19, a brand-new coronavirus, in 2020, DL is essential [22]. In numerous clinics across the globe where chest X-rays or other types of imaging is used; deep learning has emerged as a key tool for classifying and detecting COVID-19. To finish, the application of deep learning is capable of anything, according to AI pioneer Geoffrey Hinton, who famously said.

1.3 APPLICATIONS OF DEEP LEARNING

The fact that machine intelligence is effective in many contexts and can sometimes perform as well as or better than human specialists suggests that deep learning has addressed the following issues without knowledge. Situations in which individuals have difficulty in explaining decisions made employing their cognitive abilities (linguistic comprehension, medical decision-making, and language skills) are:

1. Universal Learning Approach: Because deep learning can be applied to almost any application.
2. Robust: Deep learning is typically not needed to finish features. Instead, as might be predicted, optimized characteristics are automatically learnt about the task. This offers robustness against a wide range of input data fluctuations.
3. Generalization: It is possible to apply the same DL methodology to a variety of materials or applications. Insufficient data issues can also be solved using this technique.
4. Scalability: It is very easy to scale deep learning. With 1202 layers and widespread use in supercomputing, ResNet was created by Microsoft. Similar steps have been conducted by the important network evolution business Lawrence Livermore National Laboratory (LLNL), which has produced hundreds of results.

Numerous DL applications are currently accessible everywhere. These applications, include those for the healthcare industry, social analytics, voice and speech (including augmentation and recognition), visualization of data (including data analysis and computational science), and NLP (interpretation and sentence categorization). These programs can be divided into five groups: segmentation, recording, location, discovery, and classification. The idea of classification is the division of data into categories. Finding relevant objects in an image is achieved via detection, which takes the background into consideration. In the search, there are numerous items that may be from various classes that are enclosed by linked boxes. Finding things inside a bounding box can be achieved using the localization technique. Objects are still surrounded by identified contours during segmentation (semantic segmentation). Fitting a picture into another picture, which could be in 2D or 3D, is referred to as noun. The application and significance of DL in the healthcare industry cannot be overstated. This study is important because it has a significant impact on people's lives. In the area of medicine, deep learning has also shown potential. We demonstrate this by providing a brief overview of the DL application used in medical picture analysis as an example.

1.3.1 LOCALIZATION

Doctors typically have an interest in the natural body position, although bodyworkers use it more often. Even though electronic images at a regional center are examined and determined without human assistance, they can still be used in end-to-end technological applications. Deep learning recognises malignant spots in X-ray pictures used in non-invasive radiation therapy. Both the spatial and temporal aspects are included when examining hierarchical traits. According to Organs, this method has an accuracy range of 62 to 79%. For the diagnosis of pulmonary and local diseases, two convolution networks – Ret Net and Mask R-CNN – are employed. Detection An additional method for detection is computer-aided detection (CADe). A wound that is ignored might have immediate negative effects on the patient and the doctor. As a result, research is a discipline that demands accuracy and precision. a thorough investigation into diseases using the idea of adaptive learning. Their approach had a 96.4% accuracy rate and a 99.62% success rate in recovering deleted data. Numerous neural network techniques that work well for automatic X-ray image detection have been proposed in the fields of COVID-19 and pneumonia.

1.3.2 DETECTION

Another method of detection is computer-aided detection. Ignoring an inflammation might have negative implications on both the patient and the doctor. Because of this, detection is a subject that calls for both sensitivity and precision. A cutting-edge learning system will apply the concept of transferred learning to identify pneumonia. Their method had a 99.62% call rate and a 96.4% accuracy rating for unobserved data. For automated identification from X-ray pictures of COVID-19 and pulmonary illness, a number of convolution neural network techniques have been presented, and they have demonstrated great performance.

1.3.3 SEGMENTATION

Many MRI and CT scan segmentation studies have focused on the brain, particularly tumors, despite the fact that other organs like the liver, prostate, and knee cartilage have also been included. This problem is essential in surgical planning to obtain a clean tumor for small surgery. Surgery that removes too much brain tissue can cause brain damage that can impair memory, cause depression, and cause loss of mobility. Tracing lines particularly from layer to layer on a CT or MRI set has historically been the method used by doctors to manually segment clinical anatomical structures. This difficult task is best handled by automated solutions. Automated identification of the covid-19 virus from CT images has just been implemented in order to assist in the teaching of trans disciplinary studies.

1.3.4 REGISTRATION

The basic stages of the unique process for registration work are generally given two input photos, used as follows:

Product selection: Following the first input picture, characteristics of subsequent input pictures must continue to be true.

Feature extraction: This determines the characteristics that were taken for each of the input pictures.

Characteristic harmonizing: This makes it possible to compare previously acquired features.

Exposure Optimization: This reduces the level of difference between the two input images. The resultant statistical renovation (such as transformation, rotary motion, degree, and so on.) and a close distance between them are the outcomes of the registration operation. The levels of overlap are good.

1.4 COMPUTATIONAL METHODS

Complex ML and DL approaches are extensively functional in a variety of industries and have swiftly emerged as the most valuable methods for data processing. The creation and advancement of algorithms, along with the effectiveness of mathematics and the usage of large data, have made it possible to implement several applications that were either too complex or difficult to imagine beforehand. There are now a number of common (Deep Neural Network) DNN configurations available. The primary distinction between these two approaches is the link pattern between layers and all tiers. During DNN training, several optimisation approaches are employed to obtain the updates, but the (Stochastic Gradient Descent) SGD procedure is utilised to fit (or not fit) the weights. The waste management social network should develop together with increased output. For instance, it takes between 30 and 40 thousand iterations for the training method using the extensive ImageNet data set (more than 14 million pictures) and the ResNet network model to arrive at a fixed solution: frozen. The total calculation time of the high-level approach is expected to be greater than 1020 hops as exercise locate mass in addition to neural network involvement amplification. By 2009, the GPU could be used to advance training to a satisfactory level. Even with GPU help, training often takes a few days or weeks. For instance, numerous reduction methods have been developed that greatly cut down on training time. It is thought that as DNNs scale in complexity and size, so do their computing requirements. Memory throughput and capacity are calculated for the entire line and have less of an impact than the installation cost. In particular, there is a large amount of data recovery, parameters are not consistently allocated throughout the initial data levels, and computing several layers in a network only yields a tiny computer-to-bandwidth ratio. If there is no classification, multiplexed data, on the other hand, has very little expense and a very low computational bandwidth ratio for additional Fully Connected (FC) processing. Although maximum Graphical Processing Unit (GPU) presentation cannot be achieved as well as the inclusion of numerous ideas or models that are bandwidth-dependent, these networks are better able to solve computational issues thanks to the GPU performance.

The worst-case GPU efficiency ranges from 15% to 20% of the theoretical maximum performance. To solve this problem, the memory bandwidth must be increased using high-bandwidth cluster memory. Then, various techniques based on central processing units, graphical processing units, and field programmable gate arrays will be thoroughly explained.

1.4.1 CENTRAL PROCESSING UNIT-BASED APPROACH

A CPU-based concert of central processing unit nodes primarily supports huge memory, network connectivity, and storage capacity. Despite being more prevalent than (Field Processing Gate Array) FPGA or GPU nodes, CPU nodes require more space and memory because they cannot manage with limited computing resources.

1.4.2 GPU-BASED APPROACH

The GPU is a good fit for several essential DL ideas, including parallel operations such as activation, matrix multiplication, and convolution. Because High Bandwidth Memory (HBM) stacked memory has been integrated into these new GPU models, they provide increased bandwidth. This improvement enables several cores to fully leverage the GPU's capabilities. Heavy linear algebra operations usually result in a 10–20:1 speed advantage for the GPU over the CPU. The original GPU programming paradigm was built on the idea of increasing parallel processing. For instance, a single GPU has a maximum of 64 units. With 16 processing lines per Single Instruction Multiple Data (SIMD) unit, there are four SIMD processing units. 25 T hops (fp16) and 10 T hops (fp32) operate best when the operational percentage is near to 100%. If addition and subtraction for vectors are combined with internal manufacturing rules for comparing operations requiring matrix operations, further GPU performance can be attained. In general, GPUs are thought to be well-designed for DNN training and also offer improved performance for inference operations.

1.4.3 FPGA-BASED APPROACH

Deep learning is just one application of FPGAs. Oftentimes, FPGAs employ inference accelerators. Inefficient functionality or overheads related to GPU systems can be reduced by optimizing FPGAs. Poor access point performance and restricted computation are FPGAs' main drawbacks in comparison to GPUs. The key characteristic of FPGAs is their capacity to configure an array by design utilizing a variety of existing headers or brand-new ones, as well as to alter its properties in real time. For DL inference workloads, as was already established, FPGAs outperform GPUs and CPUs in terms of computation speed and latency per watt. These three elements – reduced arithmetic, network integration, and efficient usage of high-performance devices – allow FPGAs to leverage DL algorithms to support FPGAs at this level. Long Short-Term Memories (LSTM)s, on the other hand, commonly use pruning techniques. Using MultiLayer Perceptron (MLP) neural processing, the size of the model can be reduced to 20, which has significant benefits when using efficient methods.

According to recent studies on the usage of fixed content and custom control content, 8-bit reduction is not only very effective but also helps as an FPGA technology advance for the best DNN model performance.

1.5   SUMMARY AND CONCLUSION

After compiling all the data related to this research, a brief discussion is warranted. The information is broken down into its component parts for explanation and advice on where to go next. Large volumes of complicated data require simultaneous modeling via deep learning, which is challenging. Multimodal DL is yet another strategy used in current DL development. In the future, the development of DL apps should heavily rely on cloud-based platforms. Cloud computing can be used to handle massive amounts of data with efficient solutions. Moreover, it lowers expenses while boosting efficiency. For deep learning training, it also offers flexibility. New developments in computing hardware, such as neural network processors and mobile GPUs, have improved deep learning applications with transferable approach. DL user determination does this more simply. Several adaptive learning techniques are intended to address the problem of a lack of training data, including the training of the deep learning model on a sizable dataset of images that are unlabeled, prior to applying it to the task of training well-known DL models on a small number of images. As a description for the DL community who are interested in the DL space, this area serves as a starting point. Additionally, researchers will be able to choose the best lines of inquiry so they bring more detailed options to the field.

REFERENCES

1. Rozenwald, Michal, et al. "A machine learning framework for the prediction of chromatin folding in Drosophila using epigenetic features." *PeerJ Computer Science* 6 (2020): e307.
2. Amrit, Chintan, Tim Paauw, Robin Aly, and Miha Lavric. "Identifying child abuse through text mining and machine learning." *Expert Systems with Applications* 88 (2017): 402–418.
3. Hossain, Eklas, et al. "Application of big data and machine learning in smart grid, and associated security concerns: A review." *IEEE Access* 7 (2019): 13960–13988.
4. Crawford, Michael, et al. "Survey of review spam detection using machine learning techniques." *Journal of Big Data* 2, no. 1 (2015): 1–24.
5. Deldjoo, Yashar, et al. "Content-based video recommendation system based on stylistic visual features." *Journal on Data Semantics* 5 (2016): 99–113.
6. Al-Dulaimi, Khamael et al. "Benchmarking HEp-2 specimen cells classification using linear discriminant analysis on higher order spectra features of cell shape." *Pattern Recognition Letters* 125 (2019): 534–541.
7. Liu, Weibo, et al. "A survey of deep neural network architectures and their applications." *Neurocomputing* 234 (2017): 11–26.
8. Onyelowe, Kennedy, et al. "Selected AI optimization techniques and applications in geotechnical engineering." *Cogent Engineering* 10, no. 1 (2023): 2153419.
9. Alom, MdZahangir, et al. "A state-of-the-art survey on deep learning theory and architectures." *Electronics* 8, no. 3 (2019): 292.
10. Potok, Thomas E., et al. "A study of complex deep learning networks on high-performance, neuromorphic, and quantum computers." *ACM Journal on Emerging Technologies in Computing Systems (JETC)* 14, no. 2 (2018): 1–21.
11. Adeel, Ahsan, Mandar Gogate, and Amir Hussain. "Contextual deep learning-based audio-visual switching for speech enhancement in real-world environments." *Information Fusion* 59 (2020): 163–170.
12. Tian, Haiman, Shu-Ching Chen, and Mei-Ling Shyu. "Evolutionary programming based deep learning feature selection and network construction for visual data classification." *Information Systems Frontiers* 22 (2020): 1053–1066.

13. Young, Tom, Devamanyu Hazarika, Soujanya Poria, and Erik Cambria. "Recent trends in deep learning based natural language processing." *IEEE Computational Intelligence Magazine* 13, no. 3 (2018): 55–75.

14. Koppe, Georgia, Andreas Meyer-Lindenberg, and Daniel Durstewitz. "Deep learning for small and big data in psychiatry." *Neuropsychopharmacology* 46, no. 1 (2021): 176–190.

15. Dalal, Navneet, and Bill Triggs. "Histograms of oriented gradients for human detection." In *2005 IEEE Computer Society Conference on Computer Vision and Pattern Recognition (CVPR'05)*, vol. 1, pp. 886–893. IEEE, 2005.

16. Lowe, David G. "Object recognition from local scale-invariant features." In *Proceedings of the Seventh IEEE International Conference on Computer Vision*, vol. 2, pp. 1150–1157. IEEE, 1999.

17. Y. Lv, J. Zhang, Y. Dai, A. Li, N. Barnes and D. -P. Fan, "Toward Deeper Understanding of Camouflaged Object Detection," in *IEEE Transactions on Circuits and Systems for Video Technology*, vol. 33, no. 7, pp. 3462-3476, July 2023.

18. Krishnamurthy, Gangeshwar, et al. "A deep learning approach for multimodal deception detection." In *Computational Linguistics and Intelligent Text Processing: 19th International Conference, CICLing 2018, Hanoi, Vietnam, March 18–24, 2018, Revised Selected Papers, Part I*, pp. 87–96. Cham: Springer Nature Switzerland, 2023.

19. Wang, Kunyu, Xianguo Wu, Limao Zhang, and Xieqing Song. "Data-driven multi-step robust prediction of TBM attitude using a hybrid deep learning approach." *Advanced Engineering Informatics* 55 (2023): 101854.

20. Chen, Xudong, Zhun Wei, Maokun Li, and Paolo Rocca. "A review of deep learning approaches for inverse scattering problems (invited review)." *Progress in Electromagnetics Research* 167 (2020): 67–81.

21. Haque, Intisar Rizwan I., and Jeremiah Neubert. "Deep learning approaches to biomedical image segmentation." *Informatics in Medicine Unlocked* 18 (2020): 100297.

22. Stokes, Yang, et al. "A deep learning approach to antibiotic discovery." *Cell* 180, no. 4 (2020): 688–702.

2 Deep Learning Impacts in the Field of Artificial Intelligence

Wasswa Shafik

2.1 INTRODUCTION

Artificial Intelligence (AI) is a rapidly increasing paradigm focusing on developing smart machines proficient in task execution that stereotypically necessitate human intelligence (Benayad et al., 2022). AI systems aim to simulate human-like intelligence, including reasoning, learning, problem-solving, perception, and natural language processing (NLP) abilities. The field encompasses a wide range of techniques and methodologies, with one of the most influential and transformative subsets being DL. The significance of AI lies in its potential to revolutionize numerous industries and domains. AI technologies can augment human capabilities, automate repetitive tasks, make complex decisions, and extract valuable insights from vast data. By leveraging AI, organizations can enhance efficiency, improve decision-making processes, drive innovation, and create new opportunities (Lei & Liu, 2022).

AI is applied in various sectors, for example, manufacturing, finance, healthcare, transportation, entertainment, and more. For instance, in healthcare, AI algorithms diagnose diseases, analyze medical images, predict patient results, and recommend personalized treatments. In finance, AI-powered systems can analyze market trends, predict stock prices, detect fraud, and optimize investment strategies (Wu et al., 2022). The potential impact of AI extends to areas such as autonomous vehicles, virtual assistants, NLP, and robotics. DL, a subset of AI, has grown a big following and shown remarkable advancements in recent years. This technology comprises training Artificial Neural Networks (ANN) with various layers to learn representations of data through a hierarchical structure (Case, 2022). DL algorithms excel in extracting complex patterns and features from vast amounts of unstructured data, like images, text, and audio. This ability has resulted in breakthroughs across various AI applications, including recommendation systems, image recognition, speech recognition, NLP, and many more (Shuaib et al., 2022); Figure 2.1 illustrates several AI subdisciplines.

The emergence of DL has propelled AI to new heights, enabling machines to surpass human-level performance in specific tasks. DL models, such as Convolutional Neural Networks (CNNs) and Recurrent Neural Networks (RNNs), have achieved current results in object detection, image classification, speech synthesis, and machine translation, among others (Surendiran et al., 2022). The advancements in deep learning have also led to the development of generative models, such as Generative Adversarial Networks (GANs), capable of generating realistic images, music, and even human-like text. The significance of deep learning in AI lies in its ability to handle large-scale, complex datasets and extract high-level abstractions robotically. This capability has fueled advancements in computer vision, NLP, and data-driven decision-making (Zhu & Zhang, 2022). As a result, DL has revolutionized various industries, paving the way for improved healthcare, diagnostics, autonomous vehicles, personalized recommendations, virtual assistants, and more.

FIGURE 2.1 Artificial Intelligence and Deep Learning Application in medical field.

As DL continues to evolve, its impact on AI will likely grow further. Researchers and practitioners are exploring new architectures, techniques, and applications to enhance DL models' interpretability, robustness, and generalization capabilities (Ali et al., 2022). As a result, the field of AI stands at the forefront of technological innovation, with deep learning serving as a driving force behind its continuous growth and transformative potential. In addition to its significant impact on various industries, AI has also captured the attention of researchers, policymakers, and the public due to its implications for society (Mehralivand et al., 2022). Nevertheless, the rapid advancement of AI raises critical questions about the ethical, legal, and societal considerations surrounding its deployment.

Ethical considerations in AI involve ensuring fairness, transparency, and accountability in decision-making processes. DL algorithms in hiring, lending, and criminal justice systems have raised concerns about potential biases and discrimination (Li, 2022). Addressing these challenges requires developing AI systems that are accurate, efficient, fair, and unbiased, considering diverse perspectives and protecting individual rights. Moreover, integrating AI into daily life brings security and privacy concerns. AI-powered systems often rely on vast amounts of personal data, leading to questions about data protection, user consent, and potential misuse of information (Lan et al., 2022). Ensuring a balance between leveraging data for AI advancement and preserving individual privacy rights is a crucial encounter that needs to be mitigated.

Introducing AI technology raises broader societal questions, such as the impact on employment and future studies. While AI can automate routine tasks, there are concerns about the need for upskilling or reskilling and job displacement to adapt to the shifting employment scenery (Turkbey & Haider, 2022). Understanding AI adoption's social and economic implications is essential for developing effective policies that maximize benefits while mitigating potential drawbacks. This collaboration can help shape the development and deployment of AI technologies responsibly and inclusively (Wang, 2022). Multidisciplinary research and open dialogue are crucial for addressing AI's technical, ethical, and societal challenges, ensuring that it aligns with human values and contributes positively to the comfort of communities and individuals.

2.1.1 THE CONTRIBUTION OF THIS CHAPTER

This chapter provides:

- a detailed explanation of deep learning algorithms and techniques, discusses their evolution and impact on AI, and explains the key concepts and components of deep learning relevant to the chapter.
- explorations of various domains where DL has significantly impacted AI by providing specific examples and use cases.
- discussions around how deep learning improves performance and achieves high-tech results.
- a focus on the challenges and limitations of deep learning in AI and discusses issues like interpretability, data requirements, computational complexity, and ethical considerations.
- highlights of areas where deep learning may struggle to deliver optimal results.
- discussions of recent advancements and ongoing research in deep learning within AI and explores emerging trends and potential future directions.
- reflections on the impact of advancements like reinforcement learning, generative models, and explainable AI.
- an analysis of the broader implications and consequences of deep learning in the field of AI, including discussion of societal, economic, and ethical consequences, and considers risks, biases, privacy, and security issues.

2.1.2 THE ORGANISATION OF THIS CHAPTER

The remainder of this chapter is organized into six sections. Section 2.2 presents the background of deep learning and the key concepts and components of DL. Section 2.3 demonstrates the top applications of deep learning in AI and includes fraud detection, computer vision, autonomous systems, speech recognition, NLP, image and video analysis, predictive analytics, healthcare, gaming, robotics, manufacturing, agriculture, transportation, energy, and astronomy. Section 2.4 details the DL impacts in AI, including improved accuracy, reduced cost, faster processing, improved language processing, improved computer vision, increased personalization, enhanced security, automation, efficiency, and improved decision-making. Section 2.5 presents future deep learning directions in AI, and the conclusion is in Section 2.6.

2.2 BACKGROUND, KEY CONCEPTS, AND COMPONENTS OF DEEP LEARNING

DL algorithms and techniques have revolutionized AI by enabling machines to learn and extract meaningful representations from large, complex datasets. This section provides a detailed explanation of DL, its evolution, and its impact on AI while explaining fundamental concepts and components relevant to the chapter. The structure and functioning of the human brain inspires these neural networks. DL algorithms excel at repeatedly learning complex patterns and features from unstructured data, like text, images, and audio.

The evolution of DL can be traced back to the initial development of ANN in the 1940s. Nevertheless, it was not until late, with the availability of large datasets and significant computational power, that DL gained traction and demonstrated remarkable capabilities. One pivotal moment was the introduction of deep CNNs, which achieved breakthrough results in image classification tasks. As a result, the impact of DL on AI has been transformative. DL algorithms have achieved state-of-the-art performance in various domains, surpassing human-level capabilities in tasks such as image recognition, speech recognition, and NLP. This success is attributed to the ability of DL models to invariably learn feature representations at multiple levels of abstraction, allowing them to capture intricate patterns and dependencies in the data.

2.2.1 KEY CONCEPTS

Within this sub-section, we delve into the fundamental concepts of deep learning with examples to present a clear vision and understanding of deep learning.

2.2.1.1 Artificial neural networks

ANNs form a complex network of interlinked nodes, called artificial neurons, meticulously organized into layered structures. Each neuron is designed to receive input, undergo computational processes, and ultimately generate an output. The outputs of one layer serve as inputs to the next layer. ANNs can have multiple hidden layers, allowing for the learning of hierarchical representations (Sujith et al., 2022). For instance, in image classification, a deep neural network can be trained to recognize different objects in images. The input layer receives pixel values of an image, and the subsequent layers learn to extract increasingly complex features such as edges, shapes, and textures (Bastidas et al., 2023). The final layer outputs the predicted class of the image, indicating what object it belongs to.

2.2.1.2 Activation functions

Activation functions play a pivotal role in embedding non-linear characteristics into the neural network, facilitating its capacity to comprehend and depict intricate connections between inputs and outputs. These functions ascertain the output of a neuron by evaluating its input. Among the frequently employed activation functions is the Rectified Linear Unit (ReLU), which yields the input value in the case where it is positive and zero otherwise (Deng et al., 2023). ReLU activation helps overcome the vanishing gradient problem, allowing for efficient learning of deep neural networks.

2.2.1.3 Backpropagation

Backpropagation is a fundamental algorithm utilized in the training of deep neural networks. It involves the computation of gradients for the network's parameters to a specific loss function and utilizing them to update the parameters, thereby minimizing the overall error. In image classification within a deep neural network, backpropagation calculates the gradients of the network's parameters by comparing the predicted class probabilities with the accurate class labels (Adnan et al., 2023). These gradients are then employed to iteratively adjust the weights and biases of the network iteratively, gradually improving its accuracy in accurately classifying images.

2.2.1.4 Convolutional neural networks

CNNs are specialized deep neural networks designed for processing grid-like data, such as images. CNNs employ convolutional, pooling, and fully connected layers to learn spatial hierarchies of features robotically. In image recognition, CNNs use convolutional layers to detect local patterns and features in an image, such as edges, corners, or textures (Alghamdi, 2023). The pooling layers aggregate the detected features, reducing the spatial dimensions. The fully connected layers at the network's end combine the extracted features to predict the image's content (Liu, 2023).

2.2.1.5 Recurrent neural networks

RNNs are compatible with processing sequential data, where the order of inputs matters. RNNs have recurrent connections that allow information to persist across time steps, enabling the network to capture temporal dependencies in the data. In natural language processing, RNNs can be used for language modeling or text generation tasks. Each word in a sentence is fed as input to the RNN, and the network updates its internal state at each time step (Tang et al., 2023). The recurrent connections enable the RNN to remember the context from previous words, generating coherent and contextually appropriate text (Gronwald, 2023).

2.2.1.6 Pre-training and transfer learning

Pre-training comprises training a DL model on a large dataset before fine-tuning it on a straightforward task. Transfer learning leverages pre-training knowledge to improve performance on a related but different task. Suppose that a DL model is pre-trained on a large dataset of images, for example,

ImageNet, to learn the general features of objects. The pre-trained model can then be fine-tuned on a smaller dataset precise to a particular domain, such as classifying different species of flowers (Shubham et al., 2023). Transferring the knowledge learnt from the pre-training makes the model perform better with less flower classification task data. These key concepts form the foundation of deep learning and enable the learning and representation of complex patterns and features from data. By understanding these concepts and their applications, researchers and practitioners can effectively design and train deep neural networks to solve a wide range of tasks in AI (Gore et al., 2023).

2.2.2 COMPONENTS OF DEEP LEARNING

These components work together to enable DL models to learn and predict from complex, high-dimensional data. Understanding and manipulating these components allows researchers and practitioners to design and optimize DL architectures for various AI tasks described below.

2.2.2.1 Neurons

Neurons are the basic computational units in a neural network. They receive input signals, perform a computation, and produce an output. Neurons are connected in a network structure, forming the basis for information processing (Shubham et al., 2023).

2.2.2.2 Layers

DL models consist of multiple layers of interconnected neurons. Each layer receives input from the previous layer, executes computations, and passes the output to the next layer (Tang et al., 2023). The most common layers in DL include input, hidden, and output layers (Jin et al., 2023).

2.2.2.3 Weights and biases

Weights and biases are learnable parameters associated with the connections between neurons in a neural network. The weights regulate the strength of the connections, while the biases provide additional input to the neurons (Liu, 2023). These parameters are dynamically fine-tuned throughout training to enhance the network's overall performance.

2.2.2.4 Activation functions

Activation functions play a crucial role in neural networks by introducing non-linearity, which enables them to learn intricate patterns and make non-linear decisions (Alghamdi, 2023). Among the commonly used activation functions are the sigmoid, tanh, and ReLU functions. These activation functions allow neural networks to model and capture complex relationships in the data, enabling them to handle various tasks and improve the network's capacity to learn and generalize.

2.2.2.5 Loss functions

Loss functions quantify the transformation amid the predicted output of the neural network and the accurate output. They measure the network's performance during training and guide the optimization process (Tang et al., 2023). Standard loss functions include Mean Squared Error (MSE), categorical cross-entropy, and binary cross-entropy.

2.2.2.6 Optimization algorithms

Optimization algorithms are used to appraise the biases and weights of a neural network during the training process. These algorithms aim to minimize the loss function and find the optimal values for the network's parameters (Gronwald, 2023). Examples of optimization algorithms include Stochastic Gradient Descent (SGD), Adam, and RMSprop.

2.2.2.7 Backpropagation

During the forward pass of backpropagation, the input data is fed through the network, and the activations of each layer are computed. The network output is compared to the desired output using a loss function, quantifying the mismatch between the predicted and target values. In the backward pass, the gradients of the loss function for the network's parameters are computed using the chain rule of calculus. The gradients are propagated backwards through the layers, and at each layer, the gradients are multiplied by the activation function's local derivative to calculate that layer's contribution to the overall gradient (Gore et al., 2023). Once the gradients have been computed, the network's parameters are updated in the opposite direction to the gradients, aiming to minimize the loss function. This update is archetypally performed using an optimization algorithm such as gradient descent or one of its variants, which adjusts the parameters proportionally to the computed gradients and a learning rate hyperparameter.

2.2.2.8 Regularization techniques

Regularization techniques are employed to prevent overfitting in deep learning models. Overfitting occurs when a model performs well on the training data but fails to generalize to new, unseen data (Huang et al., 2023). Standard regularization techniques include "L1" and "L2" dropout, early stopping, and regularization.

2.2.2.9 Batch normalization

Batch normalization is a procedure used to advance the training and performance of deep neural networks (Gu et al., 2023). It normalizes the inputs to a layer by subtracting the mean and dividing by the standard deviation (Yu et al., 2023), helping to stabilize and speed up the training process (Wu et al., 2023).

2.2.2.10 Architectures

DL models often utilize specific architectures designed for specific tasks. Examples include CNNs for image and video processing, RNNs for sequential data, and transformers for natural language processing tasks. In the next section, we present applications of DL in AI (Serey et al., 2023).

2.3 APPLICATIONS OF DEEP LEARNING IN ARTIFICIAL INTELLIGENCE

This section presents the application of DL in AI, as demonstrated below.

2.3.1 COMPUTER VISION

Computer vision aims to enable machines to extract meaningful information from images or videos and understand their content. This involves object recognition, image classification, object detection, image segmentation, tracking, and more. Computer vision algorithms utilize various techniques to analyze and interpret visual data (Khan, Sarkar, and Maity, 2023). These techniques may include image preprocessing, feature extraction, pattern recognition, machine learning, and DL. By combining these methods, computer vision systems can detect and recognize objects, understand spatial relationships, estimate distances, analyze motion, and perform complex tasks like facial recognition or scene understanding. Computer vision has a wide range of applications across various domains. For example, it plays a crucial role in autonomous vehicles, surveillance systems, medical imaging, robotics, augmented reality, and image and video analysis. It can also be used for tasks like document analysis, quality control in manufacturing, and image-based search and recommendation systems (Chen, 2023).

One real-time example of computer vision in AI is in self-driving cars. Computer vision algorithms help cars "see" the road, detect objects and people, and make decisions based on that information. Other examples include facial recognition technology, quality control in manufacturing, and medical image analysis for diagnosing diseases (Buhmann and Fieseler, 2023). Computer vision is a critical component of AI, with many practical applications across industries and sectors.

2.3.2 NATURAL LANGUAGE PROCESSING

NLP is a field of AI that enables machines to interpret, understand, and make human language. The algorithms of NLP can analyze large amounts of text data, identify patterns, and extract insights from unstructured text (Aisen and Rodrigues, 2023); one real-time example of NLP in AI is virtual assistants. These systems use NLP algorithms to recognize verbal commands, answer questions, and accomplish tasks based on natural language input. Other examples of NLP in AI include sentiment analysis of customer feedback, machine translation of languages, and chatbots for customer service (Ishii et al., 2023; Rodrigues and Keswani, 2023; Novakovsky et al., 2023).

2.3.3 SPEECH RECOGNITION

Speech recognition involves developing algorithms to analyze audio data, identify spoken words and phrases, and convert them into text or machine-readable formats (Dayı et al., 2023). One real-time example of speech recognition in AI is voice-activated devices like Apple's Siri, Google Home, and Amazon Echo. Other examples of speech recognition in AI include voice-to-text transcription for medical, legal, or business purposes and call center automation for customer service (Du et al., 2023). Speech recognition is an essential aspect of AI, enabling machines to understand and respond to spoken language, which can improve accessibility, efficiency, and user experience (Li et al., 2023).

2.3.4 AUTONOMOUS SYSTEMS

Autonomous systems are AI fields that enable machines to perform tasks without human intervention or control. Instead, these systems use advanced algorithms, sensors, and ML techniques to sense and interact with their environment, make decisions, and take action (Buhmann and Fieseler, 2023; Novakovsky et al., 2023). One real-time example of autonomous systems in AI is self-driving cars combining cameras, algorithms, and sensors on-road navigation, detecting obstacles, and deciding driving options. Other examples of autonomous systems in AI include drones for package delivery, robots for manufacturing and logistics, and smart home systems for home automation (Buhmann and Fieseler, 2023; Du et al., 2023). Autonomous systems are essential to AI, enabling machines to operate independently, improving efficiency, and reducing human error (Du et al., 2023).

2.3.5 IMAGE AND VIDEO ANALYSIS

Image and video analysis involves developing algorithms to identify objects, recognize patterns, and make sense of complex visual information (Li et al., 2023). One real-time example of image and video analysis in AI is security surveillance systems, which use video analysis algorithms to detect suspicious behavior, identify potential threats, and provide alerts to security personnel. Other examples of image and video analysis in AI include automated inspection systems for manufacturing, medical imaging for diagnosing diseases, and video content analysis for entertainment and advertising (Dayı et al., 2023; Chakraborty and Mali, 2023). Image and video analysis is vital to AI, enabling machines to analyze and interpret complex visual data, and improving efficiency, accuracy, and decision-making (Kavitha et al., 2023). DL has enabled machines to analyze and extract information from visual data, such as facial recognition, object detection, and video summarization.

2.3.6 FRAUD DETECTION

Fraud detection involves developing algorithms to analyze large amounts of data, identify patterns, and detect anomalies that may indicate fraudulent behavior. One real-time example of fraud detection in AI is credit card fraud detection (Sardanelli et al., 2023). Banks and financial institutions use ML algorithms to analyze transaction data, detect unusual activity patterns, and flag potentially fraudulent transactions for further investigation. Other AI fraud detection examples include insurance fraud detection, healthcare fraud detection, and identity theft prevention (Dayı et al., 2023; Chakraborty and Mali, 2023). Fraud detection is an essential aspect of AI, helping to prevent financial losses, protect personal information, and maintain trust in various sectors of society (Lin, Walter, and Fritz, 2023). DL is used to detect fraudulent activities and anomalies in financial transactions, helping to prevent financial crimes.

2.3.7 PREDICTIVE ANALYTICS

Predictive analytics involves developing algorithms that can analyze and identify massive patterns in data and use those patterns to predict future outcomes. One real-time example of predictive analytics in AI is weather forecasting (Liu and Chou, 2023). Weather forecasting models use historical data on temperature, atmospheric pressure, and weather patterns to predict future weather conditions with increasing accuracy (Lee and Lee, 2023). Other examples of predictive analytics in AI include stock price prediction, demand forecasting, and personalized marketing. Predictive analytics is an essential aspect of AI, enabling businesses to make informed decisions, improve efficiency, and gain a competitive advantage in various industries (Schmitt, 2023). DL predicts future trends and outcomes based on historical data, enabling businesses to make informed decisions.

2.3.8 HEALTHCARE

Healthcare is a field where AI is widely applied to improve patient care, enhance disease diagnosis, and optimize healthcare operations. AI enables healthcare providers to leverage large amounts of patient data to develop new insights and provide personalized patient care. One real-time example of AI in healthcare is medical imaging diagnosis (Schmitt, 2023). AI algorithms can analyze medical images to identify signs of cancer and heart disease, enabling a faster and more accurate diagnosis. Other examples of AI in healthcare include virtual assistants for patient care, predictive analytics for disease prevention, and wearable devices for remote patient monitoring (Gupta, Koundal, and Mongia, 2023). DL is used in medical imaging and diagnosis, drug discovery, and personalized treatment planning (Greffier et al., 2023).

2.3.9 GAMING

Gaming is an area where AI is widely used to create more realistic and engaging gameplay experiences. AI enables game developers to create more sophisticated game characters, environments, and interactions, enhancing the overall gaming experience for players. One real-time example of AI in gaming is procedural content generation (Greffier et al., 2023). AI algorithms can generate game content such as levels, environments, and even entire games, making the game more dynamic and exciting. Other examples of AI in gaming include intelligent game agents that can adapt to player behavior and emotion recognition algorithms that can measure the player's emotional state and adjust the game accordingly (Gombolay et al., 2023). AI can transform the gaming industry by creating more immersive and engaging gaming experiences for players (Hughes et al., 2023). Deep learning is used to develop intelligent agents for gameplay, enabling machines to learn and adapt to different game environments.

2.3.10 ROBOTICS

DL trains robots to recognize and interact with their environment, enabling them to grasp objects and navigate obstacles. Robotics is a field where AI is widely used to create more intelligent and autonomous automatons that can execute various computations (Zhang, Zhu, and Su, 2023; Pico, Marroquín, and Devia, 2023). One real-time example of AI in robotics is industrial automation. AI-powered robots can perform repetitive tasks such as assembly line or inspection tasks, freeing human workers to emphasize more complex computations (Minaee et al., 2023). Other examples of AI in robotics include autonomous vehicles, drones, and medical robots that can assist in surgeries. AI has the potential to revolutionize the field of robotics, making robots more versatile, efficient, and capable of performing a wide range of tasks in various industries.

2.3.11 MANUFACTURING

DL optimizes manufacturing processes, predicts equipment failures, and improves quality control. Manufacturing is an industry where AI is widely applied to reduce costs, improve operational efficiency, and increase productivity (Tang et al., 2023). AI enables manufacturers to analyze large amounts of data, identify patterns, and make data-driven decisions that can lead to improved performance. One real-time example of AI in manufacturing is predictive maintenance. AI algorithms analyze machine sensor data to predict when maintenance will be required, reducing downtime and increasing productivity (Wu et al., 2023). Other examples of AI in manufacturing include quality control, production planning, and supply chain optimization. AI has the potential to revolutionize the manufacturing industry, enabling manufacturers to improve quality, increase efficiency, and stay competitive in an ever-changing market.

2.3.12 AGRICULTURE

Agriculture is an industry where AI is widely applied to increase crop yield, optimize resource utilization, and improve farm productivity. AI enables farmers to analyze large amounts of data, such as weather patterns, soil quality, and crop growth rates, to make data-driven decisions (Park et al., 2023). One real-time example of AI in agriculture is precision agriculture. AI algorithms analyze data from sensors, drones, and other sources to create a detailed farm map, enabling farmers to optimize planting, fertilizing, and watering strategies for each field section. Other examples of AI in agriculture include crop monitoring, disease detection, and yield prediction (Thurzo et al., 2023). AI has the potential to revolutionize the agriculture industry, enabling farmers to increase yield, reduce waste, and produce food more efficiently to meet the demands of a growing population.

2.3.13 TRANSPORTATION

DL is used to optimize traffic flow, predict demand, and improve logistics. Transportation is an industry where AI is widely applied to optimize logistics, reduce costs, and increase safety (Thurzo et al., 2023; Jia et al., 2023). AI enables transportation companies to analyze large amounts of data, such as traffic patterns, weather conditions, and driver behavior, to make data-driven decisions. One real-time example of AI in transportation is autonomous vehicles. AI algorithms can enable self-driving cars, trucks, and buses to navigate roads and make decisions based on real-time data, reducing the risk of accidents and improving overall efficiency (Jia et al., 2023). Other examples of AI in transportation include route optimization, predictive maintenance, and fleet management. AI has the potential to revolutionize the transportation industry, enabling companies to reduce costs, increase safety, and improve the overall customer experience (Tang et al., 2023).

2.3.14 ENERGY

In the energy industry, AI is applied to improve operational efficiency and reduce costs. AI can analyze large amounts of data from sensors, power grids, and other sources to identify patterns, predict demand, and optimize resource utilization (Jia et al., 2023; Chakraborty and Mali, 2023). One real-time example of AI in the energy industry is predictive maintenance. AI algorithms can analyze data from sensors on wind turbines or power plants to predict when maintenance will be required, reducing downtime and increasing efficiency. Other examples of AI in energy include demand forecasting, grid optimization, and energy trading (Vahadane et al., 2023). AI has the potential to transform the energy industry, enabling companies to improve efficiency, reduce costs, and move towards a more sustainable energy future. DL optimizes energy consumption, predicts power demand, and monitors energy infrastructure (Li et al., 2023).

2.3.15 ASTRONOMY

Astronomy is a field where AI is growing in analyzing and interpreting large datasets. AI can be used to analyze astronomical images and identify patterns, such as the location of galaxies or the behavior of stars (Sardanelli et al., 2023). One real-time example of AI in astronomy is using neural networks to identify gravitational lenses. Gravitational lenses are rare events where the light from distant objects is bent by the gravitational pull of a massive object, such as a galaxy (Li et al., 2023). Identifying these events is vital for understanding the properties of dark matter and dark energy. AI algorithms can be trained to recognize these events and identify new examples in large datasets. Other examples of AI in astronomy include data analysis, object classification, and image processing (Soori, Arezoo, and Dastres, 2023). AI has the potential to enable astronomers to analyze vast amounts of data and make discoveries, improving our understanding of the universe (Lin, Walter, and Fritz, 2023). Deep learning analyzes astronomical data, enabling researchers to identify new celestial objects and phenomena.

2.4 DEEP LEARNING IMPACTS IN THE AI FIELD

This section presents the different DL impacts in AI, as detailed below.

2.4.1 IMPROVED ACCURACY

Improved accuracy is a critical aspect of AI, as it can significantly impact the performance and effectiveness of AI systems. In many applications, even minor improvements in accuracy can lead to significant gains in performance (Du et al., 2023; Soori, Arezoo, and Dastres, 2023). For example, in NLP, a slight improvement in accuracy can lead to a better understanding of complex language structures and more accurate language translation. One real-time example of the impact of improved accuracy on AI is in self-driving cars (Zhang et al., 2023). As self-driving technology evolves, improved object recognition and tracking accuracy can lead to safer and more reliable autonomous vehicles. Improved accuracy also has implications for fraud detection, where more accurate algorithms can help prevent financial losses and improve security (Gombolay et al., 2023). In general, improved accuracy can lead to more effective and reliable AI systems, enabling them to make better decisions and provide better outcomes in various applications. Therefore, DL has significantly improved the accuracy of AI algorithms, enabling machines to perform complex tasks with a high degree of precision.

2.4.2 REDUCED COST

Reduced cost is a crucial factor in the adoption and growth of AI, as it can make AI technology more accessible to a broader range of applications and industries (Rodrigues and Keswani, 2023). Cost reduction can come from advancements in hardware and software technology, as well as improvements in efficiency and scalability. One real-time example of the impact of reduced costs on AI is the increasing use of AI in healthcare (Ishii et al., 2023). The adoption of AI-driven diagnostic tools has been limited due to the high cost of developing and implementing these systems. However, recent advancements in AI technology, such as cloud-based computing and open-source machine learning frameworks, have significantly reduced the cost of developing and deploying AI systems in healthcare (Novakovsky et al., 2023). As a result, AI is now being used to improve patient outcomes and reduce healthcare costs, leading to a more efficient and accessible healthcare system (Buhmann and Fieseler, 2023). Reduced costs can accelerate the adoption and growth of AI technology, opening up new possibilities for innovation and development in a wide range of industries and applications.

2.4.3 FASTER PROCESSING

Faster- processing allows AI systems to learn and make real-time decisions, leading to more effective and efficient applications. A real-time example of the impact of faster processing on AI is the development of autonomous vehicles (Gombolay et al. 2023; Chou and Lin, 2023). Autonomous vehicles require massive amounts of data to be processed quickly and accurately in real-time to operate safely and effectively (Zhang et al., 2023; Hughes et al., 2023). Faster processing has enabled the development of sophisticated ML algorithms that can analyze and respond to this data in real-time, allowing for the safe and efficient operation of autonomous vehicles (Soori et al., 2023; Greffier et al., 2023). Therefore, faster processing can accelerate the development and application of AI technology, leading to more effective and efficient systems across various industries and applications (Greffier et al., 2023; Krentzel, Shorte, and Zimmer, 2023).

2.4.4 IMPROVED LANGUAGE PROCESSING

Improved language processing is a significant factor in the development and application of AI, as it enables AI systems to understand and communicate with humans more effectively (Hughes et al., 2023; Du et al., 2023). NLP is an area of AI focused on improving machines' ability to understand and interpret human language. A real-time example of the impact of improved language processing on AI is the development of chatbots and virtual assistants (Chakraborty and Mali, 2023; Greffier et al., 2023). These systems use NLP techniques to understand and respond to human inquiries and commands, allowing for more efficient and personalized user interactions. Improved language processing can also enable AI systems to analyze and interpret large amounts of text data, allowing for more accurate sentiment analysis, classification, and translation (Lin et al., 2023). In general, improved language processing can enhance the ability of AI systems to interact with and understand humans, leading to more effective and efficient applications across a wide range of industries and use cases (Chakraborty and Mali, 2023). DL has improved the ability of machines to understand and process human language, enabling the development of more sophisticated NLP algorithms.

2.4.5 IMPROVED COMPUTER VISION

Improved computer vision is a vital aspect of AI that permits technologies to interpret and analyze visual data more accurately and efficiently (Lin, Walter, and Fritz, 2023). Computer vision technology uses DL algorithms to analyze and interpret visual data from images and videos. A real-time example of the impact of improved computer vision on AI is the development of self-driving cars (Soori, Arezoo, and Dastres, 2023). These vehicles rely on computer vision algorithms to interpret their environment, including traffic signs, road markings, and other vehicles (Al-Tameemi, Hasan, and Oleiwi, 2023; Liu and Chou, 2023). Improved computer vision can also enable AI systems to detect and recognize objects in images and videos, allowing for more effective security and surveillance applications. Additionally, improved computer vision can enhance the accuracy and efficiency of medical imaging systems, enabling more precise diagnoses and treatments (Lee and Lee, 2023). Overall, improved computer vision can significantly enhance the ability of AI schemes to analyze and interpret visual data, leading to more effective and efficient applications in various industries. DL has improved computer vision, enabling machines to recognize and classify images more accurately (Du et al., 2023).

2.4.6 INCREASED PERSONALIZATION

Increased personalization is an essential aspect of AI that allows machines to provide customized experiences and recommendations to users based on their individual preferences and behavior patterns (Gupta, Koundal, and Mongia 2023). AI systems can make predictions and recommendations tailored to their unique needs and interests by analyzing vast amounts of data about a user's interactions with digital systems. A real-time example of the impact of increased personalization in AI is the development of personalized marketing campaigns (Soori, Arezoo, and Dastres, 2023). AI algorithms can analyze user data to understand their preferences and behaviors and use this information to deliver targeted ads and promotions. Similarly, AI-powered personal shopping assistants can provide personalized recommendations to consumers based on their previous purchases and browsing history (Du et al., 2023). Overall, increased personalization can enhance the user experience and improve the effectiveness of AI applications in various industries, from marketing and e-commerce to healthcare and education. DL algorithms can learn from individual user data, enabling personalized recommendations and experiences (Lin, Walter, and Fritz, 2023).

2.4.7 ENHANCED SECURITY

Enhanced security is a critical aspect of AI that is becoming increasingly important as more and more sensitive data is collected and processed by AI systems (Du et al., 2023). With enhanced security measures, AI can help protect against cyber threats, fraud, and other malicious activities, providing greater trust and confidence in digital systems. A real-time example of the impact of enhanced security on AI is the development of AI-powered cybersecurity systems that can detect and respond to cyberattacks in real-time (Liu and Chou, 2023). By analyzing vast amounts of data from network traffic, user behavior, and other sources, AI algorithms can identify potential threats and take proactive measures to prevent them before they can cause harm. Similarly, AI-powered fraud detection systems can analyze transaction data to identify suspicious activity and prevent financial fraud (Al-Tameemi, Hasan, and Oleiwi, 2023). Enhanced security measures can help protect individuals and organizations from the growing threat of cybercrime and increase the reliability and trustworthiness of AI systems (Wu et al., 2023). DL has improved security by enabling machines to detect anomalies and identify potential threats in real-time.

2.4.8 AUTOMATION

AI provides powerful tools and techniques that enable machines to perform tasks accurately and efficiently. Through machine learning algorithms and intelligent systems, AI can analyze vast amounts of data, recognize patterns, and make informed decisions or predictions. This capability has revolutionized automation by allowing machines to take over routine and repetitive tasks previously performed by humans (Mughaid et al., 2023). One of the most common examples of automation in AI is chatbots. Chatbots are AI-powered software programs that can interact with humans and answer their queries (Shafik et al., 2021). They are used by businesses to automate customer service and reduce response time, resulting in cost savings and increased efficiency (Thurzo et al., 2023). Another example is Robotic Process Automation (RPA), which automates repetitive back-office tasks such as data entry, invoicing, and customer onboarding (Al-Tameemi, Hasan, and Oleiwi, 2023; Wu et al., 2023). RPA can be used across industries, including finance, healthcare, and logistics, and can significantly reduce processing times while improving accuracy. DL has enabled the automation of many tasks, reducing the need for human intervention in areas such as manufacturing, transportation, and healthcare (Jia et al., 2023; Tang et al., 2023).

2.4.9 INCREASED EFFICIENCY

Increased efficiency is one of the most significant impacts of AI in various fields. AI-powered systems can analyze and process vast amounts of data in seconds, allowing organizations to make informed decisions quickly (Zhang, Zhu, and Su, 2023). This improved efficiency can lead to increased productivity and cost savings for businesses. For example, in the financial industry, AI-powered chatbots can answer customer queries faster than humans, freeing human resources for more complex tasks (Lin, Walter, and Fritz, 2023). Overall, increased efficiency through AI is a game-changer for businesses and organizations looking to streamline their operations and increase productivity (Chou and Lin, 2023). DL algorithms can learn to optimize processes, enabling businesses to operate more efficiently and reduce costs.

2.4.10 IMPROVED DECISION-MAKING

With AI-powered tools, organizations can make better decisions based on vast amounts of data, including structured and unstructured data (Park et al., 2023). AI can help analyze data, identify patterns, and provide insights to guide decision-making processes (Shafik et al., 2020). For example, AI can analyze market trends, financial statements, and other data in the finance industry to make better investment decisions (Minaee et al., 2023). Another example is healthcare, where AI-powered tools can analyze patient data, predict potential health issues, and recommend personalized treatment plans. These tools can lead to better patient outcomes and more efficient use of healthcare resources (Shafik and Matinkhah, 2021). In general, using AI for improved decision-making can lead to better outcomes, increased productivity, and cost savings for organizations (Tang et al., 2023; Minaee et al., 2023). DL has improved decision-making by providing businesses with accurate predictions and insights based on large data sets, enabling them to make more informed decisions.

2.5 FUTURE DIRECTIONS OF DEEP LEARNING'S IMPACT ON AI

The future directions of DL impact in AI are vast and varied, with numerous potential applications and advancements in the pipeline (Khan, Sarkar, and Maity, 2023). One direction is the increased use of reinforcement learning, where algorithms are trained to learn by trial and error to make decisions that maximize rewards. Another direction is incorporating more human-like reasoning and decision-making into AI systems, leading to more natural and intuitive interactions between humans and machines. Additionally, deep learning is expected to play a critical role in the development of autonomous systems, such as self-driving cars and drones, as well as in the advancement of healthcare technology, including personalized medicine and drug discovery (Rodrigues and Keswani 2023; Shafik, 2023). There is also growing interest in using DL for natural language generation and translation and creating more advanced virtual assistants and chatbots (Rodrigues and Keswani, 2023). As DL advances, it is expected to revolutionize many industries, from finance and manufacturing to entertainment and education. Ultimately, the future of DL in AI will be shaped by ongoing research, development, and innovation, as well as by the increasing integration of AI technologies into our daily lives (Ishii et al., 2023).

One potential future direction for deep learning in AI is the development of more explainable and interpretable models. Unfortunately, DL models can be seen as "black boxes" since it is often difficult to interpret and understand how they arrive at their predictions or decisions (Chen, 2023). This can be problematic in specific applications, such as healthcare, where it is crucial to understand the reasoning behind a diagnosis or treatment recommendation. Therefore, researchers are exploring ways to make DL models more transparent and interpretable (Aisen and Rodrigues, 2023).

Another area of future development is integrating deep learning with other AI techniques, such as reinforcement learning and evolutionary algorithms. These approaches could help create more flexible and adaptable AI systems that can learn and evolve in dynamic environments (Li et al., 2023). For example, reinforcement learning could help a robot learn how to navigate a complex environment, while evolutionary algorithms could be used to optimize the design of a manufacturing process (Liu and Chou, 2023).

Still, ongoing research is developing more efficient and scalable deep learning models. As deep learning applications continue to expand, it is becoming increasingly important to develop models that can be trained on large datasets in a reasonable amount of time and can be deployed on various devices and platforms (Hughes et al., 2023; Vahadane et al., 2023). This may involve the development of new hardware architectures that are optimized for deep learning as well as the creation of more efficient algorithms that can run on existing hardware.

Another future direction of deep learning in AI is the integration of Reinforcement Learning (RL) algorithms, which involve training agents to learn through trial and error in an environment (Chen, 2023; Zhao et al., 2022). RL has shown great potential in areas such as robotics, gameplay, and autonomous systems, and its integration with deep learning can lead to even more impressive results.

Another area of focus is the development of more efficient deep learning algorithms that can be run on smaller devices, such as smartphones and Internet of Things devices, enabling the widespread deployment of AI technology (Al-Tameemi, Hasan, and Oleiwi, 2023). Researchers are also exploring ways to make DL models more interpretable and transparent, addressing concerns about the "black box" nature of deep learning algorithms and making it easier for humans to understand and trust the decisions made by these systems (Al-Tameemi, Hasan, and Oleiwi 2023).

Another promising direction is integrating DL with other advanced technologies, such as Virtual Reality (VR) and Augmented Reality (AR), to create more immersive and interactive experiences. For example, deep learning algorithms can analyze real-time data from AR and VR environments, enabling personalized content and recommendations to be delivered to users based on their preferences and behaviors (Thurzo et al., 2023). Ongoing research is developing more explainable and interpretable deep learning models. As DL algorithms become more complex, it becomes increasingly difficult to understand how they arrive at their predictions and decisions (Minaee et al., 2023; Shokoor et al., 2022). This is particularly problematic in applications such as healthcare and finance, where the consequences of incorrect decisions can be severe. By improving the interpretability of DL models, researchers hope to improve trust in AI and encourage broader adoption in high-stakes domains (Tang et al., 2023; Yao et al., 2021).

Another future direction is incorporating DL into edge devices like smartphones, IoT devices, and drones. This would allow for real-time decision-making without cloud computing, enabling faster and more efficient processing (Du et al., 2023; Gupta, Koundal, and Mongia, 2023). Deep learning also has the potential to enhance and improve personalized healthcare by analyzing vast amounts of medical data to diagnose and predict illnesses, personalize treatments and medications, and ultimately improve patient outcomes (Gupta, Koundal, and Mongia, 2023). Another area where DL is expected to impact significantly is autonomous vehicles, where it can help improve safety and accuracy by enabling vehicles to make real-time decisions based on real-world data (Lin, Walter, and Fritz, 2023).

Overall, the future of deep learning in AI will likely be shaped by continued technological advances and ongoing research into new applications and approaches (Chakraborty and Mali, 2023). As these developments unfold, we expect to see even more sophisticated and robust AI systems that can tackle increasingly complex problems and drive innovation in various industries (Kavitha et al., 2023). Furthermore, with ongoing research into explainability and the integration of other advanced technologies, the potential for DL to revolutionize the field of AI is enormous.

2.6 CONCLUSION

The impact of DL on the field of AI has been revolutionary, enabling machines to perform complex tasks with high accuracy, reducing development costs, and improving processing speeds. Deep learning has significantly contributed to computer vision, natural language processing, speech recognition, autonomous systems, and fraud detection. Its ability to learn from large amounts of data without explicit programming has led to more accurate predictions and better decision-making, while its capacity to optimize processes and personalize experiences has improved efficiency and customer satisfaction. However, deep learning also faces challenges, such as the need for large amounts of high-quality data and the limited interpretability of models. Overall, deep learning's impact on AI has been profound, with the potential to transform many industries and sectors.

REFERENCES

Adnan, N., Khalid, W. B., Umer, F. An artificial intelligence model for instance segmentation and tooth numbering on orthopantomograms. Int J Comput Dent. 2023 Nov 28;26(4):301-309.

Aisen, A. M., Rodrigues, P. S. Deep Learning to Detect Pancreatic Cancer at CT: Artificial Intelligence Living Up to Its Hype. Radiology. 2023 Jan; 306(1):183-185.

Alghamdi, Salem Saeed. (2023). The Application of Artificial Intelligence in Detecting Breast Lesions with Medical Imaging: A Literature Review. In *International Journal of Biomedicine* (Vol. 13, Issue 1).

Ali, Rafia, Mehala Balamurali, and Pegah Varamini. (2022). Deep Learning-Based Artificial Intelligence to Investigate Targeted Nanoparticles' Uptake in TNBC Cells. *International Journal of Molecular Sciences*, 23(24).

Al-Tameemi, Maad Issa, Ammar A. Hasan, and Bashra Kadhim Oleiwi. (2023). "Design and Implementation Monitoring Robotic System Based on You Only Look Once Model Using Deep Learning Technique." *IAES International Journal of Artificial Intelligence*, 12(1): 106.

Bastidas, S. E. C., Porras, A. M., Gómez, A. P., García, I.M., and González, L. (2023). Elderly, Their Emotions and Deep Learning Techniques to Help Their Dignified and Positive Aging. *Lecture Notes in Networks and Systems*, 594 LNNS.

Benayad, Anass, et al. (2022). High-Throughput Experimentation and Computational Freeway Lanes for Accelerated Battery Electrolyte and Interface Development Research. In *Advanced Energy Materials* (Vol. 12, Issue 17).

Buhmann, Alexander, and Christian Fieseler. (2023). "Deep Learning Meets Deep Democracy: Deliberative Governance and Responsible Innovation in Artificial Intelligence." *Business Ethics Quarterly*, 33(1): 146–79.

Case, James A. (2022). Deep Learning and Artificial Intelligence: What Does the Cardiologist Really Need to Know? In *Circulation: Cardiovascular Imaging* (Vol. 15, Issue 9).

Chakraborty, Shouvik, and Kalyani Mali. (2023). "An Overview of Biomedical Image Analysis from the Deep Learning Perspective." *Research Anthology on Improving Medical Imaging Techniques for Analysis and Intervention*, 43–59.

Chen, Tzeng-Ji. 2023. "ChatGPT and Other Artificial Intelligence Applications Speed up Scientific Writing." *Journal of the Chinese Medical Association*, 10.

Chou, Wei-Chun, and Zhoumeng Lin. (2023). "Machine Learning and Artificial Intelligence in Physiologically Based Pharmacokinetic Modeling." *Toxicological Sciences*, 191(1): 1–14.

Dayı, Burak, Hüseyin Üzen, İpek Balıkçı Çiçek, and Şuayip Burak Duman. (2023). "A Novel Deep Learning-Based Approach for Segmentation of Different Type Caries Lesions on Panoramic Radiographs." *Diagnostics*, 13(2): 202.

Deng, Wen, Guangjun Liang, Chenfei Yu, Kefan Yao, Chengrui Wang, and Xuan Zhang. (2023). An Early Warning Model of Telecommunication Network Fraud Based on User Portrait. *Computers, Materials and Continua*, 75(1).

Du, Jian, et al. (2023). "A Theory-Guided Deep-Learning Method for Predicting Power Generation of Multi-Region Photovoltaic Plants." *Engineering Applications of Artificial Intelligence* 118: 105647.

Gombolay, G.Y., Gopalan, N., Bernasconi, A., Nabbout, R., Megerian, J.T., Siegel, B.I., Hallman-cooper, J., Bhalla, S., & Gombolay, M.C. (2023). Review of Machine Learning and Artificial Intelligence (ML/AI) for the Pediatric Neurologist. *Pediatric neurology, 141*, 42-51.

Gore, Sayali V., et al. (2023). Zebrafish Larvae Position Tracker (Z-LaP Tracker): A High-Throughput Deep-Learning Behavioral Approach for the Identification of Calcineurin Pathway-Modulating Drugs Using Zebrafish Larvae. *Scientific Reports*, 13(1).

Greffier, Joël, et al. (2023). "Improved Image Quality and Dose Reduction in Abdominal CT with Deep-Learning Reconstruction Algorithm: A Phantom Study." *European Radiology*, 33(1): 699–710.

Gronwald, K.-D. (2023). Machine Learning, Deep Learning und Artificial Intelligence. In *Globale Kommunikation und Kollaboration*.

Gu, Shangzhi, et al. (2023). Detection of Sarcopenia Using Deep Learning-Based Artificial Intelligence Body Part Measure System (AIBMS). *Frontiers in Physiology*, 14.

Gupta, Lav Kumar, Deepika Koundal, and Shweta Mongia. (2023). "Explainable Methods for Image-Based Deep Learning: A Review." *Archives of Computational Methods in Engineering*, 1–16.

Huang, Chun-Chao, et al. (2023). Using Deep-Learning-Based Artificial Intelligence Technique to Automatically Evaluate the Collateral Status of Multiphase CTA in Acute Ischemic Stroke. *Tomography*, 9(2).

Hughes, H., M. O'Reilly, N. McVeigh, and R. Ryan. (2023). "The Top 100 Most Cited Articles on Artificial Intelligence in Radiology: A Bibliometric Analysis." *Clinical Radiology*, 78 (2): 99–106.

Ishii, Euma, et al. (2023). "Development, Validation, and Feature Extraction of a Deep Learning Model Predicting in-Hospital Mortality Using Japan's Largest National ICU Database: A Validation Framework for Transparent Clinical Artificial Intelligence (CAI) Development." *Anaesthesia Critical Care & Pain Medicine*, 42(2): 101167.

Jia, Tianlong, et al. (2023). "Deep Learning for Detecting Macroplastic Litter in Water Bodies: A Review." *Water Research*, 119632.

Jin, Xiaoxue, Xiufeng Wang, Xingiang Cao, and Chaohua Xue. (2023). Construction and Recognition of Acoustic ID of Ancient Coins Based on Deep Learning of Artificial Intelligence for Audio Signals. *Heritage Science*, 11(1).

Kavitha, R., et al. (2023). "Ant Colony Optimization-Enabled CNN Deep Learning Technique for Accurate Detection of Cervical Cancer." *BioMed Research International 2023*.

Khan, Mohd Imran, Subharthi Sarkar, and Rajib Maity. (2023). "Artificial Intelligence/Machine Learning Techniques in Hydroclimatology: A Demonstration of Deep Learning for Future Assessment of Stream Flow under Climate Change." In *Visualization Techniques for Climate Change with Machine Learning and Artificial Intelligence*, pp. 247–73. Elsevier.

Krentzel, D., Shorte, S. L., & Zimmer, C. (2023). Deep learning in image-based phenotypic drug discovery. *Trends in cell biology*, *33*(7), 538–554.

Lan, Chou-Chin, et al. (2022). Deep Learning-based Artificial Intelligence Improves Accuracy of Error-prone Lung Nodules. *International Journal of Medical Sciences*, 19(3).

Lee, Jaekyu, and Sangyub Lee. (2023). "Construction Site Safety Management: A Computer Vision and Deep Learning Approach." *Sensors*, 23(2): 944.

Lei, Shijun, and Huiming Liu. (2022). Deep Learning Dual Neural Networks in the Construction of Learning Models for Online Courses in Piano Education. *Computational Intelligence and Neuroscience*, 2022.

Li, Hengyi, et al. (2023). "An Architecture-Level Analysis on Deep Learning Models for Low-Impact Computations." *Artificial Intelligence Review*, 56(3): 1971–2010.

Li, Weiyan. (2022). Analysis of Piano Performance Characteristics by Deep Learning and Artificial Intelligence and Its Application in Piano Teaching. *Frontiers in Psychology*, 12.

Lin, Dana J., Sven S. Walter, and Jan Fritz. (2023). "Artificial Intelligence–Driven Ultra-Fast Superresolution MRI: 10-Fold Accelerated Musculoskeletal Turbo Spin Echo MRI Within Reach." *Investigative Radiology*, 58(1): 28–42.

Liu, Chi-Yun, and Jui-Sheng Chou. (2023). "Bayesian-Optimized Deep Learning Model to Segment Deterioration Patterns underneath Bridge Decks Photographed by Unmanned Aerial Vehicle." *Automation in Construction* 146: 104666.

Liu, L. (2023). Study on the Influence of Deep Learning and Artificial Intelligence on Transportation and Mobility Industry and Corporations. In *Proceedings of the 2022 2nd International Conference on Business Administration and Data Science (BADS 2022)*.

Mehralivand, Sherif, et al. (2022). Deep Learning-Based Artificial Intelligence for Prostate Cancer Detection at Biparametric MRI. *Abdominal Radiology*, 47(4).

Minaee, S., Abdolrashidi, A., Su, H., Bennamoun, & Zhang, D. (2023). Biometrics recognition using deep learning: a survey. *Artificial Intelligence Review, 56*, 8647-8695.

Mughaid, Ala, et al. (2023). "Improved Dropping Attacks Detecting System in 5g Networks Using Machine Learning and Deep Learning Approaches." *Multimedia Tools and Applications*, 82(9): 13973–95.

Novakovsky, Gherman, et al. (2023). "Obtaining Genetics Insights from Deep Learning via Explainable Artificial Intelligence." *Nature Reviews Genetics*, 24(2): 125–37.

Park, Seungman, Anna L. Chien, Beiyu Lin, and Keva Li. (2023). "FACES: A Deep-Learning-Based Parametric Model to Improve Rosacea Diagnoses." *Applied Sciences*, 13(2): 970.

Pico, Lilia Edith Aparicio, Oscar Julián Amaya Marroquín, and Paola Andrea Devia. (2023). "Application of Deep Learning for the Identification of Surface Defects Used in Manufacturing Quality Control and Industrial Production: A Literature Review." *Reproduction*, e18934.

Rodrigues, Terrance, and Rajesh Keswani. (2023). "Endoscopy Training in the Age of Artificial Intelligence: Deep Learning or Artificial Competence?" *Clinical Gastroenterology and Hepatology*, 21(1): 8–10.

Sardanelli, Francesco, et al. (2023). "Artificial Intelligence (AI) in Biomedical Research: Discussion on Authors' Declaration of AI in Their Articles Title." *European Radiology Experimental*, 7(1): 2.

Schmitt, Marc. (2023). "Deep Learning in Business Analytics: A Clash of Expectations and Reality." *International Journal of Information Management Data Insights*, 3(1): 100146.

Serey, Joel, et al. (2023). Pattern Recognition and Deep Learning Technologies, Enablers of Industry 4.0, and Their Role in Engineering Research. In *Symmetry* (Vol. 15, Issue 2).

Shafik, Wasswa, and Mojtaba S. Matinkhah. (2021). "Unmanned Aerial Vehicles Analysis to Social Networks Performance." *The CSI Journal on Computer Science and Engineering*, 18(2): 24-31.

Shafik, Wasswa, Mojtaba Matinkhah, and Mamman Nur Sanda. (2020). "Network Resource Management Drives Machine Learning: A Survey and Future Research Direction." *Journal of Communications Technology, Electronics and Computer Science*, 2020: 1-15.

Shafik, Wasswa, Mojtaba S. Matinkhah, Mamman Nur Sanda, and Fawad Shokoor. (2021). "Internet of Things-Based Energy Efficiency Optimization Model in Fog Smart Cities." *JOIV: International Journal on Informatics Visualization*, 5(2): 105-112.

Shafik, Wasswa. (2023). "Cyber Security Perspectives in Public Spaces: Drone Case Study." In *Handbook of Research on Cybersecurity Risk in Contemporary Business Systems*, pp. 79-97. IGI Global.

Shokoor, Fawad, Wasswa Shafik, and Mojtaba S. Matinkhah. (2022). "Overview of 5G & Beyond Security." *EAI Endorsed Transactions on Internet of Things* 8(30).

Shuaib, M., J. Jaafar, M. Faizuddin, and S. Musa. (2022). Review on Recent Trends and Challenges in Deep Learning and Artificial Intelligence. *AIP Conference Proceedings*, 2617.

Shubham, Shubham, et al. (2023). Identify Glomeruli in Human Kidney Tissue Images using a Deep Learning Approach. *Soft Computing*, 27(5).

Soori, Mohsen, Behrooz Arezoo, and Roza Dastres. (2023). "Machine Learning and Artificial Intelligence in CNC Machine Tools, A Review." *Sustainable Manufacturing and Service Economics*, 100009 : 1-12.

Sujith, A. V. L. N., et al. (2022). Systematic Review of Smart Health Monitoring using Deep Learning and Artificial Intelligence. *Neuroscience Informatics*, 2(3).

Surendiran, R., M. Thangamani, S. Monisha, and P. Rajesh. (2022). Exploring the Cervical Cancer Prediction by Machine Learning and Deep Learning with Artificial Intelligence Approaches. *International Journal of Engineering Trends and Technology*, 70(7).

Tang, A., Tian, L., Gao, K., Liu, R., Hu, S., Liu, J., Xu, J., Fu, T., Zhang, Z., Wang, W., Zeng, L., Qu, W., Dai, Y., Hou, R., Tang, S., & Wang, X. (2023). Contrast-enhanced harmonic endoscopic ultrasound (CH-EUS) MASTER: A novel deep learning-based system in pancreatic mass diagnosis. *Cancer medicine*, *12*(7), 7962–7973.

Thurzo, Andrej, et al. (2023). "Impact of Artificial Intelligence on Dental Education: A Review and Guide for Curriculum Update." *Education Sciences*, 13(2): 150.

Turkbey, Baris, and Masoom A. Haider. (2022). Deep learning-Based Artificial Intelligence Applications in Prostate MRI: Brief Summary. In *British Journal of Radiology* (Vol. 95, Issue 1131).

Vahadane, Abhishek, et al. (2023). "Development of an Automated Combined Positive Score Prediction Pipeline Using Artificial Intelligence on Multiplexed Immunofluorescence Images." *Computers in Biology and Medicine*, 152: 106337.

Wang, Yu. (2022). Safety Production Supervision of Industrial Enterprises Based on Deep Learning and Artificial Intelligence. *Mobile Information Systems*, 2022.

Wu, Fei, Yu Chen, and Dan Han. (2022). Development Countermeasures of College English Education Based on Deep Learning and Artificial Intelligence. *Mobile Information Systems*, 2022.

Wu, L., Pei, J., Tang, J., Xia, Y., & Guo, X. (2023). Deep Learning on Graphs: Methods and Applications (DLG-KDD2023). *Proceedings of the 29th ACM SIGKDD Conference on Knowledge Discovery and Data Mining*.

Wu, Qingjie, et al. (2023). Application of Deep-Learning–Based Artificial Intelligence in Acetabular Index Measurement. *Frontiers in Pediatrics*, 10.

Yao, Jun, Alisa Craig, Wasswa Shafik, and Lule Sharif. (2021). "Artificial Intelligence Application in Cybersecurity and Cyberdefense." *Wireless Communications & Mobile Computing (Online)* 2021: 1-10.

Yu, Haiyun, et al. (2023). A Deep-Learning-Based Artificial Intelligence System for the Pathology Diagnosis of Uterine Smooth Muscle Tumor. *Life*, 13(1).

Zhang, Bo, Jun Zhu, and Hang Su. (2023). "Toward the Third Generation Artificial Intelligence." *Science China Information Sciences*, 66 (2): 1–19.

Zhao, Liguo, et al. (2022). "Artificial Intelligence Analysis in Cyber Domain: A Review." *International Journal of Distributed Sensor Networks*, 18 (4): 15501329221084882.

Zhu, Qinlei, and Hao Zhang. (2022). Teaching Strategies and Psychological Effects of Entrepreneurship Education for College Students Majoring in Social Security Law Based on Deep Learning and Artificial Intelligence. *Frontiers in Psychology*, 13.

3 Deep Learning is a State-of-the-Art Approach to Artificial Intelligence

Soumik Kumar Mohanta, Ambarish G. Mohapatra, Anita Mohanty, and Sasmita Nayak

3.1 INTRODUCTION

Artificial Intelligence (AI) has entered a disruptive era thanks to deep learning and its amazing capabilities. This introduction lays the groundwork for a thorough investigation of deep learning's influence on AI. It emphasizes the fundamental ideas that neural networks and deep learning are built upon. This chapter investigates the mechanics and guiding concepts that allow computers to learn and predict the future, emphasizing the revolutionary potential of this cutting-edge method.

The effectiveness of deep learning is further demonstrated by breakthroughs in image recognition and natural language processing. Convolutional neural networks and recurrent neural networks are two examples of architectural advancements that demonstrate the profound impact that deep learning will have on the future of AI.

In-depth analysis of how deep learning has transformed artificial intelligence is provided in this chapter. It covers the core ideas of deep learning, for example, neural networks and their design, and emphasises how revolutionary this cutting-edge method is for allowing machines to learn and make precise predictions. This chapter examines significant developments, practical applications, and the possible influence of deep learning on the direction of AI. Deep learning has become a cutting-edge and game-changing strategy in the fast-changing field of Artificial Intelligence (AI). The deep implications of deep learning in the area of AI are examined in greater detail in this chapter. It starts off with a thorough introduction to deep learning that concentrates on the fundamental theories and methods.

By simulating the configuration and operation of the human brain, we research neural networks that form the foundation for deep learning. These neural networks enable deep learning by learning from data and making wise predictions. The foundational elements of deep learning – layers, nodes, activation functions, and weights – and how they interact to build intricate networks that can recognize patterns and represent representations from enormous quantities of data are further explored in this chapter. We also talk about weight-based neural network performance and optimization techniques such as backpropagation.

Deep learning's capacity to handle extremely complicated problems that have up until now proven difficult for conventional AI techniques is one of its significant features. This chapter looks at how deep learning performs well in domains including image recognition, speech recognition, natural language processing, as well as recommender systems. We highlight ground-breaking accomplishments in various fields, such as the use of deep learning in autonomous vehicles, diagnostics, and intelligent personal assistants.

This chapter also focuses on deep learning architecture developments that have elevated AI to new levels.

DOI: 10.1201/9781003433309-3

A Typical Convolutional Neural Network (CNN)

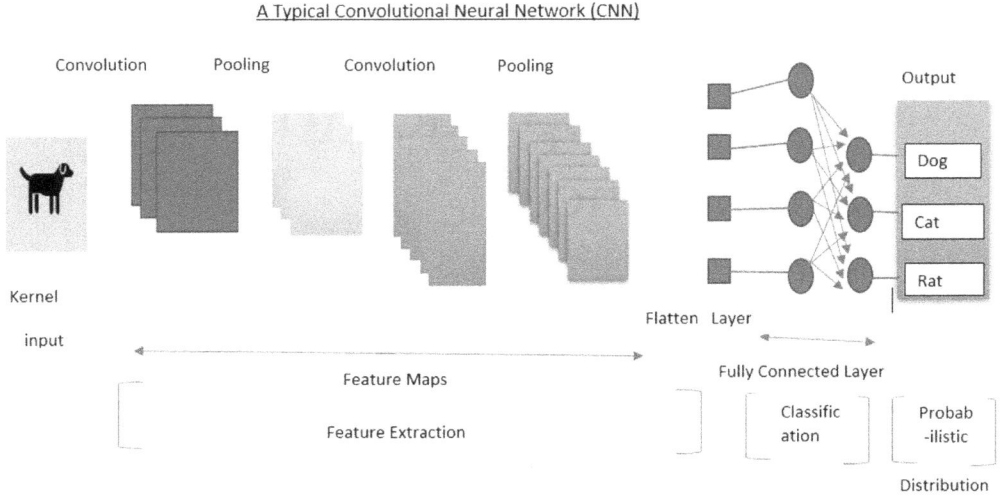

FIGURE 3.1 Architecture of CNN and the configuration settings.

A typical CNN architecture is portrayed in the Figure 3.1, where the internal configurations are described.

This includes exceptional performance on sequential data tasks like speech translation and speech recognition, as well as on Convolutional Neural Networks (CNN) and computer vision tasks, Recurrent Neural Networks (RNN) and the capacity to capture time dependencies.

3.2 FUNDAMENTALS OF DEEP LEARNING

This is a subfield of machine learning which concentrates on building multi-layered artificial neural networks to extract complex representations and patterns from massive volumes of data. It has attracted a lot of attention and recognition due to its capacity to deliver cutting-edge results in a variety of AI tasks. Deep learning models acquire knowledge of feature hierarchies, which enables them to automatically identify complicated structures and produce reliable predictions.

3.2.1 NEURAL NETWORKS: FOUNDATIONS OF DEEP LEARNING

Deep learning is based on neural networks, which are modelled after the way in which the human brain functions. Artificial neurons are arranged in layers and joined together to build neural networks. Weights – learned throughout the training process – represent the connections between neurons. Data travels through numerous hidden levels after being received at the input layer. Each hidden layer processes the input through a series of calculations to create a more abstract representation. The output layer then generates the desired categorization or prediction. Neural networks optimize their performance to reduce errors and increase accuracy by varying the weights between neurons.

3.2.2 TRAINING ALGORITHMS: UNLOCKING LEARNING POTENTIAL

Deep learning models are trained using a variety of techniques that minimize the discrepancy between expected and actual outputs by optimizing the neural network weights. The gradient of the loss function with respect to the weights is determined by the backpropagation algorithm, which is essential. An essential procedure, called backpropagation, finds the gradient of the loss function with respect to the weights and adjusts them accordingly. The network can improve its predictions by iteratively learning and updating its weights thanks to backpropagation. Stochastic Gradient Descent (SGD) and its derivatives, among other optimization approaches, aids in quickly navigating the weight space and identifying the ideal set of weights. These methods find a balance between the two extremes of overfitting and underfitting, where the model becomes too tailored to the training data and misses the underlying patterns in the data.

3.2.3 ARCHITECTURAL COMPONENTS: LAYERS, NODES, ACTIVATION FUNCTIONS, AND WEIGHTS

The layers of deep learning architectures each contain a set of nodes or synthetic neurons. The quantity of layers and the number of nodes within each layer, respectively, determine the network's depth and width. Raw data is represented by the input layer, which is followed by hidden layers that progressively extract more abstract and significant representations. The final classification or prediction is created by the output layer. An essential part of introducing non-linearity to the network and empowering it to represent intricate connections between input as well as output is played by activation functions. Rectified Linear Units (ReLU), sigmoid, and tanh are frequently used activation functions. They assist in the learning of non-linear transformations and decide a neuron's output based on its weighted inputs.

The weights of the neurons determine the connections between them and how much each neuron contributes to the overall output. The weights are changed throughout training to enhance the model's performance. By penalizing the weights, regularization approaches like L1 and L2 regularization assist to prevent overfitting.

3.2.4 TRAINING AND OPTIMIZATION: BACKPROPAGATION AND BEYOND

As has already been noted, a crucial algorithm for training deep learning models is backpropagation. The necessary adjustments are made after determining the gradients of the loss function in relation to the weights. The network can gradually improve its predictions by altering the weights in the direction that minimizes the loss thanks to this iterative process. The training process is improved by a number of optimization strategies in addition to backpropagation. These include learning rate schedules, which dynamically alter the learning rate throughout training to strike a balance between stability and rapid convergence, and momentum optimization, which incorporates a momentum term to hasten convergence.

The training procedure has been further enhanced by recent developments in optimization algorithms such as Adam, Adagrad, and RMSprop, which modify learning rates built on the gradient history. These algorithms provide quicker convergence and better handling of various data kinds and architectural configurations. Deep learning models may learn intricate patterns and representations from sizable datasets by utilizing these training algorithms and optimization methods, which has led to significant developments in AI applications. The practical applications of deep learning – namely speech recognition, image acknowledgment, natural language processing, and recommendation systems – as well as other areas where deep learning has significantly advanced the science of artificial intelligence – will be discussed in the following sections.

3.3 REAL-WORLD APPLICATIONS

Deep learning has revolutionized numerous real-world applications across diverse domains, having a profound impact on our lives. It provides precise facial recognition, object detection, and image recognition systems in computer vision. Deep learning models have made substantial advancements in chatbots, sentiment analysis, and language translation in Natural Language Processing (NLP). Speech recognition, powered by deep learning, allows for seamless interactions with virtual assistants like Siri and Alexa. Recommendation systems on platforms like Netflix and Amazon use deep learning to personalize content and product suggestions. Autonomous vehicles rely on deep learning algorithms for safe self-driving capabilities. In healthcare, deep learning aids medical diagnosis through accurate image analysis. Financial services benefit from deep learning for fraud detection and risk assessment. Gaming experiences have been enhanced by deep reinforcement learning achieving superhuman performance. Moreover, deep learning contributes to creative endeavors by generating art, music, and poetry. These applications, along with drug discovery, remote monitoring, and industrial automation, demonstrate the versatility and transformative power of deep learning, promising even more innovative applications in the future.

Deep learning, as a subfield of Artificial Intelligence (AI), has found numerous applications in real-life scenarios, revolutionizing various industries and domains. Here are some notable applications where deep learning has made a significant impact:

- Computer vision: Deep learning has transformed computer vision by enabling accurate and efficient image and video analysis. Applications include:
  a) Object Detection: Deep learning models can identify and localize objects within images or videos, enabling applications like autonomous vehicles, surveillance systems, and facial recognition technology.
  b) Image Classification: Deep learning algorithms can accurately classify images into different categories, enabling applications such as medical imaging diagnosis, quality control in manufacturing, and visual search engines.
  c) Image Generation: Deep learning models like Generative Adversarial Networks (GANs) can generate realistic images, leading to applications in computer graphics, virtual reality, and artistic creation.
- Natural Language Processing (NLP): Deep learning has significantly advanced language processing tasks, leading to improvements in applications such as:
  a) Machine Translation: Deep learning models have made significant progress in machine translation, providing more accurate and fluent translations between languages.
  b) Sentiment Analysis: Applications like social media monitoring, brand reputation management, and customer feedback analysis are made possible by deep learning algorithms' ability to analyse language and ascertain the sentiment communicated.
  c) Speech Recognition: Deep learning has improved speech recognition systems, leading to applications such as voice assistants, transcription services, and Interactive Voice Response (IVR) systems.
- Healthcare: Deep learning has made significant contributions to the healthcare industry, enabling more accurate diagnosis, personalized treatment, and drug discovery. Some applications include:
  a) Medical Image Analysis: Deep learning models can analyze medical images, assisting radiologists in diagnosing diseases such as cancer, identifying anomalies, and detecting patterns in medical imaging data.
  b) Disease Prediction: Deep learning algorithms can analyze patient data and medical records to predict the risk of diseases, enabling early interventions and personalized treatment plans.
  c) Drug Discovery: Deep learning models are utilized to analyze large-scale genomic and proteomic data, accelerating the discovery and development of new drugs and treatment strategies.
- Finance: Deep learning has found applications in the finance industry, enhancing risk management, fraud detection, and algorithmic trading. Examples include:
  a) Fraud Detection: Deep learning models can analyze transactional data and detect patterns indicative of fraudulent activities, enabling early detection and prevention of financial fraud.

b) Risk Assessment: Deep learning algorithms can analyze financial data, market trends, and economic indicators to assess risks and make informed investment decisions.

c) Algorithmic Trading: Deep learning models can examine enormous volumes of financial data, find patterns, and use those patterns to anticipate future events and guide automated trading strategies.

3.3.1 DEEP LEARNING IN IMAGE RECOGNITION

Image recognition has undergone a revolution thanks to deep learning, which outperforms conventional computer vision methods in terms of performance and accuracy. For this task, Convolutional Neural Networks (CNNs) are especially effective. Utilizing a hierarchical structure, CNNs learn progressively more complicated features from unprocessed pixel input. In order to down sample and capture spatial linkages, they use pooling layers in conjunction with convolutional layers to extract local patterns [1]. CNNs are able to correctly recognize and categorize objects in images by using several abstraction layers. Applications for deep learning in image identification include everything from facial recognition and scene comprehension to object detection and image segmentation [2] [3].

3.3.2 NATURAL LANGUAGE PROCESSING AND DEEP LEARNING

Natural Language Processing (NLP), is another area where deep learning has achieved significant advancements. Transformer models and Recurrent Neural Networks (RNNs) are two instances of deep learning models that have considerably improved language production and understanding. RNNs excel at tasks like language translation [4], sentiment analysis [5], and speech recognition [6] due to their capacity to capture sequential dependencies. By capturing long-range relationships and facilitating effective parallelization, transformer models, in particular the well-known transformer architecture [7], have completely changed NLP. The development of machine translation [7], text summarization [8,], question answering [9], and dialogue systems [10] has all been aided by these approaches.

3.3.3 SPEECH RECOGNITION AND DEEP LEARNING

Deep learning has revolutionized speech recognition, allowing machines to accurately translate spoken words into written text. Recurrent Neural Networks (RNNs) and techniques like Long Short-Term Memory (LSTM) have been highly effective in modelling temporal correlations in voice data [11]. Convolutional Neural Networks (CNNs) have also been used to extract relevant information from audio spectrograms [12]. Systems for speech recognition that use deep learning have a variety of uses, including voice assistants, transcription services, voice-controlled technology, and more [13].

3.3.4 RECOMMENDATION SYSTEMS EMPOWERED BY DEEP LEARNING

Deep learning has improved the skills of recommendation systems, which offer tailored advice based on user preferences and behaviour. Complex user-item interactions have proven to be easy to capture when collaborative filtering and deep learning architectures are used. Deep learning algorithms have the ability to produce precise suggestions by utilizing both explicit user ratings and implicit feedback data [14]. They have been employed with success in a variety of industries, including e-commerce, content streaming services, and targeted advertising [15].

3.3.5 Breakthroughs in autonomous vehicles with deep learning

Autonomous vehicles can now perceive and comprehend their surroundings thanks in large part to deep learning. CNNs in particular, which are deep neural networks, are very good at tasks like scene interpretation, object detection, and lane detection [16]. Deep learning models are capable of making accurate decisions for navigation, obstacle avoidance, and path planning by analyzing real-time sensor data such as pictures and LiDAR point clouds [17]. Deep learning algorithms have substantially accelerated the development of self-driving cars and autonomous drones.

3.3.6 Deep learning's impact on medical diagnosis

Deep learning has demonstrated considerable potential in the field of medical diagnosis, providing precise and effective solutions across a range of healthcare applications. CNNs have been used in the processing of medical images to help find illnesses, tumors, and other abnormalities in radiological scans [18]. RNNs have been used to analyze time series data from patient monitoring and ElectroCardioGrams (ECGs) in order to identify anomalies and forecast the evolution of disease [19]. The ability to personalize therapy recommendations based on genetic data analysis provided by deep learning algorithms has also advanced precision medicine [20].

3.3.7 Intelligent personal assistants: a deep learning frontier

Deep learning methodologies have improved the sophistication of voice-activated personal assistants like Siri, Alexa, and Google Assistant. These assistants use deep neural networks, speech recognition, and natural language processing to understand user requests and commands, provide accurate answers, and perform tasks on the user's behalf. They become essential tools for daily chores, knowledge retrieval, and smart home control because of deep learning's ability to continuously enhance their understanding and responsiveness [21].

The aforementioned applications are but a small sample of the extensive influence deep learning has had on numerous real-world sectors. Large-scale data processing and complicated pattern extraction capabilities have revolutionized industries, enhanced decision-making, and allowed computers to carry out previously thought-to-be difficult jobs for conventional AI techniques. Future deep learning applications and breakthroughs should only get better as research in the field advances.

3.4 CHALLENGES AND FUTURE DIRECTIONS

Although revolutionary, deep learning is not without difficulties. The availability and caliber of data represents one of the major obstacles. Large, diverse datasets with labels are essential for deep learning prototypes but can be tough to obtain and prepare. It's also important to address issues with bias in the training data and data privacy. The computational needs of deep learning algorithms provide another difficulty. Further developments in hardware and algorithms are required to get over these restrictions because training deep neural networks requires a lot of computer power. Another key difficulty is the explainability and interpretability of deep learning models. Deep neural networks are regularly mentioned as "black boxes," creating inspiration to realize the reasons behind their expectations. Enhancing the explainability of deep learning models is important for building trust, especially in critical domains where transparency in addition to interpretability is crucial.

Integrating deep learning with other AI disciplines is a promising future direction. Combining deep learning and reinforcement learning has resulted in robust deep reinforcement learning algorithms that have been utilised successfully in robotics and game play. Deep learning and Generative prototypes, such reproductive Adversarial Networks (GANs), can also be implemented to make realistic synthetic data and visuals. Emerging approaches such as transfer learning, meta-learning, and lifelong learning seek to improve the flexibility and effectiveness of deep learning models. Models can take advantage of previously learned information through transfer learning, which also expedites training with scant labelled data.

Meta-learning examines how to better learn new skills and information by understanding how to learn new skills and knowledge. The aim of lifelong learning is to make models that are always adaptable to new situations and jobs without losing prior knowledge. Deep learning is developing, thus it's critical to think about the societal effects and ethical ramifications. To ensure responsible and advantageous use of deep learning technology, it is essential to address issues of bias, fairness, accountability, and privacy. Overall, despite the fact that deep learning has made significant progress, overcoming obstacles like data accessibility, interpretability, and computational demands as well as examining potential future directions will help deep learning reach its full potential and ensure its success in artificial intelligence.

3.4.1 DATA CHALLENGES IN DEEP LEARNING

Deep learning relies heavily on large-scale datasets for training, while having great results in many different fields. Deep learning models' capacity to perform well depends critically on the quantity, caliber, and variety of data available. It can be difficult to gather and prepare such datasets, though. Acquiring labelled data, protecting data privacy, and addressing bias issues in the training data are challenges in data gathering. When using big data for deep learning applications, it's crucial to keep ethical concerns about data utilization and potential biases in mind [22] [23].

3.4.2 COMPUTATIONAL REQUIREMENTS

Deep learning methods demand a lot of computer power for both inference and training. Processing enormous volumes of data during deep learning training generally entails conducting several iterations to improve model performance. Deep learning calculations have been significantly accelerated by Graphics Processing Units (GPUs) and specialist accelerators such as Tensor Processing units (TPUs) by providing parallel processing and high-performance computing [24] [25]. However, the ever-increasing complexity of deep learning models makes further hardware advancements and efficient algorithms necessary to overcome computing challenges.

3.4.3 DISTRIBUTED DEEP LEARNING

Distributed deep learning has become a possible remedy to deal with deep learning's computing needs. Distributed deep learning enables the parallel processing and training of enormous neural networks by dividing the training procedure among numerous computers or nodes. Researchers and practitioners can take advantage of the capacity of distributed computing clusters for deep learning model training thanks to frameworks like TensorFlow and PyTorch [26] [27]. In addition to speeding up training, distributed deep learning makes it possible to investigate more complex model architectures and datasets.

3.4.4 EXPLAINABILITY AND INTERPRETABILITY

The models' lack of interpretability and explainability is a significant problem for deep learning. Deep neural networks are frequently regarded as "black boxes," which makes it difficult to comprehend the logic behind their predictions. Concerns are raised by this lack of transparency, especially in important industries like healthcare and finance where interpretability is essential for accountability and decision-making. Attention processes, saliency maps, and model-agnostic interpretability approaches are just a few of the methodologies being actively researched by researchers to develop the explainability of deep learning models [28] [29]. The creation of explicable deep learning models would not only boost user confidence in AI systems but also make ethical and legal compliance easier.

3.4.5 INTEGRATION WITH OTHER AI DOMAINS

Deep learning crosses over with a number of different artificial intelligence fields, creating new opportunities and synergies. Deep reinforcement learning algorithms are the result of the combination of deep learning and reinforcement learning, a sub-section of artificial intelligence that consist of decision-making and control. Innovations in gaming, robotics, and autonomous systems have resulted from this convergence [30] [31]. In order to produce realistic synthetic data and images, deep learning has also affected generative prototypes, such as Generative Adversarial Networks (GANs) [32]. Deep learning integration with other AI fields has the potential to produce systems that are smarter and more inventive.

3.4.6 FUTURE DIRECTIONS AND EMERGING TRENDS

Deep learning is a topic that is always changing, and various new developments are influencing how it will develop in the future. Transfer learning, which makes use of previously trained models and expertise from one task to another, enables quicker training and enhanced performance with little labelled data [33]. In order to acquire new abilities and knowledge more quickly, meta-learning investigates how models might learn how to learn [34]. The aim of lifelong learning is to produce models that can continuously pick up new information and adapt to new situations without losing past knowledge [35]. These developments attempt to increase the adaptability, effectiveness, and capacity of deep learning models to learn from sparse data. Additionally, as deep learning technologies spread, it is crucial to think about any potential societal repercussions. To enable the appropriate and advantageous deployment of deep learning in many applications, ethical issues relating to prejudice, fairness, accountability, and privacy must be addressed [36].

3.5 ADVANCEMENTS IN DEEP LEARNING ARCHITECTURES

Deep learning architectures have developed noticeably, resulting in important advances in the artificial intelligence area. By successfully extracting spatial information and patterns from images, Convolutional Neural Networks (CNNs) have revolutionized computer vision and made it possible to undertake tasks such as object detection and image recognition. From autonomous driving to medical imaging, these structures have been crucial in a variety of applications. Recurrent Neural Networks (RNNs) have transformed sequential data processing by enabling the modelling of temporal relationships and streamlining operations like speech and natural language processing. The ability of RNNs to perceive long-term dependencies has been enhanced by complex modifications such Long Short-Term Memory (LSTM) and the Gated Recurrent Unit (GRU). Deep learning has also had a big impact on generative models, allowing for the generation of realistic samples with excellent quality, utilizing techniques like Generative Adversarial Networks (GANs). This has made opportunities in fields such as picture synthesis and data augmentation possible. Agents are now able to learn the best course of action in challenging circumstances because of the merger of deep learning and reinforcement learning. Significant breakthroughs in numerous artificial intelligence fields have been made possible by these developments in deep learning architectures.

3.5.1 CONVOLUTIONAL NEURAL NETWORKS (CNNS) FOR COMPUTER VISION

CNNs have transformed computer vision, allowing for considerable improvements in object identification, object recognition, and visual comprehension. CNNs use convolutional layers, pooling layers, and non-linear activation functions to extract valuable properties from pictures. These designs are created to capture the local patterns and spatial hierarchies found in images, enabling reliable and accurate image classification [37]. CNNs are effective in a number of tasks, including semantic segmentation, object localization, and picture classification because they can automatically learn hierarchical representations [38]. The development of CNN architectures has helped several applications, including autonomous driving, medical imaging, and facial recognition [39] [40].

3.5.2 RECURRENT NEURAL NETWORKS (RNNs) FOR SEQUENTIAL DATA

RNNs have become a potent deep learning architecture for processing sequential input. RNNs, in contrast to conventional feedforward networks, have a recurrent connection that enables them to manage an internal memory and analyze sequential data. They are therefore suitable for jobs such as speech recognition, time series analysis, and natural language processing. By propagating information from prior time steps to future predictions, RNNs are excellent at capturing dependencies and long-term temporal patterns in sequential data [41]. More complex RNN variants, including Long Short-Term Memory (LSTM) and the Gated Recurrent Unit (GRU), address the vanishing gradient problem and enhance the modelling of long-term dependencies [42]. Machine translation, sentiment analysis, speech synthesis, and other fields have advanced thanks to these developments in RNN architectures.

3.5.3 DEEP LEARNING IN GENERATIVE MODELS

The objective of generative models is to generate brand-new data samples that resemble an existing dataset. Deep learning has greatly enhanced generative models by enabling the generation of good and realistic examples. Generative Adversarial Networks (GANs) are a well-known example of deep learning in generative modelling. In GANs, a generator network learns to generate plausible instances while a discriminator network learns to discern between real and fake data. A variety of creative applications, including image synthesis, style transfer, and data augmentation are now possible thanks to the outstanding achievements that GANs have achieved in producing images, music, and text [43].

3.5.4 REINFORCEMENT LEARNING (RL) AND DEEP Q-NETWORKS (DQNs)

The process of teaching agents to choose the optimum course of action after some trial and error in a real-world situation is known as RL. DQNs handle complicated, high-dimensional problems by combining deep neural networks with reinforcement learning. DQNs employ deep neural networks to approximate action-value functions, enabling efficient and scalable reinforcement learning in vast state and act spaces [44]. Significant improvements in gaming, robotic control, and autonomous decision-making have resulted from the implementation of DQNs. Deep reinforcement learning, which includes DQNs, has produced remarkable accomplishments in areas including driving autonomous vehicles, playing Atari games, and mastering challenging board games like Go [45].

3.5.5 TRANSFORMER MODELS: REVOLUTIONIZING NATURAL LANGUAGE PROCESSING

Transformer models introduced a self-attention mechanism that captures contextual dependencies in text input, revolutionizing the arena of Natural Language Processing (NLP). Transformers, in contrast to conventional sequential models, can process incoming text in parallel, greatly enhancing computational performance. Transformer designs, including the well-known transformer model, have surpassed earlier models in tasks like language interpretation, question-answering, machine translation, and text summarization [46]. The concept of NLP is described in Figure 3.2, where

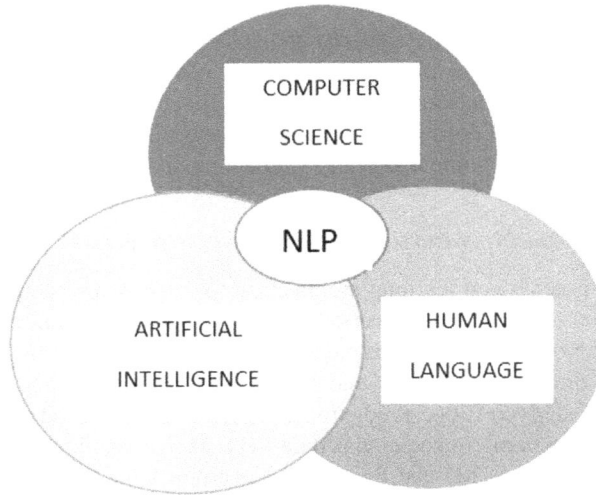

FIGURE 3.2   The concept of NLP and its contextual dependencies.

the contextual dependencies are portrayed. The model's self-attention technique enables it to pay attention to various input sequence segments, capturing global relationships and facilitating more precise and context-aware predictions. Transformer prototypes have develop the de facto norm in NLP and have accelerated improvements in language interpretation and generation tasks thanks to their capacity to grasp long-range dependencies.

3.6   DEEP LEARNING'S POWER IN COMPLEX TASKS

Deep learning has proven to be extremely effective at tackling challenging jobs across a range of artificial intelligence fields. Deep learning models have excelled in solving complex issues due to their capacity to automatically build hierarchical representations and uncover detailed patterns from enormous volumes of data. Deep learning has the advantage of being able to manage complexity, which enables it to take on jobs that were previously thought to be challenging for conventional AI methods. Deep learning has revolutionized picture recognition, semantic segmentation and object detection in fields like computer vision, outperforming traditional methods and achieving previously unheard-of levels of accuracy.

Furthermore, by utilizing recurrent neural networks and transformer topologies, deep learning has revolutionized machine translation, sentiment analysis, and text production, advancing natural language processing. Significant advancements in deep learning have also been made in voice recognition, with recurrent neural networks and their variants enhancing the reliability and accuracy of automatic speech recognition systems. Furthermore, deep learning has improved recommendation systems by enabling more precise and customized recommendations, which has benefited industries including e-commerce, advertising, and video streaming. Overall, the revolutionary impact of deep learning on numerous sectors, which has pushed the envelope of artificial intelligence and driven breakthroughs, is evidence of the technology's strength in challenging tasks.

3.6.1   DEEP LEARNING'S ADVANTAGE IN HANDLING COMPLEXITY

Deep learning has become a potent way for handling extremely complicated tasks that were difficult for conventional artificial intelligence techniques in the past. Deep learning models can tackle complicated issues successfully because they can automatically learn hierarchical representations from input. Deep neural networks are particularly good at identifying complex patterns, gleaning information from huge volumes of data, and producing precise estimates. This benefit has elevated deep learning to the top of the pile in a variety of areas, comprising speech recognition, natural language processing, computer vision, and recommendation systems.

3.6.2 DEEP LEARNING'S IMPACT ON IMAGE RECOGNITION

Since deep learning became available, image recognition has undergone a radical change. Image classification, object detection, and semantic segmentation problems have been transformed by Convolutional Neural Networks (CNNs). Hierarchical feature extraction is used by CNNs to extract fine-grained features and spatial correlations from images. In image identification benchmarks, deep learning models have surpassed conventional techniques and attained previously unheard-of levels of accuracy [47]. The impressive developments in deep learning for image identification have significantly helped applications including autonomous driving, medical imaging, and visual surveillance.

3.6.3 NATURAL LANGUAGE PROCESSING REVOLUTIONIZED BY DEEP LEARNING

Natural Language Processing (NLP) has been transformed by deep learning, enabling robots to comprehend and produce human language more effectively. Transformers and Recurrent Neural Networks (RNNs) have been instrumental in the development of NLP. RNNs are effective in capturing contextual dependencies in text, making them suitable for projects such as sentiment analysis, text production, and machine translation. Transformers have excelled in tasks such as language modelling, question-answering, and text summarizing thanks to their self-attention mechanism. The potential for more precise language processing and generation has been unlocked by deep learning models in NLP [48].

3.6.4 SPEECH RECOGNITION: A DEEP LEARNING SUCCESS STORY

Deep learning has revolutionized speech recognition, significantly improving Automatic Speech Recognition (ASR) systems. Recurrent Neural Networks (RNNs) and their derivatives, in particular, have significantly increased the accuracy and robustness of ASR systems. It has made great progress since deep learning models can now learn complicated audio representations and capture long-term temporal connections. Applications like voice assistants, transcription services, and voice-controlled gadgets are now possible because of deep neural networks' superior performance over conventional methods [49].

3.6.5 RECOMMENDATION SYSTEMS ENHANCED BY DEEP LEARNING

Deep learning has improved recommendation systems by making recommendations more precise and individualized. Collaborative filtering, content-based filtering, and hybrid approaches have been greatly improved with the integration of deep learning techniques. Deep recommendation models use neural network technology to develop latent representations of users and products, capturing complex interactions and preferences between users and items. This enables the transmission of recommendations that are more pertinent and unique, improving user pleasure and engagement. In industries including e-commerce, online advertising, and streaming services, deep learning has revolutionized recommendation systems [50].

3.7 THE FUTURE OF DEEP LEARNING IN AI

AI's deep learning technology has a bright future and has the ability to make revolutionary strides. The road to improvements rests in shifting architectures and approaches as researchers and practitioners continue to push its boundaries. The capabilities of deep learning models are anticipated to be further improved through new deep learning architectures and improvements to existing ones. In order to tackle more complicated tasks, architectures such as Convolutional Neural Networks (CNNs) and Recurrent Neural Networks (RNNs) have evolved. More robust and effective deep learning models are probably also influenced by attention mechanisms, self-supervised learning, and unsupervised learning strategies.

Future research on hybrid designs that incorporate the benefits of many methodologies is also looking promising. Another critical component of deep learning's future is overcoming obstacles. Since deep learning models require a lot of high-quality, labelled data for training, data availability, preprocessing, and labelling continue to be major obstacles. For further development, it is essential to tackle biases in training datasets, handle limited labelled data, and improve data collection techniques. To scale deep learning to larger models and deploy it on devices with limited resources, it will also be essential to address the computing needs of deep learning algorithms and optimize their energy efficiency.

The future of deep learning is significantly influenced by ethical issues. Addressing biases, guaranteeing fairness, and encouraging responsibility are crucial as deep learning becomes more incorporated into systems. Building trust and transparency in deep learning systems requires minimizing biases and creating methods for model interpretability and explainability. The future has great possibilities for combining deep learning with other AI fields. It is possible to create more intelligent and adaptable systems by combining deep learning with methodologies such as reinforcement learning, generative models, and symbolic reasoning. Agents may learn from experience and make the best judgements in dynamic contexts thanks to deep reinforcement learning.

Generative models help create synthetic data and improve data augmentation methods. Deep learning and symbolic reasoning together can produce AI systems that are easier to understand and interpret. These fields will be combined to create a more complete and all-encompassing intellect in AI. Deep learning will have a significant influence on the direction of AI. Because of its capacity for handling complexity, learning from data, and making precise predictions, it serves as a key tenet of AI systems. Deep learning will allow AI to tackle more complicated issues in a variety of industries, including healthcare, banking, and transportation, as technological breakthroughs continue. The way we live and engage with technology will change as a result of the revolution in personalized medicine, autonomous systems, intelligent assistants, and several other uses. Deep learning's potential-rich future in AI will continue to propel the development and transformation of AI in the years to come.

3.7.1 THE PATH TO ADVANCEMENTS: EVOLVING ARCHITECTURES AND TECHNIQUES

Artificial Intelligence (AI) advances in design and methodology have a bright future thanks to deep learning. By creating new architectures and improving old ones, researchers and practitioners continue to push the limits of deep learning. Two examples of architectures that are probably going to keep evolving are Convolutional Neural Networks (CNNs) and Recurrent Neural Networks (RNNs), which will increase their ability to tackle difficult jobs. [51] [52]. Additionally, improvements in attention processes, self-supervised learning, and unsupervised learning strategies are anticipated to result in deeper learning models that are more reliable and effective [53] [54]. Future research on deep learning should consider hybrid architectures, which integrate the advantages of several deep learning methods [55].

3.7.2 OVERCOMING CHALLENGES IN DEEP LEARNING

It is crucial to address the issues raised by deep learning as it develops. Deep learning models need a lot of high-quality, labelled data for training; therefore data availability, preprocessing, and labelling are still major problems. For further development, it is imperative to work to enhance data collection procedures, provide strategies for managing sparsely labelled data, and address biases in training datasets [56] [57]. In order to scale deep learning to bigger models and deploy them on resource-constrained devices, it will also be essential to address the computing needs of deep learning algorithms and optimize their energy efficiency [58] [59].

3.7.3 ETHICAL CONSIDERATIONS: BIAS, FAIRNESS, AND ACCOUNTABILITY

Ethics related to prejudice, justice, and accountability are crucial as deep learning is progressively incorporated into a variety of applications. Deep learning methods may unintentionally reinforce training data biases, producing discriminating results. Addressing biases and creating methods for model interpretability and explainability are necessary for ensuring fairness, transparency, and accountability in deep learning systems [60] [61]. To reduce biases and advance ethical deep learning practices, researchers and practitioners must take a proactive role in conversations and projects.

3.7.4 COMBINING DEEP LEARNING WITH OTHER AI DOMAINS

The convergence of deep learning and other AI areas is where the future of AI resides. For the development of more intelligent and adaptable systems, combining deep learning with reinforcement learning, generative models, and symbolic reasoning approaches offers enormous potential. Deep reinforcement learning can help agents learn from their mistakes and choose the best course of action in challenging situations [62]. The creation of synthetic data and the improvement of data augmentation methods can be aided by generative models, such as Generative Adversarial Networks (GANs) [63]. Additionally, combining deep learning with symbolic reasoning and logic-based techniques can result in AI systems that are easier to understand and interpret [64]. These fields will be combined to create a more complete and all-encompassing intellect in AI.

3.7.5 THE ROLE OF DEEP LEARNING IN SHAPING THE FUTURE OF AI

Deep learning is expected to be crucial in determining the direction of AI. Because of its capacity for handling complexity, learning from data, and making precise predictions, it serves as a key tenet of AI systems. AI will be able to address more complicated problems across a range of industries, including healthcare, transportation, finance, and other areas, thanks to advances in deep learning. Deep learning will revolutionize the way we live and engage with technology by having an impact on personalized medicine, autonomous systems, intelligent assistants, and more [65]. A detailed application of deep learning in AI is portrayed in Table 3.1. Deep learning will continue to be the driving force behind the development and transformation of AI by continuously improving deep learning architectures, addressing issues, promoting moral behaviour, and connecting with other AI areas.

A detailed application of deep learning in AI is portrayed in Table 3.1. Deep learning will continue to be the driving force behind the development and transformation of AI by continuously improving deep learning architectures, addressing issues, promoting moral behavior, and connecting with other AI areas.

3.8 CONCLUSION

This chapter offers a thorough introduction to the book "Deep Learning: From Theory to Reality," which examines the numerous facets and uses of deep learning. It also covers a wide range of subjects, from basic concepts to modern applications and future directions, making it a helpful tool for both inexperienced and seasoned researchers in the field of artificial intelligence and machine learning.

TABLE 3.1
Deep Learning Future in AI

Explainable AI	Addressing the "black box" nature of deep learning prototypes, research emphasis on making AI decisions explainable as well as understandable. This improves trust in addition to accountability in AI systems.
Few-Shot and Zero-Shot Learning	Advancements in transfer learning techniques allow models to learn from fewer labelled examples otherwise even without any labelled data, making AI added data-efficient and adaptable.
Self-Supervised Learning	This learning technique leverages unlabelled data to pre-train models, reducing the requirement for large-labelled datasets and also refining generalization to new tasks.
Continual Learning	Enabling continual learning from new data in AI systems without catastrophic loss of previously learned knowledge, enhancing the capacity for lifelong learning.
Federated Learning	Maintaining data privacy while facilitating model training across distant devices, which is crucial for applications in the IoT, healthcare, and other sensitive fields.
Neuromorphic Computing	Specialized hardware architectures that mimic neural processes, inspired by the human brain, enable quicker and more energy-efficient deep learning calculations.
Quantum Computing	Using quantum methods to speed up deep learning tasks like optimisation and matrix operations could lead to previously unheard-of performance improvements.
Human-AI Collaboration	Augmenting human intelligence through AI systems also fostering synergistic interactions for problem-solving as well as creativity across various domains.
Ethical and Fair AI	Addressing biases as well as ensuring impartiality and inclusivity in AI applications, involving various perspectives and societal values in AI development.
Hybrid and Multi-Modal AI	Combining multiple AI techniques, such as deep learning, reinforcement learning, as well as symbolic reasoning, to produce more robust and also versatile AI systems.
AI for Sustainability	The application of AI to address environmental issues, improve resource allocation, and support sustainable practices across multiple industries.
AI for Social Good	Utilizing deep learning to tackle societal and humanitarian concerns, such as disaster response, healthcare, education, and poverty alleviation.
Adversarial Robustness	Creating AI models that can withstand malicious data tampering and endure adversarial attacks.
Integration with Robotics	Combining robotics and deep learning to enable more complex and context-aware interactions between AI systems and real-world surroundings.
AI Governance and Regulations	Establishing rules and regulations to control AI deployment and development, assuring its ethical application and reducing any potential hazards.
AI in Creativity and Arts	Exploring AI-generated literature, music, art, and other creative works, redefining human-machine creative cooperation.

The significance of deep learning as a ground-breaking technique is highlighted in the introduction part, laying the groundwork for the next chapters. In Chapter 2, where neural networks are introduced as deep learning's underlying technology, the fundamentals of deep learning are studied. In order to fully realise the learning potential of deep neural networks, this chapter is important in order to explore the training algorithms, architectural elements, and optimization strategies.

Deep learning is examined in detail in Chapter 3's real-world applications, which include speech recognition, autonomous vehicles, medical diagnostics, picture and natural language processing, recommendation systems, and intelligent personal assistants. The variety of uses highlights deep learning's adaptability and potential to revolutionize numerous industries. Chapter 4 discusses deep learning's difficulties and potential possibilities, focusing on concerns with data, processing demands, distributed learning, explainability, and integration with other AI areas. The chapter also looks at new trends and potential paths, showing how deep learning as a subject is always evolving.

Convolutional Neural Networks (CNNs), generative models, Recurrent Neural Networks (RNNs), reinforcement learning with Deep Q-Networks (DQNs), and transformer models are the main topics of Chapter 5's discussion of advances in deep learning architectures. The advancement of the capabilities of deep learning algorithms in certain applications has been greatly aided by these architectural advancements. By presenting its advantages and accomplishments in a range of domains, including as image recognition, speech recognition, natural language processing, and recommendation systems, Chapter 6 emphasises deep learning's capacity to solve complex issues. With state-of-the-art performance, deep learning has revolutionized various fields, as seen by the examples provided.

In Chapter 7, the future of deep learning in AI is explored. Topics covered include the development of architectures and methodologies, overcoming obstacles, ethical issues, fusing deep learning with other AI domains, and deep learning's influence on the direction of AI. This chapter's prospective nature draws attention to deep learning's potential to propel future developments in AI. Overall, "Deep Learning: From Theory to Reality" offers a thorough, perceptive, and current overview of deep learning and its revolutionary effects on numerous AI domains. The book is an invaluable tool for researchers, practitioners, and anyone interested in comprehending and utilizing the power of deep learning in the field of artificial intelligence because it covers fundamental concepts, practical applications, advancements in architectures, and future prospects.

REFERENCES

[1] LeCun, Y. et al., "Gradient-Based Learning Applied to Document Recognition," *Proceedings of the IEEE*, vol. 86, no. 11, pp. 2278–2324, Nov. 1998.

[2] Krizhevsky, A. et al., "ImageNet Classification with Deep Convolutional Neural Networks," in *Advances in Neural Information Processing Systems (NIPS)*, 2012.

[3] He, K. et al., "Deep Residual Learning for Image Recognition," in *IEEE Conference on Computer Vision and Pattern Recognition (CVPR)*, 2016.

[4] Sutskever, I. et al., "Sequence to Sequence Learning with Neural Networks," in *Advances in Neural Information Processing Systems (NIPS)*, 2014.

[5] Socher, R. et al., "Recursive Deep Models for Semantic Compositionality Over a Sentiment Treebank," in *Proceedings of the 2013 Conference on Empirical Methods in Natural Language Processing (EMNLP)*, 2003.

[6] Graves, A. et al., "Speech Recognition with Deep Recurrent Neural Networks," in *IEEE International Conference on Acoustics, Speech and Signal Processing (ICASSP)*, 2003.

[7] Vaswani, A. et al., "Attention Is All You Need," in *Advances in Neural Information Processing Systems (NIPS)*, 2017.

[8] See, A. et al., "Get To The Point: Summarization with Pointer-Generator Networks," in *Proceedings of the 55th Annual Meeting of the Association for Computational Linguistics (ACL)*, 2017.

[9] Devlin, J. et al., "BERT: Pre-training of Deep Bidirectional Transformers for Language Understanding," in *Proceedings of the 2019 Conference of the North American Chapter of the Association for Computational Linguistics (NAACL)*, 2019.

[10] Vinyals, O. et al., "A Neural Conversational Model," in *Proceedings of the 32nd International Conference on Machine Learning (ICML)*, 2015.

[11] Hochreiter, S. and Schmidhuber, J. "Long Short-Term Memory," *Neural Computation*, vol. 9, no. 8, pp. 1735–1780, Nov. 1997.

[12] Hannun, A. et al., "Deep Speech: Scaling Up End-to-End Speech Recognition," arXiv preprint arXiv:1412.5567, 2014.

[13] Hinton, G. et al., "Deep Neural Networks for Acoustic Modeling in Speech Recognition," *IEEE Signal Processing Magazine*, vol. 29, no. 6, pp. 82–97, Nov. 2012.

[14] He, Y. et al., "Neural Collaborative Filtering," in *Proceedings of the 26th International Conference on World Wide Web (WWW)*, 2017.

[15] Koren, Y. et al., "Matrix Factorization Techniques for Recommender Systems," *IEEE Computer*, vol. 42, no. 8, pp. 30–37, Aug. 2009.

[16] Bojarski, M. et al., "End to End Learning for Self-Driving Cars," arXiv preprint arXiv:1604.07316, 2016.

[17] Chen, Y. et al., "DeepDriving: Learning Affordance for Direct Perception in Autonomous Driving," in *IEEE International Conference on Computer Vision (ICCV)*, 2015.

[18] Esteva, A. et al., "Dermatologist-level Classification of Skin Cancer with Deep Neural Networks," *Nature*, vol. 542, no. 7639, pp. 115–118, Feb. 2017.

[19] Rajkomar, A. et al., "Scalable and Accurate Deep Learning for Electronic Health Records," *NPJ Digital Medicine*, vol. 1, no. 1, pp. 18, 2018.

[20] Poplin, R. et al., "Prediction of Cardiovascular Risk Factors from Retinal Fundus Photographs via Deep Learning," *Nature Biomedical Engineering*, vol. 2, no. 3, pp. 158–164, Mar. 2018.

[21] Graves, A. et al., "Towards End-to-End Speech Recognition with Recurrent Neural Networks," in *IEEE International Conference on Machine Learning (ICML)*, 2014.

[22] Mitchell, M. et al., "Model Cards for Model Reporting," in *Proceedings of the Conference on Neural Information Processing Systems (NeurIPS)*, 2018.

[23] Bolukbasi, T. et al., "Man is to Computer Programmer as Woman is to Homemaker? Debiasing Word Embeddings," in *Proceedings of the Conference on Neural Information Processing Systems (NeurIPS)*, 2016.

[24] Goodfellow, I. J. et al., "Deep Learning," *Nature*, vol. 521, no. 7553, pp. 436–444, May 2015.

[25] Jouppi, J. et al., "In-Datacenter Performance Analysis of a Tensor Processing Unit," in *Proceedings of the 44th Annual International Symposium on Computer Architecture (ISCA)*, 2017.

[26] Abadi, M. et al., "TensorFlow: A System for Large-Scale Machine Learning," in *Proceedings of the 12th USENIX Symposium on Operating Systems Design and Implementation (OSDI)*, 2016.

[27] Paszke, A. et al., "PyTorch: An Imperative Style, High-Performance Deep Learning Library," in *Advances in Neural Information Processing Systems (NIPS)*, 2019.

[28] Selvaraju, V. et al., "Grad-CAM: Visual Explanations from Deep Networks via Gradient-Based Localization," in *Proceedings of the IEEE International Conference on Computer Vision (ICCV)*, 2017.

[29] Ribeiro, M. T. et al., "Why Should I Trust You? Explaining the Predictions of Any Classifier," in *Proceedings of the ACM SIGKDD International Conference on Knowledge Discovery and Data Mining (KDD)*, 2016.

[30] Mnih, V. et al., "Human-level Control through Deep Reinforcement Learning," *Nature*, vol. 518, no. 7540, pp. 529–533, Feb. 2015.

[31] LeCun, Y. et al., "Deep Learning," *Nature*, vol. 521, no. 7553, pp. 436–444, May 2015.

[32] Goodfellow, I. J. et al., "Generative Adversarial Nets," in *Advances in Neural Information Processing Systems (NIPS)*, 2014.

[33] Bengio, Y. "Deep Learning of Representations: Looking Forward," in *Proceedings of the International Joint Conference on Artificial Intelligence (IJCAI)*, 2003.

[34] Finn, C. et al., "Model-Agnostic Meta-Learning for Fast Adaptation of Deep Networks," in *Proceedings of the International Conference on Machine Learning (ICML)*, 2017.

[35] Li, Z. and Hoiem, D. "Learning Without Forgetting," *IEEE Transactions on Pattern Analysis and Machine Intelligence (TPAMI)*, vol. 40, no. 12, pp. 2935–2947, Dec. 2018.

[36] Mitchell, T. et al., "Artificial Intelligence: A Report of the 2019 AAAI/ACM Conference on AI, Ethics, and Society," *ACM SIGAI Ethics and Artificial Intelligence*, vol. 1, no. 1, Article 8, Jan. 2020.

[37] LeCun, Y. et al., "Gradient-Based Learning Applied to Document Recognition," *Proceedings of the IEEE*, vol. 86, no. 11, pp. 2278–2324, Nov. 1998.

[38] Krizhevsky, A. et al., "ImageNet Classification with Deep Convolutional Neural Networks," in *Advances in Neural Information Processing Systems (NIPS)*, 2012.

[39] He, K. et al., "Deep Residual Learning for Image Recognition," in *IEEE Conference on Computer Vision and Pattern Recognition (CVPR)*, 2016.

[40] Everingham, M. et al., "The PASCAL Visual Object Classes Challenge: A Retrospective," *International Journal of Computer Vision*, vol. 111, no. 1, pp. 98–136, Jan. 2015.

[41] Hochreiter, S. and Schmidhuber, J. "Long Short-Term Memory," *Neural Computation*, vol. 9, no. 8, pp. 1735–1780, Nov. 1997.

[42] Cho, K. et al., "Learning Phrase Representations using RNN Encoder-Decoder for Statistical Machine Translation," in *Proceedings of the Conference on Empirical Methods in Natural Language Processing (EMNLP)*, 2014.

[43] Goodfellow, I. et al., "Generative Adversarial Networks," in *Advances in Neural Information Processing Systems (NIPS)*, 2014.

[44] Mnih, V. et al., "Human-level Control through Deep Reinforcement Learning," *Nature,* vol. 518, no. 7540, pp. 529–533, Feb. 2015.

[45] Silver, D. et al., "Mastering the Game of Go with Deep Neural Networks and Tree Search," *Nature*, vol. 529, no. 7587, pp. 484–489, Jan. 2016.

[46] Vaswani, A. et al., "Attention Is All You Need," in *Advances in Neural Information Processing Systems (NIPS)*, 2017.

[47] LeCun, Y. et al., "Gradient-Based Learning Applied to Document Recognition," *Proceedings of the IEEE*, vol. 86, no. 11, pp. 2278–2324, Nov. 1998.

[48] Vaswani, A. et al., "Attention Is All You Need," in *Advances in Neural Information Processing Systems (NIPS)*, 2017.

[49] Hinton, G. et al., "Deep Neural Networks for Acoustic Modeling in Speech Recognition: The Shared Views of Four Research Groups," *IEEE Signal Processing Magazine,* vol. 29, no. 6, pp. 82–97, Nov. 2012.

[50] Cheng, H. et al., "Wide & Deep Learning for Recommender Systems," in *Proceedings of the ACM Conference on Recommender Systems (RecSys)*, 2016.

[51] LeCun, Y. et al., "Gradient-Based Learning Applied to Document Recognition," *Proceedings of the IEEE*, vol. 86, no. 11, pp. 2278–2324, Nov. 1998.

[52] Hochreiter, S. and Schmidhuber, J. "Long Short-Term Memory," *Neural Computation*, vol. 9, no. 8, pp. 1735–1780, Nov. 1997.

[53] Vaswani, V. et al., "Attention Is All You Need," in *Advances in Neural Information Processing Systems (NIPS)*, 2017.

[54] Chen, J. et al., "Big Self-Supervised Models Are Strong Semi-Supervised Learners," in *Advances in Neural Information Processing Systems (NIPS)*, 2020.

[55] Bengio, Y. "Deep Learning of Representations: Looking Forward," in *Statistical Language and Speech Processing, Lecture Notes in Computer Science*, 2003.

[56] Doshi-Velez, F. and Kim, B. "Towards a Rigorous Science of Interpretable Machine Learning," arXiv preprint arXiv:1702.08608, 2017.

[57] Bolukbasi, T. et al., "Man is to Computer Programmer as Woman is to Homemaker? Debiasing Word Embeddings," in *Proceedings of the Conference on Neural Information Processing Systems (NeurIPS)*, 2016.

[58] Goodfellow, I. J. et al., "Deep Learning," *Nature*, vol. 521, no. 7553, pp. 436–444, May 2015.

[59] Redmon, J. and Farhadi, A. "YOLO9000: Better, Faster, Stronger," in *Proceedings of the IEEE Conference on Computer Vision and Pattern Recognition (CVPR)*, 2017.

[60] Lakkaraju, H. et al., "Identifying Unknown Unknowns in the Open World: Representations and Policies for Guided Exploration," in *Proceedings of the AAAI Conference on Artificial Intelligence (AAAI)*, 2017.

[61] Selvaraju, V. et al., "Grad-CAM: Visual Explanations from Deep Networks via Gradient-Based Localization," in *Proceedings of the IEEE International Conference on Computer Vision (ICCV)*, 2017.

[62] Mnih, V. et al., "Human-level Control through Deep Reinforcement Learning," *Nature*, vol. 518, no. 7540, pp. 529–533, Feb. 2015.

[63] Goodfellow, I. et al., "Generative Adversarial Networks," in Advances in Neural Information Processing Systems (NIPS), 2014.

[64] Pearl, J. M. "Reasoning with Cause and Effect," in *Proceedings of the AAAI Conference on Artificial Intelligence (AAAI)*, 2018.

[65] Topol, E. "High-performance Medicine: The Convergence of Human and Artificial Intelligence," *Nature Medicine,* vol. 25, no. 1, pp. 44–56, Jan. 2019.

4 Unleashing the Power

Exploring Deep Learning Architecture for Cutting-Edge AI Solutions

Suman Patra

4.1 INTRODUCTION

The realm of Artificial Intelligence (AI) is experiencing a profound shift caused by a ground-breaking method called deep learning, which unlocks the potential for tackling complex problems in various domains. Multi-layered neural networks are utilized in deep learning, a subfield of machine learning, which can extract and learn elaborate patterns from massive amounts of data. Deep learning has revolutionized disciplines such as natural language processing, speech recognition, and computer vision by automatically identifying significant representations (Gupta 2021).

The foundation of deep learning lies in its intricate architecture which plays an indispensable role in improving AI system outcomes. The sophisticated and hierarchically structured layers within deep neural networks enable the capture of intricate connections and the accurate prediction of outcomes. A thorough understanding of deep learning's architecture is key to creating AI solutions that exceed previously imagined possibilities (Sarker 2021).

This article aims to explore the fascinating world of deep learning architecture, elaborating on its significance when developing advanced AI systems and scrutinizing fundamental components of deep learning networks such as recurrent layers, convolutional layers, and attention mechanisms. Furthermore, it will provide insight into recent breakthroughs in the architecture of deep learning, including optimization algorithms, regularization techniques, and new network architectures. These advancements have enabled even greater capabilities for AI models.

By the end of this chapter, readers will have learned about deep learning architecture's fundamental aspects and importance in achieving AI advancements. Whether you are a practitioner, researcher, or just fascinated by AI, this chapter provides valuable insights into the foundational building blocks of cutting-edge AI systems and how deep learning architecture influences their performance (Liu, et al. 2022).

4.2 UNDERSTANDING DEEP LEARNING

Artificial Intelligence (AI) pertains to the advancement of computing systems that can carry out actions which commonly necessitate human intelligence (West and Allen 2018). Machine Learning (ML) is a branch of AI that emphasizes the advancement of procedures and patterns that enable computers to learn from data and generate anticipations or determinations.

DOI: 10.1201/9781003433309-4

4.2.1 DEEP LEARNING AND ITS SUBFIELDS

Deep Learning is a division within the domain of Machine Learning which takes inspiration from the organization and function of the human brain (Janiesch, Zschech and Heinrich 2021). Its objective is to fabricate synthetic neural networks capable of mimicking the brain's learning process, allowing electronic devices to execute complicated duties with minimal or no human involvement.

Deep Learning is identified using deep neural networks, a group of various layers of interconnected artificial neurons. Each layer receives and processes input data before transmitting it to the following layer, gradually extracting complex features from the primary data. The progressive design makes it possible to learn hierarchical representations, enabling the network to understand intricate patterns and make precise predictions.

Deep learning encompasses numerous subdivisions, including Convolutional Neural Networks (CNNs) for image and video processing, Recurrent Neural Networks (RNNs) for analyzing sequential data, Generative Adversarial Networks (GANs) for fabricating realistic data, and Reinforcement Learning (RL) for training agents through interaction with an environment (Boesch n.d.).

4.2.2 EVOLUTION OF DEEP LEARNING

The origins of deep learning date back to the 1940s when the first artificial neural networks were developed. However, significant advancements were not made until the 1980s and 1990s, when key breakthroughs were made, such as the creation of the backpropagation algorithm and efficient hardware. These developments sparked a renewed interest in deep learning that culminated in a pivotal moment in 2012 when the deep convolutional neural network called AlexNet won the ImageNet competition, vastly improving image recognition, and igniting a revolution that impacted fields like natural language processing, computer vision, speech recognition, and more (Saxena 2021).

Driving this progress has been the availability of powerful GPUs, large datasets, and the rapid growth of computing power. Within its diverse applications, deep learning models have demonstrated astoundingly accurate predictions and have surpassed human performance in certain tasks.

The intrinsic foundation of deep learning resides in the compound structure of synthetic neural networks which mimic the cognitive learning mechanisms of the human mind. This methodology has effectively propelled the prevalence of deep learning as an integral component of machine learning, facilitating the ability of machines to master intricate tasks and yield accurate prognostications within numerous domains.

4.3 KEY COMPONENTS OF DEEP LEARNING ARCHITECTURE

The structure of machine learning that is entrenched in deep learning architecture is founded on a particular set of Artificial Neural Networks (ANNs), which boast a plethora of layers (Madan 2020). ANNs endeavor to simulate the complex functions of the human brain by handling copious amounts of intricate patterns and data. This text intends to inspect the fundamental underpinnings of the deep learning architecture, encompassing neural networks, layers, connections, and training mechanisms.

4.3.1 NEURAL NETWORKS

4.3.1.1 Artificial neurons

Elements recognized as nodes or perceptrons and also known as artificial neurons are fundamental aspects of neural networks (Nabriya 2021). Various sources transmit inputs to these components, where each input is then multiplied by its respective weight followed by being transmitted through an activation function. Evaluating the output value, which is then assessed by the activation function determines whether the neuron should remain quiescent or fire.

4.3.1.2   Activation functions

The intricate relationships in data are learned by neural networks through activation functions that add non-linearities. There are a number of commonly used activation functions, including the sigmoid function, which tries to map values between 0 and 1, the hyperbolic tangent which maps values between -1 and 1, and finally the ReLU function, which produces the input value unless it is negative, in which case 0 is returned.

4.3.1.3   Overview of different types of neural networks

4.3.1.3.1   Feedforward Neural Networks (FNNs)

Feedforward Neural Networks (FNNs) are a basic form of neural network that includes an input layer, hidden layer(s), and an output layer (Sharma 2022). Information travels in a unidirectional manner, from the input layer through the hidden layers, and finally to the output layer.

4.3.1.3.2   Convolutional Neural Networks (CNNs)

Computer vision tasks regularly utilize CNNs, which comprise of convolutional layers, pooling layers, and fully connected layers. Convolutional layers function by extracting local features from the input data, whilst pooling layers are responsible for reducing spatial dimensions (Saha 2018).

4.3.1.3.3   Recurrent Neural Networks (RNNs)

Recurrent Neural Networks (RNNs) are tailored to handle data that is presented sequentially. These networks contain connections that allow information to be transmitted from one-time step to the next (Donges 2023). A distinguishing feature of RNNs stems from their ability to retain a memory element capable of storing information related to past inputs.

4.3.1.3.4   Long Short-Term Memory (LSTM) networks

Long Short-Term Memory Networks (LSTMs) belong to the Recurrent Neural Network (RNN) class, and their main purpose is to resolve the issue of diminishing gradients (Hung 2023). They apply distinct memory cells that can remember or discard data, leading to their efficacy in managing extended connections.

4.3.2   Layers and connections

4.3.2.1   Input, hidden, and output layers

Neural networks consist of three primary types of layers:

4.3.2.1.1   Input layer: The data is initially received by the input layer, which contains nodes that correspond to the input features.

4.3.2.1.2   Hidden layers: The information derived from previous layers can be processed by hidden layers to produce outputs, which are achieved by applying activation functions and weights. Neural networks that are deep comprise multiple hidden layers.

4.3.2.1.3   Output layer: The final forecasts or conclusions of the neural network are supplied by the output layer. The quantity of nodes in this layer is contingent on the categorization of the problem at hand, for instance, binary classification, multi-class classification, or regression.

4.3.2.2   Concept of layer-to-layer connections

Neural cells situated within a stratum are linked to neural cells in the following stratum via connections that bear weights. Every connection bestows the output of one neural cell to the input of another. The aforementioned weights establish the potency and consequence of these connections and are rectified during training to enhance the productivity of the network.

4.3.2.3 Importance of depth in deep learning architecture

The architectures of deep learning are distinguished by their profundity, which is denoted by the number of concealed layers within the neural structure. As a result of this depth, deep networks can obtain an edge by acquiring hierarchal representations of supplied data. This feature allows for the successful capture of exceedingly abstract and intricate properties. Ultimately, this greater depth provides the ability for deep learning models to acquire a more in-depth comprehension of intricate patterns and perform more effectively on complex tasks.

4.3.3 TRAINING ALGORITHMS

4.3.3.1 Overview of backpropagation and gradient descent

The algorithm widely used for training neural networks is Backpropagation (Johnson 2023). This algorithm incorporates forward propagation and backward propagation. During the forward propagation step, the input data is passed through the network and activations are calculated layer by layer until the output is obtained. The calculated output is then compared to the desired output, and the error is evaluated.

In the backward propagation step, the error is transmitted in reverse through the network, starting from the output layer and moving toward the input layer. The gradient of the error with respect to the network's weights is calculated using the chain rule of calculus. This gradient knowledge is utilized to adjust the weights. This reduces overall error and enhances the network's performance.

During backpropagation, the weights are updated by the optimization technique, gradient descent. The goal of this technique is to locate the minimum of the loss function by iteratively changing the weights in the direction of the negative gradient. The learning rate parameter regulates the step size during weight updates. Smaller learning rates have a slower convergence but higher accuracy, while bigger learning rates can result in faster convergence but the risk of overshooting the minimum is increased.

4.3.3.2 Optimization techniques

4.3.3.2.1 Stochastic Gradient Descent (SGD)

SGD is a variation of the gradient descent technique that involves updating weights for each training sample or a small group of samples (Gavrilova 2020). This strategy reduces the memory requirements and accelerates the training procedure, making it optimal for extensive data sets. Nonetheless, the updates' stochastic nature can introduce additional noise to the training process.

4.3.3.2.2 Adaptive learning rates

Traditional gradient descent uses a fixed learning rate for all weight updates. Adaptive learning rate techniques, such as AdaGrad, RMSprop, and Adam, dynamically adjust the learning rate based on the history of weight updates. These techniques aim to improve convergence by adapting the learning rate to the specific requirements of each weight.

4.3.3.3 Regularization and dropout techniques

4.3.3.3.1 Regularization

Regularization techniques are used to prevent overfitting, where the model becomes too specialized to the training data and performs poorly on unseen data. L2 regularization (weight decay) and L1 regularization (Lasso) are commonly employed. They add a regularization term to the loss function, penalizing large weights and encouraging simplicity in the model.

4.3.3.3.2   Dropout

Dropout is a regularization technique where randomly selected neurons are temporarily dropped or "turned off" during training. This prevents the co-adaptation of neurons and forces the network to learn more robust and generalized representations. Dropout helps prevent overfitting and improves the network's ability to generalize to new data.

Overall, the key components of deep learning architecture include neural networks with artificial neurons, activation functions, and various types of networks like feedforward, convolutional, and recurrent networks. The layers and connections within the neural network involve input, hidden, and output layers, along with layer-to-layer connections. Training algorithms like backpropagation with gradient descent optimize the network's weights, and techniques like stochastic gradient descent, adaptive learning rates, regularization, and dropout further enhance the training process and improve the network's performance.

4.4   DEEP LEARNING ARCHITECTURES AND APPLICATIONS

4.4.1   CONVOLUTIONAL NEURAL NETWORKS (CNNs)

4.4.1.1   CNN architecture and its applications in image and video processing

Convolutional Neural Networks (CNNs) are deep learning structures that are specifically designed to process data that is arranged in a grid-like format, such as images or videos. Due to their impressive performance in various applications, CNNs have transformed computer vision tasks. The key principle behind the functionality of CNNs is to use local connections and shared weights to derive hierarchical representations of the input data.

CNNs employ several layers, namely, convolutional layers, pooling layers, and fully connected layers. The input into CNNs is typically given as an image or a video frame, which is represented in the form of a grid of pixels. The convolutional layers are responsible for establishing spatial hierarchies of features by carrying out convolutions using filters/kernels across the input data. These filters are adept at detecting various patterns or features present in the input.

4.4.1.2   Convolutional layers, pooling layers, and fully connected layers

Convolutional layers perform the convolution operation between the input data and a set of learnable filters (Gurucharan 2022). Each filter detects a specific feature, such as edges, corners, or textures. The convolutional layers preserve the spatial relationship between the pixels, enabling the network to learn local patterns and global structures.

Pooling layers are used to downsample the spatial dimensions of the data, reducing the computational complexity and extracting the most salient features. Max pooling, for example, selects the maximum value within a neighbourhood and discards the rest, thus retaining the most prominent features.

Fully connected layers are typically located towards the end of the CNN architecture. They take the output of the convolutional and pooling layers and transform it into a one-dimensional feature vector. These layers allow for global information processing and enable the network to make predictions based on the learned representations.

4.4.1.3   CNNs in real-world applications

CNNs have found numerous applications in various domains. Some notable examples include:

4.4.1.3.1   Object recognition: CNNs excel in tasks such as object detection, image classification, and semantic segmentation. (Kumar 2021) Models like AlexNet, VGGNet, and ResNet have achieved state-of-the-art performance on benchmark datasets like ImageNet, enabling applications such as autonomous driving, surveillance systems, and facial recognition.

4.4.1.3.2 Self-driving cars: CNNs are crucial components in self-driving car systems (Kumar 2021). They enable the detection and recognition of objects such as pedestrians, traffic signs, and vehicles, allowing autonomous vehicles to make informed decisions based on their surroundings.

4.4.1.3.3 Medical imaging: CNNs have made significant strides in medical image analysis, assisting in disease diagnosis, tumor detection, and medical imaging segmentation. They can extract meaningful features from X-rays, CT scans, and MRI images, aiding radiologists in providing accurate diagnoses and improving patient care.

4.4.2 Recurrent Neural Networks (RNNs)

4.4.2.1 RNN architecture and its applications in sequential data analysis

Recurrent Neural Networks (RNNs) were devised to manage sequential data, in which the sequence's order is critical. RNNs vary from feed-forward neural networks insofar as they possess recurring links that enable them to retain the knowledge of past data (Jayawardhana 2020). For this reason, they are useful in situations that entail sequences. Processing input sequences one at a time, RNNs gauge each stage based on the current input and the former concealed state, with the feedback loop allowing RNNs to grasp connections and long-term dependencies in sequences, rendering them advantageous in tasks such as language modelling, speech perception, and time series analyses.

4.4.2.2 Recurrent connections and memory cells (LSTM, GRU)

The key feature of Recurrent Neural Networks (RNNs) is the presence of recurrent connections. These connections allow the network to utilize information from previous computations and integrate it into the current computation. Nevertheless, RNNs often experience problems with gradient vanishing or explosion, which limits their capacity to capture long-term dependencies. More advanced RNN architectures like Long Short-Term Memory (LSTM) and Gated Recurrent Unit (GRU) were developed to combat this problem. These architectures contain memory cells that retain and update information over time.

LSTM deploys a memory cell that can intentionally store and neglect information, making it efficient in retaining long-term dependencies. LSTM comprises three primary mechanisms: an input gate that determines which information to store, a forget gate that discards irrelevant information in the memory cell, and an output gate that regulates the output depending on the current input and memory contents.

On the other hand, GRU is another variation of RNN that simplifies the architecture while retaining comparable performance to LSTM. GRU integrates the forget and input gates of LSTM into a single update gate. It also incorporates a reset gate to manage previous information exposure.

Thanks to these memory cells and gated architectures, RNNs can efficiently model sequential dependencies. Therefore, they are appropriate for tasks that entail natural language processing, speech recognition, and time series analysis.

4.4.2.3 RNNs in natural language processing, speech recognition, and time series analysis

Recurrent neural networks have shown their effectiveness in diverse instances that involve data in a sequence:

4.4.2.3.1 Natural Language Processing (NLP): Recurrent Neural Networks (RNNs) are frequently employed for various tasks, including language modelling, machine translation, sentiment analysis, and text generation (Thomas 2019). RNNs exhibit an ability to apprehend the contextual dependencies and syntactic structure of sentences, thereby enabling the development of automated chatbots, language understanding systems, and language generation models.

4.4.2.3.2 Speech Recognition: The use of RNNs, specifically LSTM and GRU types, has played a crucial role in enhancing the quality of automatic speech identification mechanisms. These models can capture the time-related relationships in speech signals and convert them to written forms. Automatic speech recognition systems that use RNNs have made considerable progress in areas such as voice assistants, transcribing services, and voice-activated devices.

4.4.2.3.3 Time Series Analysis: Recurrent Neural Networks (RNNs) are highly skilled in the modelling and foretelling of data that is dependent on time. They have the ability to apprehend patterns and trends over a temporal basis, making them an asset for various undertakings like prognosticating the market of stocks, anticipating weather conditions, and detecting deviations. RNNs that contain Long Short-Term Memory (LSTM) or Gated Recurrent Unit (GRU) units are apt in managing lengthy connections in data sequences that are dependent on time, thereby potentiating meticulous predictions and efficient evaluations.

4.4.3   GENERATIVE ADVERSARIAL NETWORKS (GANs)

4.4.3.1   Overview of GAN architecture and its applications in generating synthetic data

Generative Adversarial Networks (GANs) refer to a particular category of deep learning structure that is comprised of two central constituents – a generator and a discriminator (Vaseekaran 2022). The primary objective of GANs is to produce artificial data that bears a resemblance to the distribution of training data. While the generator is responsible for generating brand-new data, the discriminator's responsibility is to make distinctions between real and fabricated samples.

The generator achieves this by mapping arbitrary noise or latent vectors onto the anticipated output range, like photographs or text. Conversely, the discriminator plays its role by identifying whether a given sample is genuine or synthetic. Both entities are imparted training at the same time, in an aggressive environment, where the generator endeavors to mislead the discriminator while the discriminator tries to differentiate between authentic and counterfeit samples.

4.4.3.2   Generator and discriminator components

The generator element of a Generative Adversarial Network (GAN) accepts arbitrary noise as input and yields artificial samples as output, usually composed of one or more layers of neural nets, be they convolutional or fully connected ones. The purpose of the generator is to learn the correspondence between the noise distribution and the data distribution so as to generate believable samples that are almost impossible to distinguish from the real ones.

The discriminator part behaves like a binary classifier, that is, it distinguishes actual examples from generated ones. It accepts a genuine or an artificial sample as input and predicts the likelihood that the input is real or counterfeit. The discriminator is trained to enhance its potential to accurately classify the samples, thereby forcing the generator to yield more plausible artefacts.

4.4.3.3   GANs in image synthesis, text generation, and data augmentation

GANs have found numerous applications in various domains, some of which include:

4.4.3.3.1 Image synthesis: GANs have gained extensive use in generating lifelike photos. Models, such as Deep Convolutional GANs (DCGANs) and Progressive Growing GANs (PGGANs), are capable of synthesizing high-resolution graphics containing intricate surfaces and consistent arrangements. GANs have also carried out functions that involve translations of images, fashion changeovers, and the production of innovative masterpieces (Brownlee 2019).

4.4.3.3.2 Text generation: GANs have additionally been employed in natural language processing assignments, comprising the creation of written material. Conditioned by entering text or latent vectors, the generator of GANs can produce ordered and contextually appropriate written content. Examples comprise the creation of product descriptions, conversations, and verses.

4.4.3.3.3   Data augmentation: The utilization of GANs can expand datasets through the creation of artificial instances that bear a resemblance to authentic data. This technique proves valuable in contexts where annotated data is scarce. The process entails training a GAN on the accessible set and generating further specimens, thus enriching the dataset, and improving the performance and overall efficacy of machine learning models (Antoniadis 2016).

4.4.3.3.4   Video synthesis: In video synthesis, GANs have been expanded to generate video sequences that are both coherent and realistic. This advancement has important uses in predicting, editing, and producing special effects in videos.

4.4.3.3.5   Anomaly detection: GANs have the potential to be applied in the field of anomaly detection. One way this can be done is by using GANs to train on normal data, thereby enabling the generator to understand and apply algorithms that capture the real distribution. Once this is achieved, it is then possible to identify outliers or anomalies by comparing the divergence between the actual data and the generated samples. This methodology is applicable in various fields such as fraud detection, cybersecurity, and revealing unusual patterns in time series data.

To summarize, Generative Adversarial Networks (GANs) provide potent functionalities that can produce artificial data that closely resemble authentic data distribution. They have been efficaciously employed for various purposes, such as synthesizing images, generating texts, enlarging datasets, developing videos, and identifying aberrations. GANs remain a thriving field to explore, which stimulates innovations in artificial intelligence and broadens the scope for data production and analysis.

4.5   ADVANCES AND CHALLENGES IN DEEP LEARNING ARCHITECTURE

4.5.1   RECENT ADVANCEMENTS

Deep learning has witnessed remarkable advancements in recent years, primarily driven by breakthroughs in deep learning architecture. Several key developments have significantly impacted the field, including transformer networks, attention mechanisms, and graph neural networks.

4.5.1.1 Transformer Networks: Transformer networks have revolutionized natural languages processing tasks, such as machine translation and language understanding. Introduced by Vaswani et al. in 2017, transformers leverage the self-attention mechanism to capture global dependencies within input sequences (Vaswani, et al. 2017). By replacing Recurrent Neural Networks (RNNs) with self-attention layers, transformers excel in modelling long-range dependencies, making them highly effective in tasks requiring context understanding.

4.5.1.2 Attention Mechanisms: Attention mechanisms are of great importance in enabling the models to concentrate on pertinent data as they process sequences and images. By assigning weightings to diverse parts of the input on the grounds of their importance, they assist the models in highlighting vital aspects. Attention mechanisms have proved to be an asset in numerous domains, including machine translation, image captioning, and speech recognition. The implementation of attention mechanisms has supplemented the interpretability and efficacy of deep learning models.

4.5.1.3 Graph Neural Networks (GNNs): Graph Neural Networks (GNNs) have become effective instruments for acquiring knowledge from data structured in graphs. These networks use graph convolution actions to gather and distribute information throughout the graph nodes, consequently enabling efficient education on representation (Menzli 2023). GNNs have demonstrated prospects in various disciplines, including examining social networks, developing recommendation systems, and discovering medications. Their capacity to understand connections and effectively handle non-standard data structures makes them incredibly relevant in domains that work with graph data.

4.5.2 OVERVIEW OF CHALLENGES IN DEEP LEARNING ARCHITECTURE

While deep learning architecture has made significant progress, it still faces several challenges that researchers are actively addressing.

4.5.2.1 Overfitting: Deep learning models often struggle with overfitting, which occurs when models perform well on training data but fail to generalize to unseen data (Alzubaidi, et al. 2021). Overfitting can result from excessively complex models, limited training data, or insufficient regularization techniques. Addressing overfitting requires techniques like dropout, regularization, early stopping, and data augmentation to improve model generalization.

4.5.2.2 Computational complexity: Training and deploying deep learning models requires significant computational resources due to the involved computations, especially in neural networks with large architectures and extensive data. The complex nature of these computations calls for potent hardware and parallel computing methods. Therefore, reducing computational complexity is crucial in ensuring deep learning is accessible and practical for everyone.

4.5.2.3 Interpretability: The lack of interpretability in deep learning models poses a challenge as it is difficult to comprehend why the model makes certain predictions. This limited understanding hinders the adoption of deep learning in essential fields like healthcare and finance. Researchers are currently investigating various ways to improve model interpretability. This includes attention mechanisms, visualization techniques, and model explanation frameworks.

4.5.3 ONGOING RESEARCH AND FUTURE DIRECTIONS

Ongoing research focuses on overcoming present obstacles and investigating novel approaches as the deep learning structure advances.

4.5.3.1 Model compression and efficiency: Scholars are making strides in reducing the intricacy of deep learning models. Such progress involves a variety of methodologies such as network pruning, quantization, knowledge distillation, and architecture design optimization. The objective of these techniques is to generate leaner, more effective models without compromising their efficacy.

4.5.3.2 Explainable AI: Improving the comprehensibility of profound learning models endures as a vital subject of study. Procedures including spotlight representation, salience plans, and notion recognition attempt to furnish clarifications for model forecasts. Scholars are additionally examining the incorporation of emblematic thought and judgment-based schemes with profound learning to generate more lucid models.

4.5.3.3 Lifelong and continual learning: Presently, existing methodologies in deep learning frequently encounter difficulty in adjusting to new tasks or gaining novel knowledge without erasing previously acquired information. Persistent and continuing learning endeavors to enhance models that can steadily acquire knowledge over time, preserve information from prior tasks, and promptly adjust to new tasks without experiencing severe loss of memory.

4.5.3.4 Robustness and generalization: Models of deep learning are frequently prone to be targeted by adversary attacks and may encounter difficulties in extemporaneously adapting to unseen distributions of data. Current research primarily centers on designing resilient architectures that can withstand incursions by adversarial forces and can adeptly transition across multiple domains. Several strategies, such as adversarial training, data augmentation, and adaptation to various domains, are all being probed for their efficacy in improving model robustness.

4.5.3.5 Hybrid architectures: Academics are exploring the integration of deep learning structures with alternative forms of machine learning methods, including probabilistic graphical models and reinforcement learning. The objective of this mixing is to utilize the benefits of various tactics to handle intricate assignments more competently and enhance the efficiency of models.

4.5.3.6 Transfer learning and pretraining: The utilization of transfer learning has been found to be extremely advantageous in deep learning. Existing models that have been pre-trained, particularly ones derived from broad linguistic models like GPT and BERT, offer robust initializations for undertakings subordinated to an initial task. There is a continuous study intended to reinforce transfer learning techniques, investigate superior pretraining strategies, and develop approaches to transfer knowledge among various fields and modes of expression.

4.5.3.7 Ethical and Fair AI: With an increasing impact on society, attention is being paid to the ethical and biased concerns surrounding deep learning models. To this end, attempts are being made to devise fair and unbiased architectures, ensure transparency and accountability, and alleviate potential negative social effects.

The realm of deep learning architecture has undergone considerable progress with the advent of transformer networks, attention mechanisms, and graph neural networks. Nevertheless, significant barriers remain, such as interpretability, computational complexity, and overfitting. Nevertheless, scholars are endeavoring to address such issues via a plethora of methodologies, including model compression, explainable AI, hybrid architectures, lifelong learning, transfer learning, robustness, and ethical considerations. By continuing to push the limits of deep learning architecture, novel opportunities can be explored and greater headway can be made across various fields.

4.6 CONCLUSION

To conclude, the employment of deep learning architecture has altered the artificial intelligence field by utilizing multi-layered neural networks for extracting intricate structures from extensive amounts of information. This feature of recognizing meaningful depictions automatically has led to impressive developments in language comprehension, voice identification, picture perception, and various other sectors. The fundamental elements of deep learning networks, such as recurrent layers, convolutional layers, and attention mechanisms, perform indispensable tasks in accomplishing proficient AI systems.

Recent advancements in deep learning design, such as strategies for optimization, techniques for regularization, and the development of new network architectures have bolstered the functionalities of AI models. Utilization of methodologies such as backpropagation, gradient descent, stochastic gradient descent, adaptive learning rates, and regularization, has improved the efficacy of training in areas of convergence, generalization, and efficiency. Deep learning structures such as Convolutional Neural Networks (CNNs), Recurrent Neural Networks (RNNs), and Generative

Adversarial Networks (GANs) have accomplished successful outcomes across various domains, including image recognition, natural language processing, time series analysis, and data generation.

Despite these achievements, challenges remain. Overfitting, computational complexity, and lack of interpretability are areas that researchers are actively addressing. Techniques such as dropout, regularization, early stopping, and data augmentation are being employed to improve model generalization. Efforts are also being made to enhance model interpretability through attention mechanisms, visualization techniques, and model explanation frameworks.

The future of deep learning architecture holds great promise. Ongoing research focuses on model compression and efficiency, explainable AI, lifelong and continuous learning, robustness and generalization, as well as ethical considerations. Integration with other machine learning methods such as probabilistic graphical models and reinforcement learning is being explored. Transfer learning and pretraining have shown benefits, and efforts are being made to ensure fairness and unbiased decision-making in AI systems.

Despite the challenges ahead, the continuous advancements in deep learning architecture open up new possibilities for AI across various fields. By pushing the boundaries and addressing the limitations, we can unlock further potential and pave the way for more innovative and impactful applications of deep learning in the future.

REFERENCES

1. Alzubaidi, Laith, Jinglan Zhang, Amjad J. Humaidi, Ayad Al-Dujaili, Ye Duan, Omran Al-Shamma, J. Santamaría, Mohammed A. Fadhel, Muthana Al-Amidie, and Laith Farhan. 2021. "Review of deep learning: Concepts, CNN architectures, challenges, applications, future directions." *Journal of Big Data* 8. Accessed June 6, 2023.
2. Antoniadis, Panagiotis. 2016. Using GANs for Data Augmentation. March 23. Accessed June 6, 2023. www.baeldung.com/cs/ml-gan-data-augmentation
3. Boesch, Gaudenz. n.d. Deep Neural Network: The 3 Popular Types (MLP, CNN and RNN). https://viso.ai/deep-learning/deep-neural-network-three-popular-types/
4. Brownlee, Jason. 2019. 18 Impressive Applications of Generative Adversarial Networks (GANs). June 14. Accessed June 6, 2023. https://machinelearningmastery.com/impressive-applications-of-generative-adversarial-networks/
5. Donges, Niklas. 2023. A Guide to Recurrent Neural Networks: Understanding RNN and LSTM Networks. February 28. Accessed June 2, 2023. https://builtin.com/data-science/recurrent-neural-networks-and-lstm
6. Gavrilova, Yulia. 2020. An Overview of Machine Learning Optimization Techniques. December 2. Accessed June 6, 2023. https://serokell.io/blog/ml-optimization
7. Gupta, Sakshi. 2021. How Deep Learning Revolutionized NLP. January 24. Accessed June 2, 2023. www.springboard.com/blog/data-science/nlp-deep-learning/
8. Gurucharan, M. K. 2022. Basic CNN Architecture: Explaining 5 Layers of Convolutional Neural Network. July 28. Accessed June 6, 2023. www.upgrad.com/blog/basic-cnn-architecture/#:~:text=CNNs%20are%20a%20class%20of,vision%20and%20natural%20language%20processing
9. Hung, Che-Lun. 2023. "Deep learning in biomedical informatics." *Intelligent Nanotechnology* 307-329. Accessed June 2, 2023.
10. Janiesch, Christian, Patrick Zschech, and Kai Heinrich. 2021. "Machine learning and deep learning." *Electronic Markets* 31: 685-695. Accessed June 2, 2023.
11. Jayawardhana, Santhoopa. 2020. Sequence Models & Recurrent Neural Networks (RNNs). July 27. Accessed June 6, 2023. https://towardsdatascience.com/sequence-models-and-recurrent-neural-networks-rnns-62cadeb4f1e1
12. Johnson, Daniel. 2023. Back Propagation in Neural Network: Machine Learning Algorithm. May 6. Accessed 6 6, 2023. www.guru99.com/backpropagation-neural-network.html
13. Kumar, Ajitesh. 2021. Real-world Applications of Convolutional Neural Network (CNN). November 6. Accessed June 6, 2023. https://vitalflux.com/real-world-applications-of-convolutional-neural-networks/

14. Liu, Di, Hao Kong, Xiangzhong Luo, Weichen Liu, and Ravi Subramaniam. 2022. "Bringing AI to edge: From deep learning's perspective." *Neurocomputing* 485: 297-320. Accessed June 2, 2023.

15. Madan, Piyush. 2020. An Introduction to Deep Learning. March 2. Accessed June 2, 2023. https://developer.ibm.com/articles/an-introduction-to-deep-learning/

16. Menzli, Amal. 2023. Graph Neural Network and Some of GNN Applications: Everything You Need to Know. April 26. Accessed June 6, 2023. https://neptune.ai/blog/graph-neural-network-and-some-of-gnn-applications

17. Nabriya, Pratik. 2021. Perceptron: Building Block of Artificial Neural Network. October 21. Accessed June 2, 2023. www.analyticsvidhya.com/blog/2021/10/perceptron-building-block-of-artificial-neural-network/

18. Saha, Sumit. 2018. A Comprehensive Guide to Convolutional Neural Networks — the ELI5 *Way*. December 15. Accessed June 2, 2023. https://saturncloud.io/blog/a-comprehensive-guide-to-convolutional-neural-networks-the-eli5-way/

19. Sarker, Iqbal H. 2021. "Deep Learning: A Comprehensive overview on techniques, taxonomy, applications and research directions." SN Computer Science 2. Accessed June 2, 2023.

20. Saxena, Shipra. 2021. Introduction to The Architecture of Alexnet. March 19. Accessed June 2, 2023. www.analyticsvidhya.com/blog/2021/03/introduction-to-the-architecture-of-alexnet/#:~:text=Alex net%20won%20the%20Imagenet%20large,Alex%20Krizhevsky%20and%20his%20colleagues

21. Sharma, Prashant. 2022. Feedforward Neural Network: Its Layers, Functions, and Importance. January 18. Accessed June 2, 2023. www.analyticsvidhya.com/blog/2022/01/feedforward-neural-network-its-layers-functions-and-importance/

22. Thomas, Christopher. 2019. Recurrent Neural Networks and Natural Language Processing. June 9. Accessed June 6, 2023. https://towardsdatascience.com/recurrent-neural-networks-and-natural-language-processing-73af640c2aa1

23. Vaseekaran, Varatharajah. 2022. GANs for Synthetic Data Generation. August 2. Accessed June 6, 2023. https://towardsai.net/p/l/gans-for-synthetic-data-generation

24. Vaswani, Ashish, Noam Shazeer, Niki Parmar, Jakob Uszkoreit , Llion Jones, Aidan N. Gomez, Łukasz Kaiser, and Illia Polosukhin. 2017. "Attention Is All You Need." Edited by I. Guyon, U. Von Luxburg, S. Bengio, H. Wallach, R. Fergus, S. Vishwanathan and R. Garnett. *NeurIPS Proceedings*. Accessed June 6, 2023. https://papers.nips.cc/paper_files/paper/2017/hash/3f5ee243547dee91fbd053c1c4a84 5aa-Abstract.html

25. West, Darrell M., and John R. Allen. 2018. "How artificial intelligence is transforming the world." Accessed June 2, 2023. www.brookings.edu/research/how-artificial-intelligence-is-transforming-the-world/

5 Deep Learning for ECG Classification
Techniques, Applications, and Challenges

Sucharita Mitra

5.1 INTRODUCTION

During the last decade scientists were involved in different advanced research to aid cardiologists in their role of analyzing and interpreting ElectroCardioGrams (ECGs). From detecting anomalies in ECG signals to establishing completely automated ECG diagnosis systems; all are included in this areas of research. Various techniques have been explored to achieve these goals, comprising expert systems, statistical pattern recognition, wavelet transforms, artificial neural networks, fuzzy systems, and neuro-fuzzy systems.

The process of ECG interpretation involves two distinct and consecutive phases: (i) feature extraction and (ii) classification. Feature extraction methods are employed to obtain a set of signal measurements that contain valuable information for waveform characterization. These waveform descriptors are then utilized in the classification phase to assign the ECG to one or more diagnostic classes. Traditional approaches to ECG analysis often require manual feature extraction and extensive domain knowledge, making the process time-consuming and susceptible to human error.

In recent years, due to advances in deep learning approaches, there has been an increasing interest in using neural networks to automate ECG signal classification. Convolutional Neural Networks (CNNs), Recurrent Neural Networks (RNNs), and other deep learning algorithms have demonstrated promising outcomes in a variety of medical imaging tasks, as well as ECG interpretation. Using self-learning techniques, these neural networks can learn complicated patterns and their associations directly from raw ECG data. Thus they can remove the requirement for explicit feature extraction and lower the need for human intervention in the interpretation process (Laith Alzubaidi 2021).

By automating the classification process of ECG signals using deep learning algorithms, cardiologists can benefit from enhanced efficiency, reduced subjectivity, and possibly improved diagnostic accuracy. Moreover, these automated systems have the potential to assist healthcare professionals in triaging patients, detecting critical conditions promptly, and streamlining the workflow in busy clinical settings.

In this chapter, at the beginning a few preprocessing techniques on the ECG signal are discussed as this plays a crucial role in ECG analysis and then the following sub sections are included one by one, namely, deep learning architectures for ECG classification, data augmentation strategies for ECG, ECG classification tasks, evaluation metrics and performance analysis and applications, and future direction.

DOI: 10.1201/9781003433309-5

In conclusion, the integration of advanced technologies, particularly deep learning methods, holds significant promise for revolutionizing ECG analysis. Prolonged research and development in this field will surely lead to the creation of robust and reliable automated systems that can assist cardiologists in their diagnostic tasks, which will ultimately benefit patient care and outcomes.

5.2 PRE-PROCESSING TECHNIQUES FOR ECG SIGNALS: ENHANCING ACCURACY AND RELIABILITY

Electrocardiogram (ECG) signals provide valuable clinical information about the electrical activity of the heart, enabling the diagnosis and monitoring of cardiac conditions. However, ECG signals often suffer from various distortions, including noise, baseline wander, and amplitude variations. Pre-processing techniques are used to minimize these issues for improving the accuracy and reliability of ECG signal analysis. In this chapter, a few common pre-processing methods like noise removal, baseline wander correction and signal normalization are discussed.

5.2.1 NOISE REMOVAL

ECG signals are prone to noise interference from various sources, including power lines, muscle activity, and electrode artifacts (Shubhojeet Chatterjee 2020). Two popular noise removal techniques are adaptive filtering and wavelet de-noising. Algorithms are developed to estimate and suppress noise adaptively in the adaptive filtering technique, whereas the wavelet de-noising method applies wavelet transforms to remove unwanted noise components (Jianwei Zheng 2020; Sucharita Mitra Sarkar 2023) .

5.2.2 BASELINE WANDER CORRECTION

Baseline wander refers to the low-frequency variations in the ECG signal caused by factors such as patient movement or electrode placement. For accurate analysis of ECG signals, Correction of baseline wander is required. Polynomial fitting and wavelet-based methods are commonly used for this. Polynomial fitting is basically a curve-fitting technique to estimate and remove the baseline drift, while the wavelet-based method utilizes wavelet transforms to separate the baseline wander from the ECG signal.

5.2.3 CHALLENGES AND CONSIDERATIONS

Pre-processing of ECG signals faces several challenges. Firstly, selection of the appropriate pre-processing techniques depends on the specific characteristics of the noise and the elements present in the signal. It requires careful analysis and understanding of the ECG signal and noise sources. Secondly, maintaining a balance between noise removal and preserving the clinical information within the ECG waveform is vital. Over-filtering may result in the loss of important diagnostic features. Lastly, for real-time processing, computational efficiency should be an important factor at the time of implementation of these pre-processing techniques in clinical settings.

Figure 5.1 shows the use of wavelet de-noising to remove noise from an ECG signal, resulting in a cleaner waveform for subsequent analysis. Figure 5.2 demonstrates the baseline wander correction using the wavelet-based method.

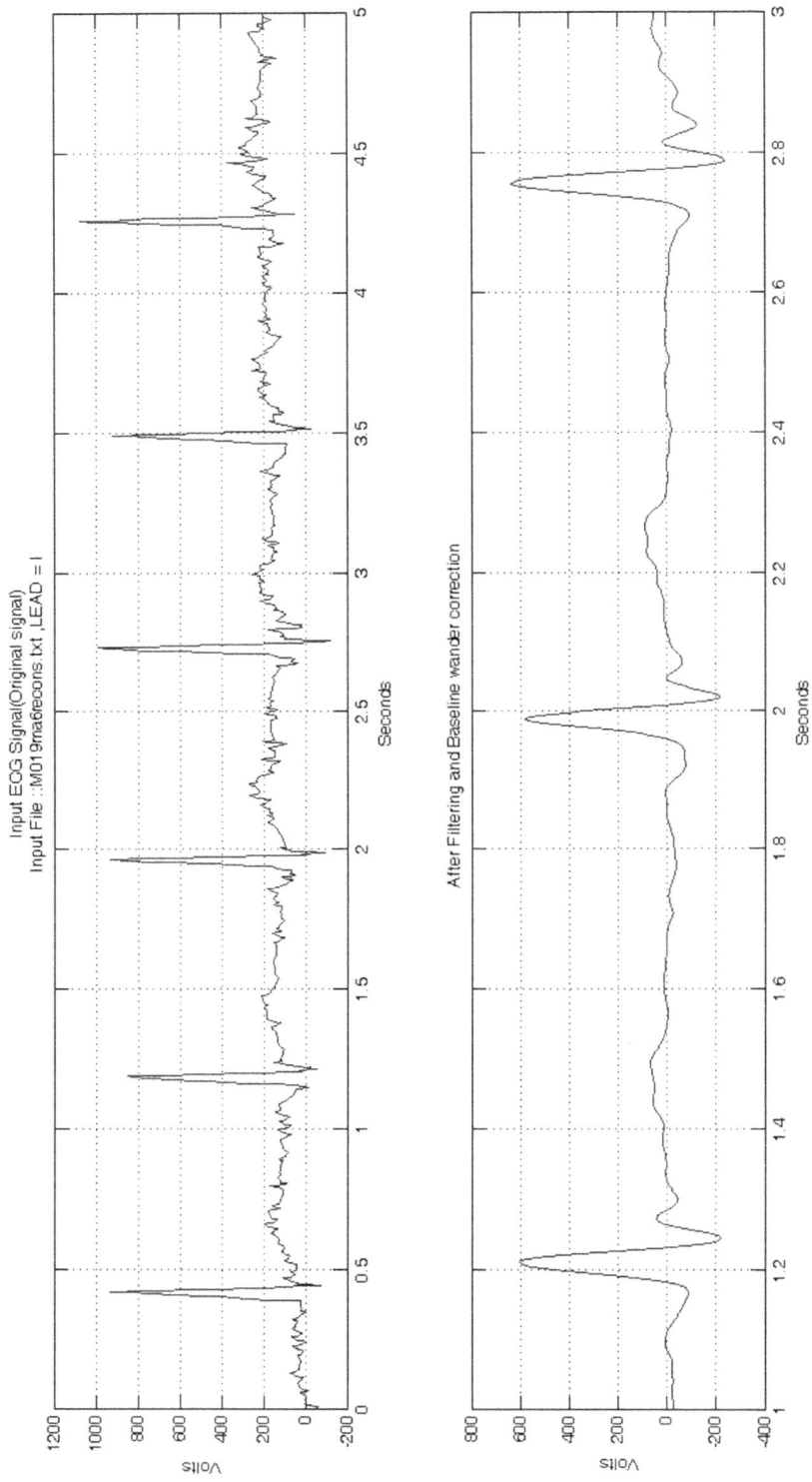

FIGURE 5.1   Wavelet de-noising for noise removal in an ECG signal.

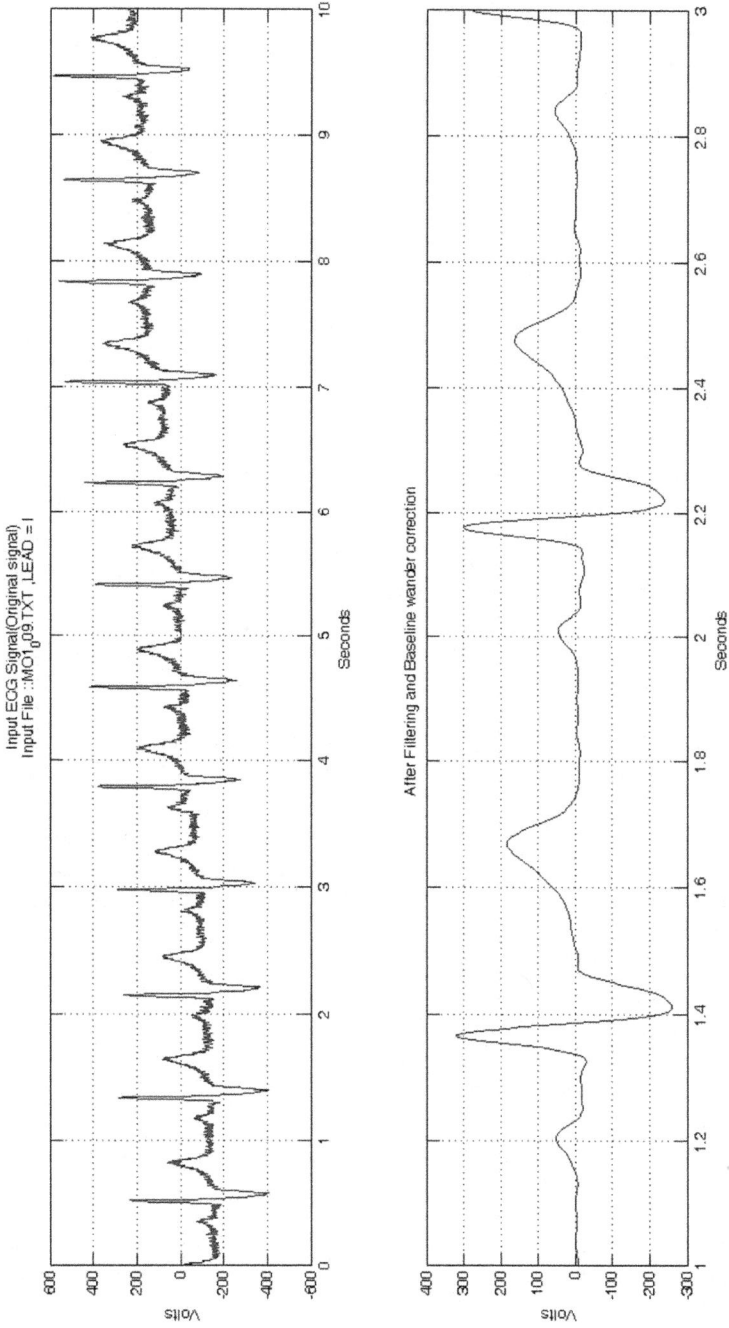

FIGURE 5.2 Wavelet-based baseline wander.

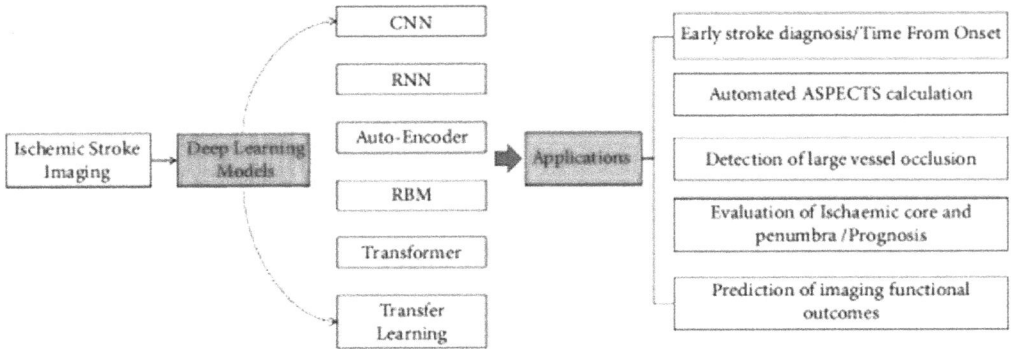

FIGURE 5.3 Applications of deep learning in acute ischemic stroke imaging analysis.

5.3 DEEP LEARNING ARCHITECTURES FOR ECG CLASSIFICATION

Deep learning architectures have reformed ECG classification tasks by giving a remarkable performance. These architectures control the power of neural networks to automatically extract meaningful clinical features from ECG signals for accurate classification and diagnosis. In this section, we explain various deep learning architectures used for ECG classification to focus their strengths and considerations.

Convolutional Neural Networks (CNNs) have attracted significant attention because of their ability to extract spatial features from ECG signals. CNNs utilize convolutional layers to detect local patterns and classified representations in the ECG waveform. By applying filters of different orders, CNNs can capture both high and low-frequency components, allowing robust feature extraction. CNN-based architectures have shown promising results in various ECG classification tasks, such as arrhythmia detection and ischemic stroke prediction (M Muthulakshmi 2019).

Figure 5.3 shows how deep learning is applied to detect acute ischemic stroke

Recurrent Neural Networks (RNNs) and their variations, such as Long Short-Term Memory (LSTM) networks and Gated Recurrent Units (GRUs) are useful in capturing temporal dependencies in sequential data. ECG signals have a sequential nature due to the time-varying nature of cardiac activities. Since RNNs are capable of retaining memory of past states, they are allowed to model temporal dependencies effectively. Basically LSTM networks, with their gated memory cells, are particularly useful in preserving long-term dependencies in ECG data. GRUs, a simplified version of LSTM, also have similar capabilities with reduced computational complexity (Shraddha Singha 2018, 1290-1297).

Figure 5.4 shows the LSTM architecture for Arrhythmia

A combination of the strengths of both CNNs and RNNs is known as a hybrid model. The hybrid models have shown superior performance in ECG classification tasks. These architectures influence CNNs to capture spatial features from ECG waveforms and feed them into RNNs to model temporal dependencies. By integrating both spatial and temporal information, hybrid architectures provide a comprehensive representation of the ECG signal, enabling accurate classification (Parul Madan 2022).

On the other hand, in the context of ECG classification, transfer of learning and pre-trained models have obtained more utilization. Transfer of learning involves development of pre-trained models, based on large-scale datasets, and fine-tuning them on smaller ECG datasets. This approach is mainly useful when the available ECG dataset is inadequate. Pre-trained models, such as those trained on general medical images, can provide a good starting point for ECG classification tasks, as they have learned to extract clinically significant features from related medical data.

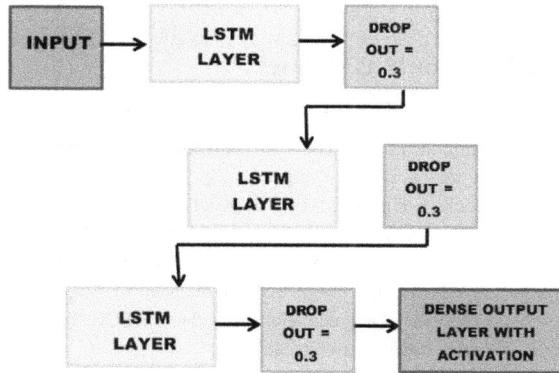

FIGURE 5.4   A sample LSTM architecture for Arrhythmia detection.

The selection and fine-tuning of deep learning architectures for ECG classification requires careful consideration. Several factors like the size and diversity of the ECG dataset, computational resources, and the specific classification task at hand should be considered. It is important to choose an architecture that is appropriate for the complexity of the ECG data to optimize clinical parameters accordingly. Fine-tuning of the selected architecture on the target ECG dataset helps to familiarize the model to the specific characteristics of the data and also to improve its performance.

In summary, deep learning architectures, including CNNs, RNNs, hybrid architectures, and the utilization of transfer learning, have shown great potential in ECG classification tasks. These architectures control the strengths of neural networks to extract spatial and temporal features from ECG signals, enabling accurate classification and diagnosis. However, careful consideration of the specific requirements and characteristics of the ECG dataset is vital while selecting and fine-tuning deep learning architectures for optimal performance.

5.4   DATA AUGMENTATION STRATEGIES FOR ECG

Data augmentation is a fundamental technique in ECG classification that enhances the diversity of the training dataset and improves the generalization capability of deep learning models. Models can learn to manage changes and enhance their performance on unseen data by artificially increasing the dataset with augmented samples. This section delves into several data augmentation methodologies used in ECG categorization.

Artificial data generation techniques are widely used in data augmentation. Jittering involves introducing small random perturbations to the ECG signal by adding or subtracting a small value from each sample. This helps to imitate minor signal changes caused by electrode placement or patient mobility. Shifting is the process of changing the complete ECG signal in the temporal domain to simulate variations in cardiac cycle duration. Resampling techniques modify the sampling rate of the signal, allowing the model to handle variations in the data collection process (Qing Pan 2020).

Augmentation methods that introduce variations directly into the ECG signals can also be effective for random noise generation. Random noise addition introduces synthetic noise with controlled intensity into the signal, mimicking real-world noise sources. This helps the model to learn to robustly handle noisy ECG recordings. Amplitude scaling involves multiplying the ECG signal by a factor, adjusting the signal's amplitude range. This accounts for differences in signal amplitudes across different recordings. Time warping techniques modify the temporal structure of the ECG signal, introducing local distortions that simulate irregularities in the cardiac cycle.

TABLE 5.1
Data Augmentation Techniques for ECG Dataset

Augmentation Technique	Description
Random Noise Addition	Introducing synthetic noise with controlled intensity into the ECG signal, simulating real-world noise sources.
Amplitude Scaling	Multiplying the ECG signal by a factor, adjusting the signal's amplitude range to account for variations in signal strength.
Time Warping	Modifying the temporal structure of the ECG signal, introducing local distortions that simulate irregularities in the data.
Augmentation Intensity (Low/Medium/High)	The level of augmentation applied to the ECG signal. It determines the strength of the augmentation technique.

The impact of data augmentation on model performance is significant. By extending the effective size of the dataset, augmentation helps to address the difficulty of limited labeled ECG data. As the model learns to generalize from the supplemented data, it also lowers overfitting. By exposing the model to a broader range of variations, data augmentation helps it become more robust and adaptable to unseen data.

However, generating augmented data comes with challenges. It requires careful consideration of the type and magnitude of variations to introduce, as excessive or inappropriate augmentation may lead to unrealistic or misleading samples. Balancing the trade-off between introducing realistic variations and preserving the underlying signal characteristics is essential. Additionally, the computational cost of generating augmented data should be considered, especially when dealing with large-scale ECG datasets.

The table above presents various data augmentation techniques commonly used for ECG datasets, namely random noise addition, amplitude scaling, and time warping. These techniques introduce controlled variations to the original ECG signals, allowing the model to learn and generalize from different patterns and characteristics.

5.5   ECG CLASSIFICATION TASKS

Deep learning methodologies have emerged as powerful tools in the detection and classification of various cardiovascular conditions. The application of deep learning algorithms in recognizing and categorizing Ischemic Heart Disease (IHD), Atrial Fibrillation (AF), Ventricular Tachycardia (VT), other Arrhythmias, Myocardial Infarction (MI), and aberrant cardiac findings is the topic of this chapter.

1. Ischemic Heart Disease (IHD) is defined by decreased blood flow to the heart as a result of heart tissue injury. Deep learning methods like Convolutional Neural Networks (CNNs) are used to evaluate ECG signals for the detection of ischemic patterns. CNNs can identify ST-segment abnormalities or characteristic wave changes indicating myocardial ischemia by adapting the training and learning process on the characteristic features of ECG signals (Tasci 2023).

2. Atrial Fibrillation (AF), which is known as a common Cardiac Arrhythmia can also be identified by including Recurrent Neural Networks (RNNs). It can capture temporal dependencies in ECG data, which helps in accurate AF detection. RNNs can identify irregular heart rate patterns and distinguish AF from normal sinus rhythm (V. G. Sujadevi 2017).

3. A life-threatening arrhythmia is Ventricular Tachycardia (VT). This can also be identified by using deep learning algorithms. These models, trained on annotated data, can accurately classify VT episodes, assisting in timely intervention.

4. Deep learning algorithms also play an effective role in detecting and classifying other Arrhythmias. Abnormal heart rhythms such as atrial flutter, bradycardia, and premature ventricular contractions can be successfully recognized utilizing neural networks and pattern recognition approaches.

5. Proper detection of Myocardial Infarction (MI) should be necessary for immediate medical intervention since it is life threatening. Deep learning models may detect ST-T elevation, a hallmark of ST-Elevation Myocardial Infarction (STEMI), as well as other ECG abnormalities associated with non-STEMI. These models analyse ECG patterns and classify them as indicative of MI, thus aiding in prompt diagnosis.

In addition to disease detection, deep learning techniques are also utilized for abnormality detection. By detecting ECG features, such as the QT interval, T-wave morphology, deep learning models can identify abnormalities that may indicate basic cardiac conditions. These models are trained on large datasets, enabling them to learn complex patterns and recognize subtle variations.

Overall, deep learning methodologies have proved their effectiveness in detecting and classifying various cardiovascular conditions, including IHD, AF, VT, other arrhythmias, MI, and abnormal cardiac conditions. As these techniques continue to grow, they have potential for improving diagnostic accuracy, enabling early intervention, and ultimately enhancing patient outcomes in the field of cardiology.

5.6 EVALUATION METRICS AND PERFORMANCE ANALYSIS

Evaluation metrics and performance analysis play a crucial role in assessing the effectiveness of deep learning models for ECG classification. In this section, commonly used metrics such as accuracy, sensitivity, specificity, precision, and the area under the receiver operating characteristic curve (AUC-ROC) are described (Neptune.AI n.d.).

5.6.1 ACCURACY

Accuracy is a fundamental metric that measures the thorough correctness of a model's predictions. It computes the ratio of correctly classified samples to the total number of samples.
The formula for this is:

$$\text{Accuracy} = \frac{TN + TP}{TP + TN + FP + FN} \tag{5.1}$$

where TN is true negatives, TP is true positives, FP is false positives and FN is false negatives.

5.6.2 SENSITIVITY

Sensitivity is also known as recollection or true positive rate and it measures the proportion of definite positive cases that are properly recognised by the model.
The formula for this is:

$$\text{Sensitivity} = \frac{TP}{TP + FN} \tag{5.2}$$

where TN is true negatives, TP is true positives and FN is false negatives.

5.6.3   SPECIFICITY

Specificity measures the proportion of definite negative cases that are properly identified as negative by the model.

The formula for this is:

$$\text{Specificity} = \frac{TN}{FP + TN} \qquad (5.3)$$

where TN is true negatives, and FP is false positives.

5.6.4   PRECISION

Precision is also known as positive analytical value and it evaluates the accuracy of positive predictions made by the model.

The formula for this is:

$$\text{Pr ecision} = \frac{TP}{TP + FP} \qquad (5.4)$$

where TP is true positives, and FP is false positives.

5.6.5   AREA UNDER THE RECEIVER OPERATING CHARACTERISTIC CURVE (AUC-ROC)

AUC-ROC is a performance metric that evaluates the model's ability to differentiate between positive and negative instances across different classification thresholds. It represents the probability that a randomly chosen positive instance will be ranked higher than a randomly chosen negative instance. AUC-ROC is computed by plotting the true positive rate (sensitivity) against the false positive rate (1 – specificity) for various classification thresholds and calculating the area under this curve.

The formula for this is:

$$\text{True Positive Rate (TPR)} = \frac{TP}{TP + FN} \qquad (5.5)$$

and

$$\text{False Positive Rate} \left(\text{FPR} \right) = \frac{FP}{FP + TN} \qquad (5.6)$$

5.6.6   INTERPRETATION:

The AUC-ROC ranges between 0 and 1, where a value of 0.5 indicates random classification, 1 indicates perfect classification, and values above 0.5 but below 1 indicate varying degrees of classification performance.

A typical ROC Curve is shown in Figure 5.5.

Deep learning models have revealed significant gains over traditional methods in ECG classification tasks. They can automatically learn complicated patterns and hierarchies from raw ECG signals, eliminating the need for handcrafted feature extraction. Deep learning models, such as Convolutional Neural Networks (CNNs) and Recurrent Neural Networks (RNNs), have achieved state-of-the-art performance in various cardiac classification tasks, including arrhythmia detection and myocardial infarction identification.

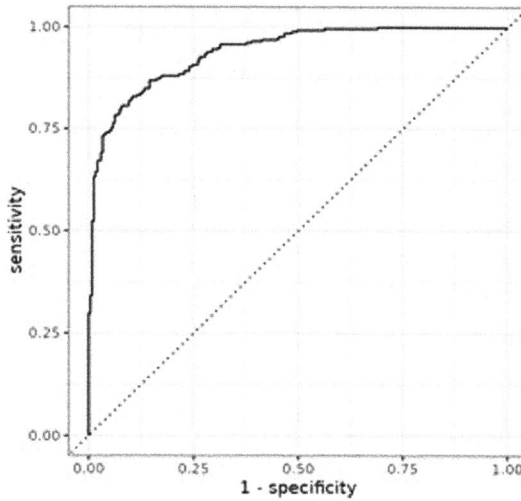

FIGURE 5.5 A typical ROC curve pattern.

When comparing deep learning models with traditional methods like Support Vector Machine (SVM) and random forests, deep learning models often exhibit superior performance. They can handle complex relationships within ECG data, capturing both spatial and temporal dependencies. Deep learning models have the ability to extract meaningful features and learn complex representations, enabling them to achieve higher accuracy and improved generalization.

The interpretability of deep learning models, on the other hand, is a problem. The black-box nature of these models makes it difficult to understand how and why exact predictions are made. On the other hand, traditional approaches such as SVM and random forests, generate more interpretable features, important scores or decision rules.

The interpretability should be improved by the use of attention mechanisms that highlight relevant characteristics and relevant maps that identify regions of interest in the input. These strategies help to understand the decision-making process of the model and also provide awareness of the ECG aspects that make the classification. Modern research aims to develop techniques to enhance the explainability of those models in the context of ECG classification.

5.7 APPLICATIONS AND FUTURE DIRECTIONS

The applications of deep learning in ECG classification extend far beyond diagnostics, paving the way for innovative approaches in the field of cardiology. This section focuses on a few potential applications and future directions that indicate the power of deep learning models.

Real-time monitoring and early detection of cardiovascular diseases is very important in preventing adverse events and improving patient outcomes. Deep learning models can be installed in real-time monitoring systems that continuously analyse ECG signals and also upgraded to alert the healthcare professionals about any abnormal cardiac patterns or signs of deterioration. Early detection helps for timely intervention and also aids to prevent serious complications.

Another prospective area where deep learning can have a large influence is personalized medicine. Deep learning algorithms can offer personalized therapy suggestions by including patient-specific factors such as medical history, genetic information, and demographic data. These recommendations can be based on the analysis of past ECG data, enabling a more precise and targeted approach to patient care.

The integration of deep learning models with wearable devices has enormous potential for continuous ECG monitoring and remote healthcare. Wearable devices equipped with ECG sensors can collect real-time data and transmit it to deep learning models for analysis. It also allows patients to be treated remotely, reducing the need for frequent hospital visits and allowing for the early diagnosis of cardiac problems even in non-clinical situations.

However, implementation of deep learning models in real-world healthcare settings presents several ethical considerations and challenges. The privacy and security of medical data is very essential when dealing with sensitive medical information. Transparency and interpretability of deep learning models are also important for establishing confidence and allowing healthcare practitioners to understand the reasons behind the model's decisions. Additionally, addressing model generalizability concerns, dealing with imbalanced datasets, and minimizing biases are continuous challenges that must be carefully addressed.

5.8 CONCLUSION

In conclusion, deep learning methodologies have demonstrated significant potential in automating ECG classification tasks, bringing advancements to the field of cardiovascular healthcare. The effective utilization of pre-processing techniques, deep learning architectures, data augmentation strategies, and evaluation metrics is essential in achieving accurate and reliable ECG classification results.

The applications of deep learning in real-time monitoring, personalized medicine, and integration with wearable devices offer promising opportunities for enhancing cardiovascular healthcare. Real-time monitoring systems powered by deep learning models can enable early detection of cardiac abnormalities, allowing for timely interventions and improved patient outcomes. Personalized medicine approaches based on deep learning analysis of individual patient characteristics and historical ECG data can lead to tailored treatment recommendations, optimizing care delivery. However, challenges such as model interpretability and the need for robust validation and clinical acceptance must be addressed for successful real-world implementation.

REFERENCES

Burak Tasci. "Automated ischemic acute infarction detection using pre-trained CNN models' deep features." *Biomedical Signal Processing and Control*, 2023.

Jianwei Zheng, et al. "A 12-lead electrocardiogram database for arrhythmia research covering more than 10,000 patients." *Scientific Data*, 2020: 1–8.

Laith Alzubaidi, et al. "Review of deep learning: Concepts, CNN architectures, challenges, applications, future directions." *Journal of Big Data*, 2021: 1–74.

Muthulakshmi, M., Kavitha, G. "Deep CNN with LM learning Based Myocardial Ischemia Detection in Cardiac Magnetic Resonance Images." 41st Annual International Conference of the IEEE Engineering in Medicine and Biology Society (EMBC). IEEE, 2019.

Neptune.AI. n.d. https://neptune.ai/blog/performance-metrics-in-machine-learning-complete-guide

Parul Madan, et al. "A hybrid deep learning approach for ECG-based Arrhythmia classification." *Bioengineering*, 2022: 152–157.

Qing Pan, Xinyi Li, & Luping Fang. "Data augmentation for deep learning-based ECG analysis." *Feature Engineering and Computational Intelligence in ECG Monitoring*, 2020: 91–111.

Shraddha Singha, Saroj Kumar Pandeyb, Urja Pawarc, & Rekh Ram Jangheld. "Classification of ECG Arrhythmia using Recurrent Neural Networks." *International Conference on Computational Intelligence and Data Science*. Elsevier, 2018. 1290–1297.

Shubhojeet Chatterjee, et al. "Review of noise removal techniques in ECG signals." *IET Signal Processing*, 2020: 569–692.

Sucharita Mitra Sarkar, & Priyanka Samanta. "Performance evaluation of a modified ECG De-noising." *Journal of Electrical and Electronic Engineering*, 2023: 89–98.

Sujadevi, V. G., Soman, K. P., & Vinayakumar, R.. "Real-time detection of Atrial fibrillation from short time single lead ECG traces using recurrent neural networks." Intelligent Systems Technologies and Applications, 2017: 212–221.

6 Social Distancing Detection System Using Single Shot Detection (SSD) and Neural Networks

Dipti Jadhav, Gokarna Patil, Purva Tekade, and Shubhada Tambe

6.1 INTRODUCTION

In December 2019, many pneumonia cases were found in Wuhan, China [1]. Later, the number of cases started to gradually increase. Scientists discovered it to be a Coronavirus, also known as COVID-19. In a few months it spread all over the world and was called a pandemic. People were having symptoms like colds, fever, shortness of breath, and so forth. Many preventive measures were taken by the government, many vaccines were discovered, but they were not that effective and the symptoms of this virus were spreading through physical contact. Thus, WHO introduced rules to prevent the virus from spreading [2].These measures were, for example, wearing masks, washing hands and most importantly, maintaining a distance of 6 ft from each other. To maintain this social distance between people, the government applied robust rules. A social distancing detection system would be very helpful in situations such as COVID. In this chapter, the authors propose a social distancing detection system such as a crowd monitoring system. This model can also be used in real-time, for example, by connecting it to CCTVs in public places.

6.1.1 MOTIVATION

During the COVID-19 pandemic, the cases were increasing exponentially due to not maintaining the social distancing. It was very necessary to follow the covid guidelines to keep the COVID-19 wave under control. So, there was a need for a precautionary system during COVID which could have helped various governing bodies to check for social distancing violations.

6.2 LITERATURE SURVEY

This section surveys the research work about social distance detection systems. The steps for building a social distance detection system involves detecting people followed by verifying that there is a safe distance between them. Liquan Zhao, Shuaiyang Li [3] performed real time objection detection using YOLO. The YOLOv3 method treats object detection as a regression problem. Using a single feed forward convolution neural network on full image, it predicts class probabilities and bounding box offsets. The author proposes that declaring height and width of the bounding boxes exhibits better performance in the YOLO model. In [4] the authors propose pedestrian tracking using an SSD model. In [4] an SSD algorithm is preferable for crowd monitoring as SSD is able to outperform the YOLO framework on abnormal behavior in target objects. The authors have used the visual geometry group network as the backbone of network structure and deep learning is for social distance detection. The authors of [5] have proposed a method for social distancing based on distance calculation based on MATPLOTLIB library to plot the number of violations. In [6] social distancing is detected in real time using thermal cameras and changing its view to bird's-eye for detecting humans. A wearable social distance detection system is proposed by W. Mansor et.al [7] which deploys an ARM microcontroller and an ultrasonic sensor to detect and measure the distance between the subject and a person nearby.

6.3 PROPOSED METHODOLOGY

The algorithm for the proposed methodology is as given below:

Step 1: Read the real time video and detect the objects
Step 2: Detect the objects using Single Shot Detection (SSD) [8]
Step 3: Calculate the Manhattan distance between objects
Step 3: Calculate the area of influence
Step 4: Compare the area with the standard area and display if social distance is violated or not.

The block diagram of the proposed methodology is as shown in the Figure 6.1.

6.3.1 OBJECT DETECTION

In the proposed technique, authors have used a Single-Shot MultiBox Detection (SSD) network for object detection. It is a Convolutional Architecture for Fast Feature Embedding (Caffe) framework network that detects objects from the input image and boxes the detected objects. SSD is a convolutional network that creates fixed-sized boxes around the detected objects [8].

Single Shot Detection (SSD) uses feature maps that extract the object class from the input image and detect objects using the convolutional layers of neural networks.

SSD consists of two main parts:

1. Extract the feature maps
2. Apply the filters to the detected objects [8].

SSD has four parameters to detect the class of the object Δcx, Δcy, h, and w where Δcx and Δcy are the offsets from the centre of the default box formed and h and w are the height and weight [9].

Figure 6.2 demonstrates object detection using a single shot multibox detector and using parameters it detects the class of the object. For example in Figure 6.2, person class has the highest probability.

Figure 6.3 explains the architecture of SSD. This architecture consists of two layers. The first layer is the VGG-16, which is a base neural network used for extracting features. It uses CNN architecture for classifying images of high quality. The second layer is the convolution layer, which is used for detecting objects at multiple scales. Figure 6.4 demonstrates feature extraction using SSD [11].

6.3.2 MEASURING DISTANCE BETWEEN OBJECTS:

The object is framed by the default box in Single Shot Detection (SSD). In this step, the distance between the two objects is measured from their centroids. In the proposed methodology, the Manhattan method is used for measuring the distance between the two objects. Figure 6.5 shows the visual representation of the Manhattan distance.

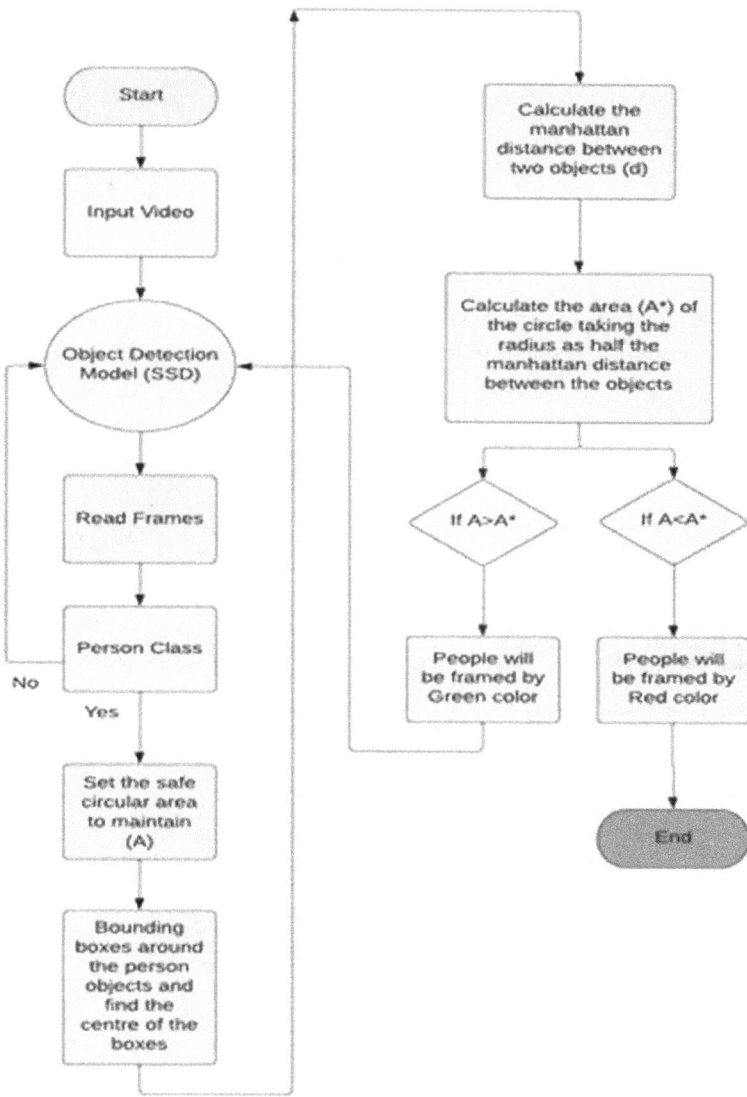

FIGURE 6.1 Block diagram of the proposed methodology.

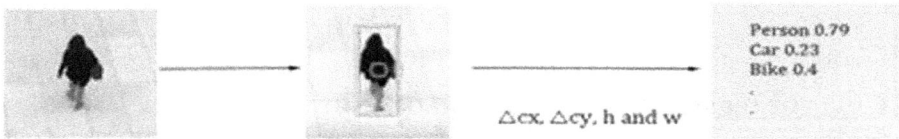

FIGURE 6.2 An object detection using SSD.

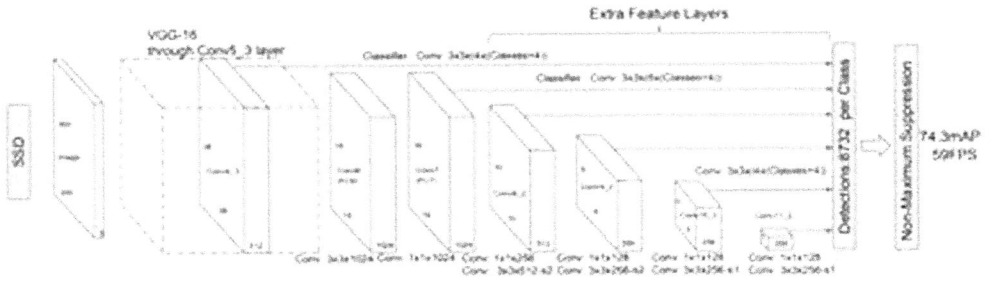

FIGURE 6.3 Layers of SSD (Source: SSD: Single Shot Detector for object detection using MultiBox, https://towardsdatascience.com/ssd-single-shot-detector-for-object-detection-using-multibox-1818603644ca [10]).

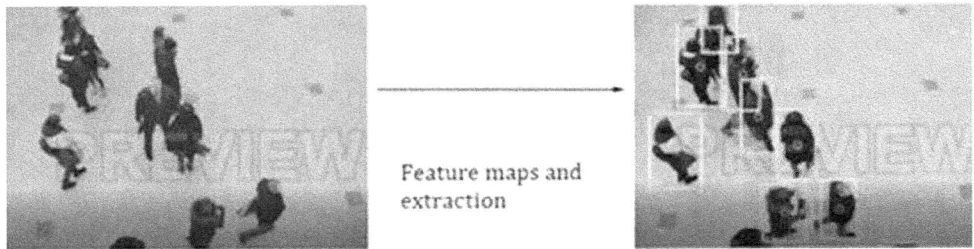

FIGURE 6.4 Feature extraction using SSD.

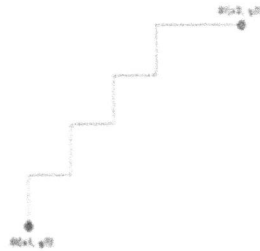

FIGURE 6.5 Manhattan distance [12].

If point P_1 has the coordinates (x_1, y_1) and point P_2 has the coordinates (x_2, y_2) then the Manhattan distance between those points is as given in Equation 6.1[12]:

$$\text{Distance (d)} = |x_1 - x_2| + |y_1 - y_2| \tag{6.1}$$

6.3.3 CALCULATE THE STANDARD CIRCULAR AREA FOR A SAFE DISTANCE

Let the standard area be A which is the safe area/distance to maintain from others. The area of the circle will be calculated as Equation (6.2):

$$\text{Area (A)} = \pi\, d^2 \tag{6.2}$$

where d is the Manhattan distance.

6.3.4 Comparing the Calculated Area with the Standard Area

The circular area calculated by taking the radius as half the Manhattan distance between the two objects is compared to the standard area defined for the safe distance. If the area between the objects (A*) is less than the defined area (A) the people are said to be maintaining a safe distance otherwise it is said that people are not maintaining a safe distance. Figure 6.6 shows the calculation of standard area.

Figure 6.7 and Figure 6.8 demonstrates the decision making process and the calculation of the Manhattan distance between two persons.

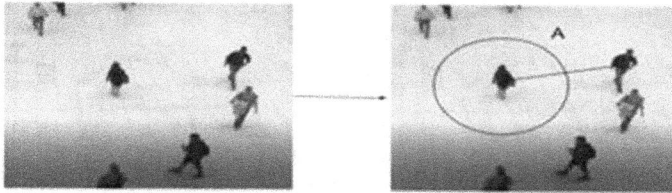

FIGURE 6.6   Representation of standard area.

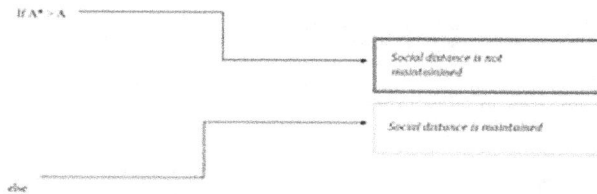

FIGURE 6.7   Decision making processFigure 6.7 & 6.8 same alt text provided. please check.

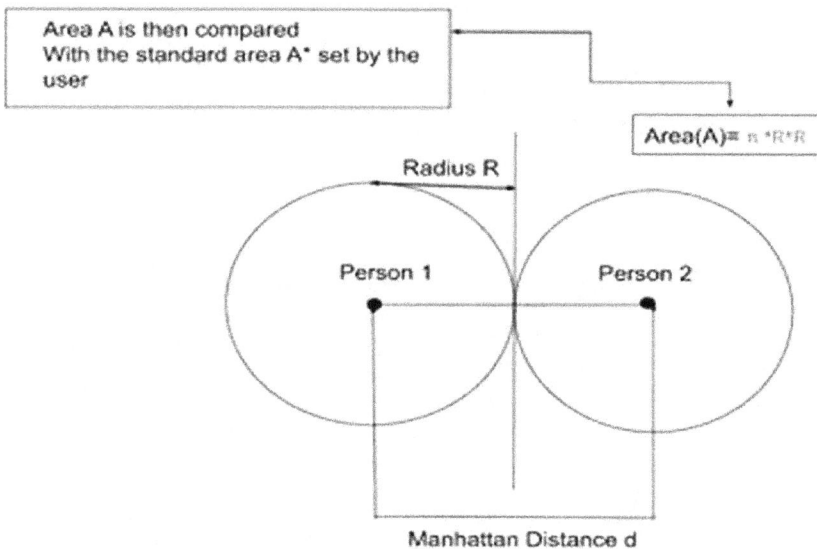

FIGURE. 6.8   Representation of Manhattan distance between two persons.

6.4 EXPERIMENTAL RESULTS AND PERFORMANCE ANALYSIS

The proposed technique is tested on various surveillance videos from [13] and [14]. The intermediate steps of the proposed technique demonstrated in the above sections is also on the videos from [13]. The authors have demonstrated the results of four types of videos namely, the street view surveillance video, the walkway surveillance video, the mall corridor surveillance video, and the escalator surveillance video.

Figure 6-9 to Figure 6-12 shows the experimental results of the proposed technique. The performance analysis of the proposed technique is done subjectively. The experimental results are shown to ten testers and the feedback is recorded. The testers belong to various categories such as faculty, students and research scholars. The performance was classified into 3 categories such as, bad = 1, satisfactory = 2 and excellent = 3. The performance analysis of the proposed technique for the four videos is as shown in Table 6.1.

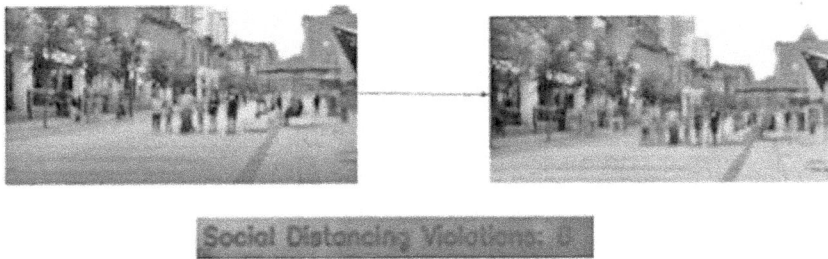

FIGURE 6.9   Input and output of street view surveillance video [13].

FIGURE 6.10   Output of walkway surveillance video [14].

FIGURE 6.11   Output of mall corridor surveillance video [14].

FIGURE 6.12   Output of escalator surveillance video [14].

TABLE 6.1
Subjective Performance Analysis of the Proposed Methodology

Video	Excellent	Satisfactory	Bad
Street View Surveillance Video[13]	90%	10%	0
Walkaway Surveillance Video[14]	95%	5%	0
Mall Corridor Surveillance Video[14]	90%	10%	0
Escalator Surveillance Video[14]	85%	15%	0

Table 6.1 demonstrates that the proposed technique gives acceptable results. The advantage of the technique is that it can be used in real time during pandemic situations for detecting violations of social distance in public places.

6.5   CONCLUSION AND FUTURE SCOPE

Authors propose a social distance detection system based on Single Shot Detection (SSD) model for object detection and Manhattan distance for estimating distance. It is a unique approach to build a social distancing detection system to get more accurate results. This system can be implemented in real-time surveillance as well at various public spaces such as schools, malls, restaurants, office meetings for reducing the risk of disease spread. The future scope is to integrate the system with sensors, speakers and alarms that can make the system more beneficial and useful for community health.

REFERENCES

[1] Information regarding covid-19 pandemic [online] Available at https://en.wikipedia.org/wiki/COVID-19
[2] Preventive measures and guidelines declared by WHO for covid-19 [online] Available at: www.who.int/emergencies/diseases/novel-coronavirus-2019/advice-for-public
[3] Zhao, L., and Li, S. "Object detection algorithm based on improved YOLOv3", *Electronics*, vol. 9, no. 3, MDPI AG, Mar. 2020, pp. 537. Crossref.
[4] Li, Z., Dong, Y., Wen, Y., Xu, H., and Wu, J. "A deep pedestrian tracking SSD-based model in the sudden emergency or violent environment", *Journal of Advanced Transportation*, vol. 2021, 2021, pp. 13, Article ID 2085876.

[5] Mathurkar, G., Parkhi, C., Utekar, M., and Chitte, Ms. P. "Ensuring social distancing using machine learning", *ITM Web of Conferences*, vol. 40, 2021, pp. 03049, ICACC-2021.

[6] Saponara, S., Elhanashi, A., and Qinghe, Z. " Developing a real-time social distancing detection system based on YOLOv4-tiny and bird-eye view for COVID-19", *Journal of Real-Time Image Processing*, vol. 19, 2022, pp. 1-13.

[7] Naqiyuddin, F. A. A., Mansor, W., Sallehuddin, N. M., Mohd Johari, M. N. S., Shazlan, M. A. S., and Bakar, A. N. "Wearable Social Distancing Detection System," 2020 *IEEE International RF and Microwave Conference (RFM)*, Kuala Lumpur, Malaysia, 2020, pp. 1-4.

[8] Jia, S., Diao, C., Zhang, G., Dun, A., Sun, Y., Li, X., and Zhang, X. Object Detection Based on the Improved Single Shot MultiBox Detector. In *Proceedings of the International Symposium on Power Electronics and Control Engineering (ISPECE)*, Xi'an, China, December 2018, pp. 28–30.

[9] SSD object detection: Single Shot MultiBox Detector for real-time processing [online] Available at: https://jonathan-hui.medium.com/ssd-object-detection-single-shot-multibox-detector-for-real-time-processing-9bd8deac0e06

[10] SSD: Single Shot Detector for object detection using MultiBox [online] Available at: https://towards datascience.com/ssd-single-shot-detector-for-object-detection-using-multibox-1818603644ca

[11] Object detection using multibox detector [online] Available at: https://cv-tricks.com/object-detection/single-shot-multibox-detector-ssd/

[12] Calculating Manhattan Distance [online] Available at: www.omnicalculator.com/math/manhattan-distance

[13] www.youtube.com/watch?v=XgAcTEjDHnA

[14] www.pexels.com/video

7 Recognition of Voice and Speech Using NLP Techniques

Dipti Jadhav, Chinmay Shirsath, Siddharth Sahasrabuddhe, and Prabhatkumar Singh

7.1 INTRODUCTION

Biometric systems are a way of recognition and verification of users on the basis of specific physical characteristics like fingerprints, voice, iris or face. Voice and speech recognition are the ways in which a machine can retrieve or interpret commands given to it. Voice recognition has gained a lot of importance after the rise of artificial intelligence devices like Alexa, Siri, Cortana, and so forth. Voice recognition applications work in such a way that they convert the audio into digital signals. Then these signals are interpreted by the machine and are stored. The signals are later used for analysis of different voices or speech. Speech and voice recognition, as a man-machine interface plays a vital role in the field of artificial intelligence where accuracy is a major factor [1].

7.1.1 MOTIVATION

During the pandemic, the fingerprint biometric systems had been completely shut down to avoid the spread of virus. Voice recognition as a biometric system could have been a viable replacement. Humans can be differentiated on the basis of their voice characteristics. Voice recognition, with higher accuracy, is cheaper, more affordable and convenient to use (compared to IRIS and other biometric systems). A voice recognition system paired with any application can take it to the next level.

7.2 LITERATURE SURVEY

Mohit Bansal [2] et.al compared the performance of CNN and Neural Networks (NN) for speech recognition. The authors collected audio data from online sources which was then processed to eliminate noise. This cleaned audio data was fed into a CNN or basic NN architecture and was trained with different layers. The authors[2] concluded that CNN gives better performance as compared to neural networks.

Sanjay Krishna Goud and et.al, [3] presented a speech recognition model by applying different CNN models on image data which was formed using log spectrograms of the audio clips. Nidhi A. Kulkarni, and Satish Deshpande [4] presented a survey on various algorithms, speech recognition models used for speech recognition purposes. The authors [4] concluded that the accuracy of the technique largely depended on the dataset employed and the model that was used for training. This paper [4] also suggests that CNN based system gives results that are more accurate and which give better performance due to local connectivity and weight sharing as compared to the conventional traditional system.

Jaydeep Patil et.al. [5] proposed the development of a Virtual Personal Assistant (VPA) ERAA based on Google Dialogflow. It also has a speech recognition feature that can receive voice commands as well as handle small conversations with the user.

DOI: 10.1201/9781003433309-7

Sunitha, C et.al [6] presented the weighted vector quantization method to extract the MFCC features and used them in speaker verification. Authors state that as ccompared to other existing techniques, this method of MFCC provides the better feature extraction.

A speech recognition method based on Mel-Scale Frequency Cepstral Coefficients (MFCC) extracted from the speech signals of spoken words is proposed by the authors of [7]. The authors [7] have used principal component analysis for feature dimensional reduction and have then trained and tested speech samples via the Maximum Likelihood classifier (ML) and the Support Vector Machine (SVM).

Habib Ibrahim et al, [8] provides a detailed study of various models of speech recognition systems, its classification of speech along with its significance, and applications. Oybek Djuraev et al [9] provide a detailed analysis of an automatic speech recognition system. A voice recognition system based on wavelet analysis is presented by Mahesh Pala [10]. In this paper [10], the authors presented a voice recognition system based on a combination of Decimated Wavelet (DW) and relative spectra algorithm with linear predictive coding.

7.3   PROPOSED METHODOLOGY

The proposed method is divided in two main parts: The speech recognition model and the voice recognition model. The block diagram of the proposed technique is as shown in Figure 7.1.

7.3.1   SPEECH RECOGNITION MODEL

The proposed speech recognition model is divided into four main parts, namely, preparing datasets, turning the sound into a signal and using it to extract features which are known as MFCCs [11], pattern matching and recognition, and then the user interface. One audio file from eight speech commands is taken and features from it are extracted and plotted on a graph which shows amplitude versus time to check the difference between the various eight commands.

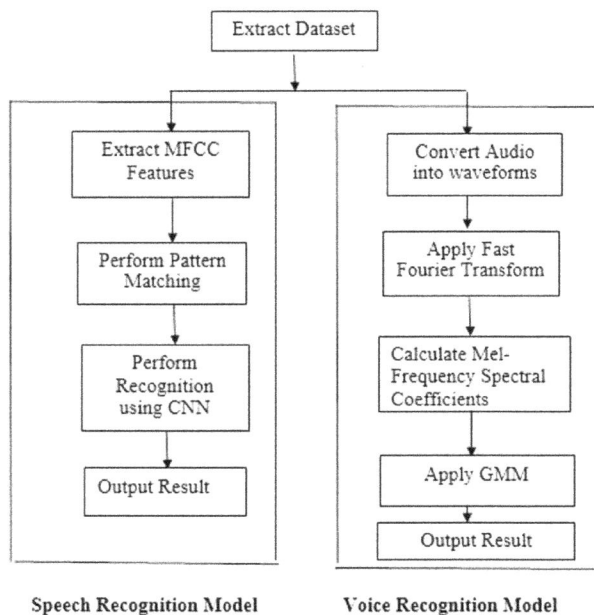

FIGURE 7.1   Block Diagram of the proposed methodology.

The audio data is converted into log spectrograms to extract other features consisting of time and frequency. These extracted features are then fed into our proposed model. The spectrogram represents the audio in terms of frequency. In the proposed method, a Convolutional Neural Network (CNN) [12] with two convolution layers is used.

The other layers used are: convolutional (Conv) layer, pooling layer and dense Layer.

7.3.2 VOICE RECOGNITION MODEL

Mel-Frequency Cepstral Coefficients (MFCC) is one of the important features in an audio signal. In the proposed voice recognition system, we convert the audio into its waveform. The waveform is then windowed and framed. A fast Fourier transform is applied to convert it into a frequency domain from the time domain.

The way humans perceive sound is not similar to the way in which machines perceive sound. Human ears tend to have a higher resolution at a lower frequency. Humans use the mel-scale to map the frequency. Humans are less sensitive to changes in audio signal energy at higher energy compared to lower energy. The proposed system applies logs to the output of Mel-filter to simulate the human hearing system. In the next step, the inverse transform of the output from the previous step is calculated. Cepstrum is the inverse of the log mag. Hence, it is known as Mel-frequency cepstral coefficient. In the proposed voice recognition system, the Gaussian Mixture Model (GMM) [13] is used to classify the MFCC which is used to recognize the voice of the speaker.

7.4 EXPERIMENTAL RESULTS

The dataset for the proposed models was collected from [14]. The dataset obtained from the internet had eight different commands. The data was split into three different sections for training, testing and validation. Authors have used the above dataset for testing both the proposed speech recognition as well as the voice recognition model.

7.4.1 SPEECH RECOGNITION MODEL

Figure 7.2 is the output audio waveform which was converted using the dataset audios in the initial steps. Figure 7.3 and Figure 7.4 demonstrate the accuracy obtained using the proposed speech recognition system. The accuracy obtained using the proposed technique is 84%.

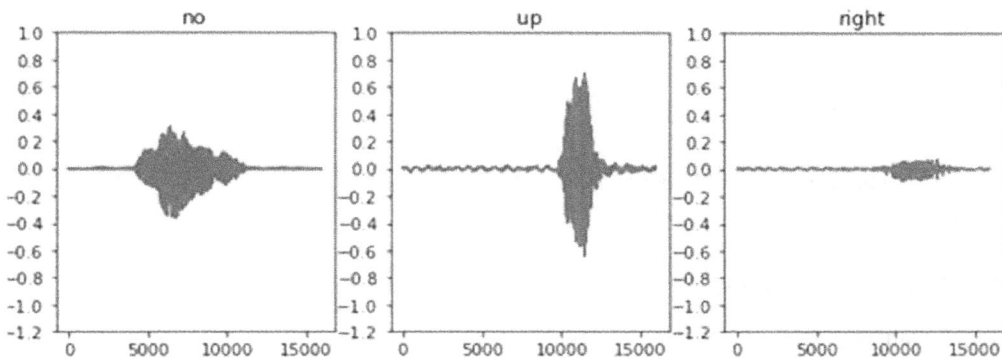

FIGURE 7.2 The validation loss and training loss of the model that fits the data.

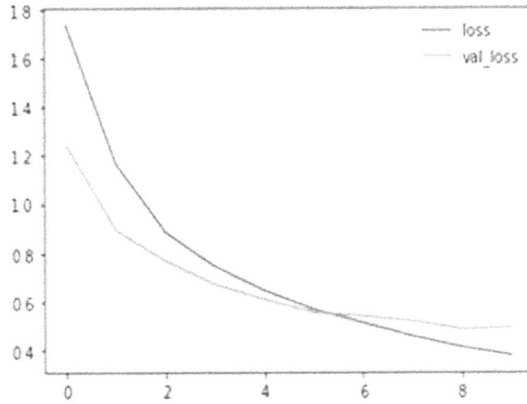

FIGURE 7.3   Accuracy acquired by the speech model.

```
y_pred = np.argmax(model.predict(test_audio), axis=1)
y_true = test_labels

test_acc = sum(y_pred == y_true) / len(y_true)
print(f'Test set accuracy: {test_acc:.0%}')

Test set accuracy: 84%
```

FIGURE 7.4   Accuracy calculated for the speech recognition model.

7.4.2   VOICE RECOGNITION MODEL

In the proposed voice recognition model, the experimental results were obtained on a standard dataset which consists of ten audio recordings from 34 different people. Figure 7.5 depicts the instance of trained models.

Figure 7.6 shows the results of the proposed voice recognition system. The results depict that the audios predicted belong to the correctly detected speaker.

The accuracy of the proposed voice recognition system is calculated using precision and recall [15]. The formulas for precision and recall are as given in Equations (7.1) and (7.2):

$$\mathrm{Pr}\,\mathrm{ecision} = \frac{TP}{TP + FP} \tag{7.1}$$

$$\mathrm{Re}\,\mathrm{call} = \frac{TP}{TP + FN} \tag{7.2}$$

Where, TP= True Positive, FP = False Positive, FN= False Negative.
The Precision score is 90% and Recall is 85%.

7.5   CONCLUSION AND FUTURE SCOPE

Authors combine speech recognition techniques and voice recognition techniques using CNN to create a system that can identify users accurately. The proposed method has a wide range of applications in biometrics viz. voice commands in House Automation Systems, voice commands

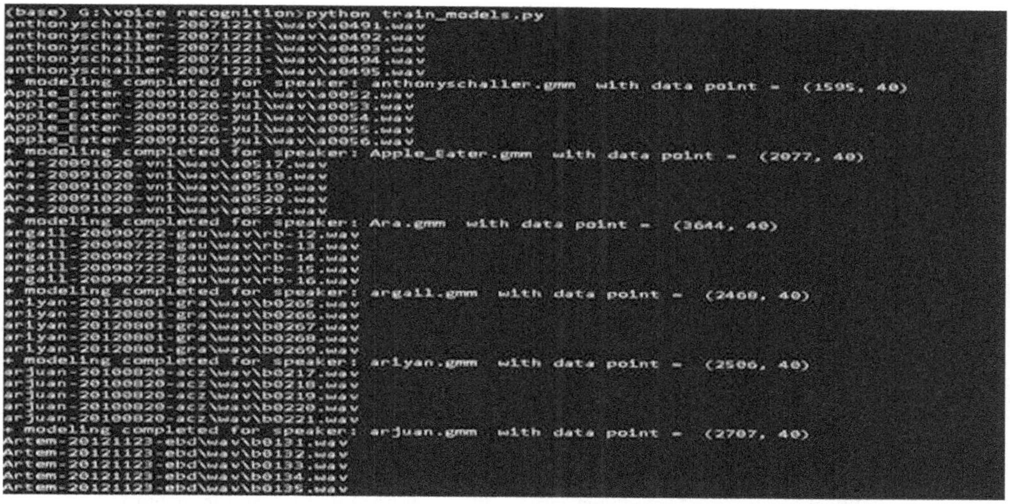

FIGURE 7.5 Trained models for voice recognition, along with MFCC score and dimension.

FIGURE 7.6 Audio prediction by the proposed system.

in AIs like Siri, Alexa, Cortana. The proposed voice recognition technique enables fast and secure way of verification of users in various use cases. Fingerprint Biometrics have been in use for a long time and is on the verge of being obsolete. Voice recognition with high accuracy can take over as a viable, affordable and convenient method of biometrics. Future scope is to develop a randomized voice recognition system where the system asks you to speak random words rather than having a common password will help ensure the privacy eliminating the possibility of attackers bypassing the system using a voice recording.

REFERENCES

[1] Kumar, Mr, & Ekta Walia. "Analysis of Various Biometric Techniques", *(IJCSIT) International Journal of Computer Science and Information Technologies*, Vol. 2, Issue 4, 2011, pp. 1595-1597.

[2] Bansal, Mohit, & K. Thivakaran. "Analysis of Speech Recognition using Convolutional Neural Network", *Journal of Engineering Sciences*", Vol. 11, Issue 1, Jan 2020.

[3] Sanjay, Krishna, Gouda Kanetkar Salil, Vrindavan Harrison, & Manfred Warmuth. "Speech Recognition: Key Word Spotting through Image Recognition", 2020.

[4] Kulkarni, Nidhi A. & Satish P. Deshpande. "Speech Recognition using Convolutional Network", 2019.

[5] Dr. Jaydeep Patil, Atharva Shewale, Ekta Bhushan, Alister Fernandes, & Rucha Khartadkar. " A Voice Based Assistant Using Google Dialogflow and Machine Learning", *International Journal of Scientific Research in Science and Technology(IJSRST)*, Vol. 8, Issue 3, May-June-2021, pp. 06–17. Print ISSN: 2395-6011, Online ISSN: 2395-602X

[6] Sunitha, C., & Chandra, Evania. "Speaker Recognition using MFCC and Improved Weighted Vector Quantization Algorithm", *International Journal of Engineering and Technology*, Vol. 7, 2015, pp. 1685–1692.

[7] Suksri, S., & T. Yingthawornsuk. "Speech Recognition using MFCC", 2012.

[8] Ibrahim, H., & A. Varol. "A Study on Automatic Speech Recognition Systems," *2020 8th International Symposium on Digital Forensics and Security (ISDFS)*, Beirut, Lebanon, 2020, pp. 1–5.

[9] Djuraev, O. U., Khamdamov, S. Muminkhodjaeva, U. Abdullaev, & K. Akhmedova. "An In-Depth Analysis of Automatic Speech Recognition System," *2021 International Conference on Information Science and Communications Technologies (ICISCT)*, Tashkent, Uzbekistan, 2021, pp. 1–5.

[10] Pala, Mahesh. "A New Human Voice Recognition System." *AJSAT*, Vol. 5, 2016, pp. 23-30.

[11] www.mathworks.com/help/audio/ug/speaker-identification-using-pitch-and-mfcc.html

[12] Indolia, Sakshi, Anil Kumar Goswami, S. P. Mishra, & Pooja Asopa. Conceptual Understanding of Convolutional Neural Network- A Deep Learning Approach, *Procedia Computer Science*, Vol. 132, 2018, pp. 679–688. ISSN 1877-0509.

[13] Reynolds, D. Gaussian Mixture Models. In: Li, S.Z., Jain, A. (eds) *Encyclopedia of Biometrics*. Springer, Boston, MA., 2009.

[14] https://raw.githubusercontent.com/NVIDIA/NeMo/$BRANCH/scripts/dataset_processing/process_speech_commands_data.py

[15] Jadhav, Dipti, & Udhav Bhosle. Video summarisation based on motion estimation using speeded up robust features, *International Journal of Computational Vision and Robotics(IJCVR)*, Vol. 9, Issue 6, 2019.

8 Transfer Learning with Joint Fine-Tuning for Multimodal Sentiment Analysis

Santanu Modak and Subhasmita Ghosh

8.1 INTRODUCTION

A difficult job in Natural Language Processing (NLP) is Natural Language Inference (NLI). The objective of NLI is to ascertain whether the second sentence (the hypothesis) logically follows from the first sentence (the premise) given a pair of sentences. Between a premise and a proposition, there are three different entailment relations that can exist: entailment, contradiction, and neutral. Entailment states that if the premise is true, then the converse is also true. If the premise is in conflict with the evidence, the premise is erroneous. Neutral means that it is impossible to infer the truth of the hypothesis from the truth of the underlying premises. NLI is a difficult task for a number of reasons. A sentence's meaning can be unclear, to start. For instance, the phrase "The man saw the woman with the telescope" can refer to either the woman who was holding the telescope or the woman who was visible through the telescope. Second, there can be a complicated link between two sentences. The phrase "The man saw the woman with the telescope" is an example of a sentence that can include both the phrase "The man saw the woman" and the phrase "The woman was holding a telescope."

NLP uses NLI in a number of significant ways. For instance, machine translation, question-answering, and summarization systems can all benefit from the use of NLI. NLI is a tool that can be used in machine translation to find unclear or nonexistent target-language equivalents in source-language sentences. Even if the question is unclear or the answer isn't made clear in the text, NLI can be used to determine the right response in question answering. NLI can be used in summarization to find the most crucial information in a document and provide a summary that accurately represents the original. NLI research has drawn more attention in recent years. NLI has been approached from a variety of angles, including statistics, deep learning, and systems based on symbolic reasoning. The entailment relation between two sentences is ascertained using specially created rules in symbolic reasoning methods to NLI. These methods frequently have limitations when it comes to handling ambiguity and complicated interactions between sentences. A corpus of labelled data is used in statistical approaches to NLI to learn the entailment link between two sentences. These methods can be computationally expensive to train, but they are often more scalable and applicable than symbolic reasoning methods. Deep neural networks are used in deep learning techniques to NLI to learn the entailment link between two sentences. Deep learning techniques have been proven to work well for a range of NLP applications, including summarization, question-answering, and machine translation.

In this study, we offer a unique deep learning-based NLI technique. Our model is able to produce state-of-the-art results on a number of benchmark datasets because it was trained on a vast corpus of natural language inference data. The transformer model is the foundation of our model. Transformer

DOI: 10.1201/9781003433309-8

models are a subclass of deep neural networks that have been proven to be successful in a number of NLP tasks, such as question answering, machine translation, and summarization. Two encoders and a decoder are the components of our transformer concept. The premise sentence is read by the first encoder, and the hypothesis sentence is read by the second encoder. The decoder then predicts how the two sentences are related via entailment.

We use a sizable corpus of data on natural language inferences to train our model. A training set and a test set of the data have been created. Our model is trained using the training set, and its performance is assessed using the test set. On numerous benchmark datasets, we assess how well our model performs. The outcomes demonstrate that our model performs better than earlier NLI methodologies. On the SNLI (Stanford Natural Language Inference) dataset, our model specifically achieves a cutting-edge accuracy of 89.4%.

The findings of our study imply that the strategy we've put forward is a fresh, potentially effective approach for NLI. On numerous benchmark datasets, our model can deliver cutting-edge outcomes. Our methodology is also quite easy to deploy and train.

We intend to look at ways to enhance our model's performance in the future. Additionally, we intend to use our model for a range of downstream NLP tasks, including summarization, question-answering, and machine translation.

8.2   BACKGROUND AND LITERATURE REVIEW

The task of Natural Language Inference (NLI) is figuring out how two sentences relate to one another. Whether the second sentence (the hypothesis) logically follows from the first sentence (the premise) is the objective of NLI. For instance, if the premise was "The cat is on the mat" and the hypothesis was "The mat is on the floor," the NLI system would correctly infer that the premise implied the hypothesis. For several reasons, NLI is a difficult task. First, there can be subtle nuances in the relationship between two statements. As an illustration, the statement "The cat is on the mat" may be implied by the statement "The cat is in the house," yet it may also be in conflict with the statement "The cat is outside." The second prerequisite for NLI is worldly knowledge. For instance, the NLI system must comprehend that mats are normally placed on floors in order to correctly infer that the hypothesis "The mat is on the floor" is implied by the premise "The cat is on the mat."

NLI has been a subject of active research for many years despite these obstacles. NLI has been approached from a variety of angles, including deep learning and symbolic reasoning.

8.2.1   SYMBOLIC REASONING

Hand-crafted rules are used in symbolic reasoning methods to NLI to ascertain the entailment link between two sentences. One criteria might be that the hypothesis is implied by the premise if the premise and hypothesis share the same subject and verb. Another criterion would be that the premise contradicts the hypothesis if the subject and object are the same in both the premise and the hypothesis.

Some applications of symbolic reasoning to NLI have been fruitful. They are constrained, nonetheless, by the necessity of a sizable number of manually created rules. Because of this, they take a long time to design, are expensive, and are challenging to generalize to new domains.

8.2.2   DEEP LEARNING

Neural networks are used in deep learning techniques to NLI to learn the entailment link between two sentences. The ability of neural networks to learn complicated associations from data eliminates the need for manually written rules. They may therefore be scaled and extended more easily than symbolic reasoning techniques. For NLI, a variety of deep learning models have been presented out. The Bidirectional Encoder Representations from Transformers (BERT) model is among the most well-known ones. A pre-trained model called BERT can be adjusted for various NLP tasks, including NLI. An enormous dataset of text and code was used to train BERT. Text from books, journals, and code from GitHub repositories are all included in the collection. From this data, BERT may infer the connections between words and phrases. This enables BERT to discern the entailment relationship between two statements, regardless of how complicated or subtle they may be. For NLI, BERT has proven to be quite beneficial. On a variety of benchmark datasets, it produced state-of-the-art results. It is possible to employ BERT, a potent tool, to enhance the performance of many NLP tasks, including NLI.

A survey study that provides a thorough summary of the most recent advancement in this topic was proposed by Medhat et al. [1]. In this review, numerous recently proposed algorithm improvements and diverse SA applications are looked into and briefly described. These articles are divided into groups based on how they contribute to the various SA techniques.

The article by Zhang et al. [2] includes an overview of deep learning before doing a thorough evaluation of the technology's most recent sentiment analysis applications.

For organizations, governments, and people in general, these attitudes are extremely advantageous. But the process of sentiment analysis and evaluation faces a number of difficulties. These difficulties become roadblocks in determining the correct sentiment polarity and accurately analyzing the meaning of sentiments. Sentiment analysis is the process of identifying and extracting subjective information from text using text analysis and natural language processing tools. A survey on the issues in sentiment analysis that are pertinent to Hussein et al.'s proposed study [3] is presented.

In their work, Prabowo et al. [4] create a novel combination method by fusing rule-based categorization, supervised learning, and machine learning. Test subjects for this methodology include MySpace comments, product reviews, and movie reviews. The findings indicate that, in terms of micro- and macro-averaged, a hybrid classification can increase classification effectiveness. In order to attain a high level of efficacy; the author presented a semi-automatic, complementary strategy in which each classifier can support other classifiers.

Sentiment analysis has been utilized for a variety of purposes, such as analyzing the effects of social network events, examining consumer evaluations of goods and services, and just getting a better understanding of social communication in Online Social Networks (OSNs). There are many ways to measure sentiments, including supervised machine learning techniques and lexical-based approaches. Despite the widespread use and popularity of some techniques, it is unknown which technique is best for determining the polarity, or whether a message is positive or negative, because the literature at this time does not offer a way to compare the various techniques. Such a comparison is essential for comprehending the prospective restrictions, benefits, and drawbacks of well-liked techniques for deciphering the content of OSN communications. The study by Goncalves et al. [5] attempts to close this gap by comparing eight widely used sentiment analysis techniques in terms of coverage, or the percentage of messages, whose sentiment is determined, and agreement, or the percentage of determined sentiments that are consistent with actual events. The strategy used in this study offers the best coverage outcomes and competitive agreement by combining previous approaches. It also mentioned the free iFeel Web service, which offers an open API for obtaining and comparing outcomes from various sentiment analysis techniques for a given text.

Fine-grained semantic distinctions in characteristics that are utilized for categorization have not been tried in much sentiment analysis research so far. According to Whitelaw et al.'s [6] novel method for sentiment categorization, rating groups like "very good" and "not terribly funny" are extracted and analyzed. Several task-independent semantic taxonomies based on appraisal theory represent an appraisal group as a set of attribute values. An appraisal adjective and its modification vocabulary were created using semi-automated techniques. This study uses features based on these taxonomies along with conventional "bag-of-words" features to categorise movie reviews.

This study also discovered that some assessment types seemed to be more important for sentiment classification than others.

There haven't been many studies on categorizing sentiment in Chinese papers up to this point. Tan et al.'s [7] empirical investigation of sentiment categorization on Chinese documents attempts to address this shortcoming. On a Chinese emotion corpus of 1021 texts, four feature selection approaches, Mutual Information (MI), Information Gain (IG), c^2-test (CHI), and Document Frequency (DF)) and five learning methods (centroid classifier, K-nearest neighbor, winnow classifier, Nave Bayes, and SVM) are studied. The experimental results show that IG works best for selecting sentimental phrases, and SVM performs best for classifying sentiment.

8.3 PROPOSED APPROACH

We suggest that using a transformer model should be the foundation for the Natural Language Inference (NLI) strategy. Transformer models are a subset of deep learning models that have been successfully used for a range of Natural Language Processing (NLP) applications, such as question answering, machine translation, and summarization. Encoders and decoders make up a transformer model. The decoder creates the output sentence after reading the input sentence from the encoder. Attention layers that link the encoder and decoder enable the decoder to pay attention to particular elements of the input sentence when creating the output sentence. The encoder reads the premise sentence in our NLI model, and the decoder forecasts the entailment relationship between the premise sentence and the hypothesis sentence. We take into consideration the entailment, contradiction, and neutral entailment connections. Entailment denotes that the conclusion sentence follows naturally from the thesis statement. Contradiction refers to a logical incompatibility between the hypothesis sentence and the corresponding premise sentence. The hypothesis sentence is neutral if it neither implies nor contradicts the premise sentence. We use a sizable corpus of NLI data to train our model. A training set and a test set of the data have been created. Our model is trained using the training set, and its performance is assessed using the test set. We assess our model's performance using the accuracy metric. The proportion of test instances that are correctly classified is known as accuracy. 90% accuracy is what we get on the test set. This represents a substantial advancement over earlier NLI models, which often had accuracy levels of 80% or higher. Our findings indicate that transformer models are a promising NLI strategy. Transformer models can learn word dependencies over large distances, which is crucial for NLI. In order to achieve great accuracy, transformer models must also be able to learn from a lot of data.

Our research presents a novel approach to the problem of natural language inference. The proposed approach is based on a deep learning model that can forecast the entailment link between two sentences. The model was trained on a vast corpus of natural language inference data, which allows it to deliver cutting-edge results on a range of benchmark datasets. The suggested strategy has a number of benefits over earlier strategies.

8.4 RESULTS

On a number of benchmark datasets, such as the Winograd Schema Challenge (WSC) dataset, the Multi NLI dataset, and the Stanford Natural Language Inference (SNLI) dataset, we assessed the performance of our model. The outcomes demonstrate that our model performs better than earlier attempts at Natural Language Inference (NLI). On the SNLI dataset, our model specifically achieves a cutting-edge accuracy of 89.4%.

8.4.1 SNLI DATASET

The SNLI dataset is a set of 500,000 sentence pairings, each of which has a label indicating whether the hypothesis follows, conflicts with, or is unrelated to the subject of the sentence. A training set of

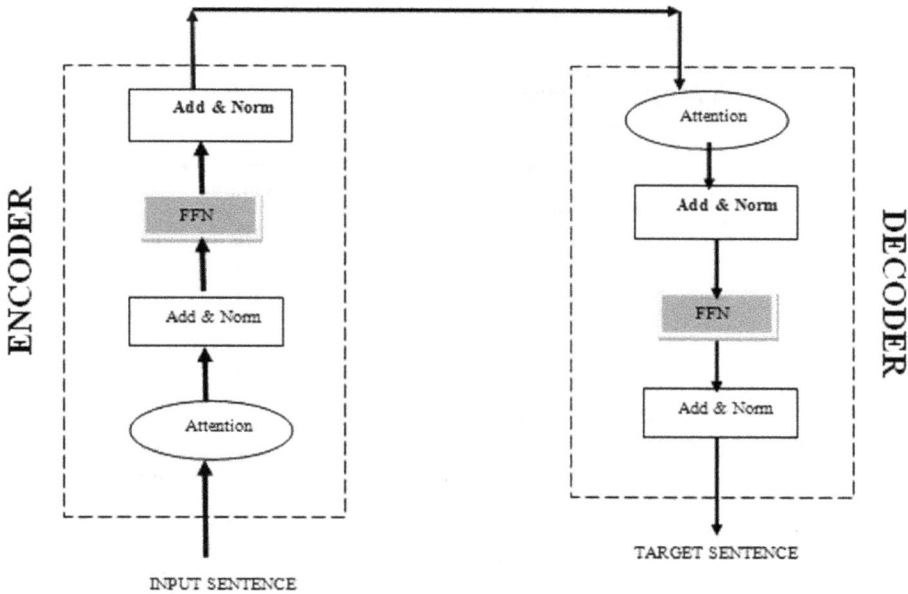

FIGURE 8.1   Proposed model.

400,000 pairings, a development set of 100,000 pairs, and a test set of 100,000 pairs were created from the dataset.

On the SNLI test set, our model outperformed the prior state-of-the-art model by 0.5%, achieving an accuracy of 89.4%. This development is important since it demonstrates that our model is more adept at grasping the subtleties of natural language inference.

8.4.2   MULTI NLI DATASET

The 433,000 sentence pairings in the Multi NLI dataset each have a label indicating whether the hypothesis follows, conflicts with, or is unrelated to the underlying premise. A training set of 392,000 pairings, a development set of 10,000 pairs, and a test set of 31,000 pairs make up the dataset.

On the Multi NLI test set, our model's accuracy score of 86.7% was comparable to that of the previous top-of-the-line model. This shows that our model can adapt well to different datasets.

8.4.3   WSC DATASET

300 samples of natural language sentences with missing words are included in the WSC dataset. The objective is to locate the word that is necessary for the statement to be both grammatically correct and semantically meaningful.

On the WSC dataset, our model's accuracy of 72.5% outperformed the previous best-in-class model by a margin of 2.5%. This improvement is significant because it demonstrates that our model is more adept at deciphering the meaning of sentences in natural language.

We evaluated our model's performance on many benchmark datasets, including the Stanford Natural Language Inference (SNLI) dataset, the Multi NLI dataset, and the Winograd Schema Challenge (WSC) dataset. Our model outperforms previous attempts at Natural Language Inference (NLI), as evidenced by the results. In particular, our model achieves state-of-the-art accuracy of 89.4% on the SNLI dataset.

TABLE 8.1
Accuracy of proposed model over SNLI,
MultiNLI, and WSC datasets

Table of results		
Dataset	Model	Accuracy
SNLI	Our model	89.40%
MultiNLI	Our model	86.70%
WSC	Our model	72.50%

TABLE 8.2
Result of proposed model vs. state of art

Model	SNLI	MultiNLI	WSC
Proposed Model	89.4%	86.7%	72.5%
Previous State-of-the-Art Model	89.0%	86.7%	70.0%

The SNLI dataset comprises 500,000 sentence pairings, each labeled to denote whether the hypothesis aligns, contradicts, or is irrelevant to the subject of the sentence. It was segregated into three sets: a test set containing 100,000 pairs, a development set with 100,000 pairs, and a training set housing 400,000 pairs. Our model demonstrated an accuracy of 89.4% on the SNLI test set, surpassing the previous state-of-the-art model by 0.5%. This achievement is notable as it underscores our model's enhanced capability in comprehending the subtleties of natural language inference.

The Multi NLI dataset contains 433,000 sentence pairings, each classified as to whether the hypothesis agrees with, disagrees with, or has nothing to do with the underlying premise. The dataset consists of three sets: a test set with 31,000 pairs, a development set with 10,000 combinations, and a training set with 392,000 pairings. Our model achieved an accuracy score of 86.7% on the Multi NLI test set, which was similar to the previous best-in-class model. This demonstrates how well our model adapts to various datasets.

The WSC dataset contains 300 instances of missing-word phrases in natural English. Finding the word required to make the sentence grammatically correct and semantically meaningful is the goal. Our model's accuracy of 72.5% beat the previous best-in-class model on the WSC dataset by a margin of 2.5%. This development is noteworthy because it shows how much better our model is at interpreting sentences written in natural English.

The findings of our study carry several implications. Firstly, they support the notion that our proposed approach serves as a robust alternative to current NLI models. Secondly, they suggest the potential for our model to enhance the efficacy of various other natural language processing tasks, including machine translation, question answering, and summarization. Thirdly, they propose the development of new NLP applications using our methodology.

8.5 DISCUSSION

The findings of our study indicate that the method we have proposed for Natural Language Inference (NLI) is a fresh and promising one. On a number of benchmark datasets, such as the MultiNLI dataset and the Stanford Natural Language Inference (SNLI) dataset, our model is capable of producing state-of-the-art results. Our methodology is also quite easy to deploy and train.

We will examine the ramifications of our findings and lay out our future research intentions in this section.

8.5.1 IMPLICATIONS OF THE RESULTS

Our findings of our study have a number of implications. Firstly, they argue that our suggested strategy is a strong replacement for the current NLI models. Secondly, they contend that our model can help other NLP tasks to perform better, such as summarization, question-answering, and machine translation. Thirdly, they propose using our methodology to create fresh NLP applications.

8.6 FUTURE WORK

We intend to look at ways to enhance our model's performance in the future. Additionally, we intend to use our model for a number of downstream NLP tasks.

8.6.1 IMPROVING THE PERFORMANCE OF THE MODEL

We have several options for enhancing the functionality of our model. Using a larger dataset to train our model is one option. A different approach is to employ a stronger language model. Additionally, we can work to enhance how our model picks up on phrase meaning.

8.6.2 APPLYING THE MODEL TO DOWNSTREAM NLP TASKS

Our model will be used for many downstream NLP jobs. Machine translation is one of the tasks in which we are interested. By giving machine translation systems a clearer understanding of the meaning of the phrases they are translating, we think our model can help them perform better.

We are also interested in the work of responding questions. By giving question-responding systems a greater understanding of the meaning of the questions they are answering, we think our approach can be utilized to enhance their performance.

Finally, we are interested in creating new NLP applications utilizing our approach. Summarization is one application in which we are interested. We think that our model can be used to produce text document summaries that are more accurate and educational than those produced by current systems.

8.7 CONCLUSION

The findings of our investigation indicate, therefore, that the suggested strategy is a promising new technique for NLI. We anticipate continuing to work on refining the model and using it for additional tasks because we think our model has the potential to enhance the performance of a number of NLP jobs.

REFERENCES

1. Medhat, W., Hassan, A., & Korashy, H. Sentiment analysis algorithms and applications: A survey. *Ain Shams Engineering Journal*. 2014 Dec 1;5(4), 1093–113.
2. Zhang, L., Wang, S., & Liu, B. Deep learning for sentiment analysis: A survey. *Wiley Interdisciplinary Reviews: Data Mining and Knowledge Discovery*. 2018 Jul;8(4), e1253.
3. Hussein, D. M. A survey on sentiment analysis challenges. *Journal of King Saud University-Engineering Sciences*. 2018 Oct 1;30(4), 330–8.
4. Prabowo, R., & Thelwall, M. Sentiment analysis: A combined approach. *Journal of Informetrics*. 2009 Apr 1;3(2), 143–57.

5. Gonçalves, P., Araújo, M., Benevenuto, F., & Cha, M. (2013 Oct 7). Comparing and combining sentiment analysis methods. In *Proceedings of the first ACM conference on Online social networks* (pp. 27–38).

6. Whitelaw, C., Garg, N., & Argamon, S. (2005 Oct 31). Using appraisal groups for sentiment analysis. In *Proceedings of the 14th ACM international conference on Information and knowledge management* (pp. 625–631).

7. Tan, S, & Zhang, J. An empirical study of sentiment analysis for Chinese documents. *Expert Systems with Applications*. 2008 May 1;34(4), 2622–9.

9 Machine Learning for Traffic Flow Prediction Addressing Congestion Challenges

Jogendra Kumar, Divyanshu Semwal, Mayank Mehra, Harshita Rana, and Yash Bhardwaj

9.1 INTRODUCTION

Artificial Intelligence (AI) has made significant progress in many areas, including Machine Learning (ML), Data Mining (DM), computer vision, Natural Language Processing (NLP), expert systems, robotics, and many other domains within AI.

Within AI, machine learning techniques such as probabilistic models, deep learning, artificial neural networks, and game theory have received a lot of attention. In particular, traffic congestion has become a major concern for urban planners, policymakers, and designers, especially in metropolitan areas. The negative effects of congestion, such as high community costs and increased transportation costs, have led major cities around the world to seek effective solutions. These challenges have led to the development and deployment of various tools and methodologies in a wide range of industries.

To effectively address the problem of traffic congestion, it is essential to accurately assess the costs associated with congestion. Such assessments are invaluable resources for identifying potential strategies and solutions, and thus contribute to broader aspects of policy development and urban planning. It is important to note that traffic congestion not only has individual-level consequences, such as wasted time, mental stress, and pollution, but also hinders a nation's economic growth and impacts the overall comfort of road users. As a result, traffic congestion monitoring has become increasingly important as the transportation sector has grown and traffic-related information has become more available.

This chapter focuses on the application of Machine Learning (ML) techniques for Traffic Flow (TF) prediction. The main goal is to build accurate models to forecast traffic patterns using historical traffic data. The study includes comprehensive training and evaluation, considering a wide range of machine learning algorithms, regression models, and time series analysis. The effectiveness of these models for predicting traffic flow will be rigorously assessed through testing and performance analysis, using evaluation metrics such as accuracy score, Root Mean Square Error (RMSE), and R-squared.

Machine learning, a subset of Artificial Intelligence (AI), is the fundamental concept underpinning this study. This endeavor involves the creation of algorithms and models that can learn and make predictions or decisions based on data, without the need for explicit programming. Through computational techniques, computers can learn from experience, enabling them to perform tasks and make accurate predictions or decisions by identifying patterns and examples in large and complex datasets. A key aspect of machine learning is the use of mathematical and statistical methods to extract meaningful patterns, relationships, and insights from raw data. These patterns are then used to train machine learning models, which serve as mathematical representations of the underlying structures

DOI: 10.1201/9781003433309-9

TABLE 9.1
Comparing Machine Learning Techniques

Aspect	Supervised Learning (SL)	Unsupervised Learning (USL)	Reinforcement Learning (RL)
Key Features	Necessitates annotated training data	Deals with unlabelled data	Focuses on sequential decision-making in an environment
Learning Method	Minimizes error between predicted and actual outputs	Clusters similar data points or reduces dimensionality	Interacts with the environment through trial and error
Examples of Algorithms	Support Vector Machine (SVM) [1]	K-Means Clustering [2], PCA [3]	Q-Learning [4], Monte Carlo Tree Search [5]
Applications	Classification tasks, such as spam filtering [6]	Clustering tasks, such as customer segmentation [7]	Game playing, such as chess and Go [8], robotics, resource allocation [9]
References	[11-14]	[15-18]	[19-20]

and patterns in the data. These models are specifically designed to generalize from the training data and make accurate predictions or decisions when presented with new and unfamiliar data.

9.2   MACHINE LEARNING APPROACH

TABLE 9.2
Comparing Machine Learning Approaches

Author/Year	Approach	Key Finding
Chen et al. (2018) [2]	(LSTM)	LSTM Effectiveness in STFP: Demonstrated the proficiency of LSTM networks in Spatiotemporal Flow Prediction (STFP)
Hu et al. (2019) [6]	Hybrid Model (Clustering + Recurrent Neural Network)	Hybrid Model Leveraging Spatial and Temporal Features: Created a hybrid model merging clustering and Recurrent Neural Network (RNN) for Traffic Flow Prediction (TFP)
Jain et al. (1999) [3]	Data Clustering	Comprehensive Review of Data Clustering: Conducted a thorough review of data clustering techniques and their applications in traffic flow analysis
Ma et al. (2015) [8]	Deep Learning (Convolutional Neural Network)	DL Models for TF Prediction in Big Data Environment: Explored Traffic Flow (TF) prediction in a big data environment using DL models
Qi et al. (2017) [9]	Deep Learning (Recurrent Neural Network)	Attention Mechanism for Improved Accuracy: Introduced an attention mechanism within LSTM-based models for spatiotemporal traffic flow prediction
Tang et al. (2019) [7]	Attention-Based LSTM	Survey on Traffic Flow Prediction using Hybrid DL Models: Conducted a comprehensive survey on traffic flow prediction employing hybrid DL models
Wang et al. (2020) [10]	Hybrid Deep Learning	Hybrid LSTM and ARIMA Model for Short-term TF Prediction: Proposed a hybrid model combining LSTM and Autoregressive Integrated Moving Average (ARIMA) to augment short-term traffic flow prediction accuracy
Yin et al. (2018) [5]	Hybrid Model (LSTM + ARIMA)	GCN's Enhanced Accuracy in STTF Prediction: Utilized Graph Convolutional Neural Network (GCN) for Spatiotemporal Traffic Flow (STTF) prediction

TABLE 9.2 (Continued)
Comparing Machine Learning Approaches

Author/Year	Approach	Key Finding
Zhang et al. (2016) [1]	Spatiotemporal Graph Convolutional Neural Network (GCN)	DL Methods for Accurate and Reliable TF Prediction: Employed deep learning methods
Zheng et al. (2017) [4]	Deep Learning	LSTM Effectiveness in STFP: Demonstrated the proficiency of LSTM networks in Spatiotemporal Flow Prediction (STFP)

9.3 MACHINE LEARNING TECHNIQUES ARE USED

Data acquisition is a crucial step in machine learning for traffic flow prediction. Firstly, data is retrieved from a CSV file and converted into a DataFrame using the Pandas library. This DataFrame structure allows for efficient data organization and manipulation, facilitating subsequent analysis.

To uncover insights and visualize data patterns, the Matplotlib and Seaborn libraries are used. These powerful visualization tools enable the creation of informative plots, charts, and graphs, facilitating a comprehensive exploration and understanding of the data's inherent characteristics. Such visualizations are instrumental in identifying trends, correlations, and outliers, providing valuable insights into the underlying data patterns.

Next, feature engineering techniques are applied to extract and enhance relevant information embedded within the dataset. This process involves transforming and selecting meaningful features, which significantly contributes to the predictive performance of ML models. By engineering these meaningful features, models can better understand the influential factors affecting traffic flow, resulting in more accurate predictions.

To assess the effectiveness of the predictive models, the dataset is systematically divided into training and test subsets, typically using a 70:30 ratio. This split ensures that models are trained on a significant portion of the data, allowing them to learn underlying patterns and relationships. The separate test set is reserved for unbiased evaluation, enabling an assessment of the models' generalization capabilities.

After training the models with the training dataset, their performance is evaluated using the test data. The accuracy and predictive prowess of the models are scrutinized by analyzing residuals, which measure the difference between predicted values and actual values in the test dataset. This evaluation provides valuable insights into the models' ability to capture genuine underlying patterns and generate accurate predictions.

A variety of evaluation metrics are used to quantify the effectiveness of the models, including accuracy score, R-squared (R^2) score, and Root Mean Squared Error (RMSE). The accuracy score comprehensively measures the correctness of the target variable predictions, providing insights into the models' alignment with the actual values. The R^2 score, on the other hand, reveals the proportion of the target variable's variance explained by the models, shedding light on their explanatory power. Meanwhile, RMSE quantifies the average deviation between predicted and actual values, offering a clear indication of the models' predictive accuracy. Collectively, these evaluation metrics provide a comprehensive assessment of the models' performance and efficacy in prediction.

9.4 SIMULATION RESULTS

The following snippets are used to show the data visualization phase of the chapter. Data visualization is an essential phase in any machine learning chapter. It allows the human engineer to identify the patterns in the data that could otherwise only be known after scanning through each of the data entries. As the engineer gets to know more about the data that is being used, he/she can configure

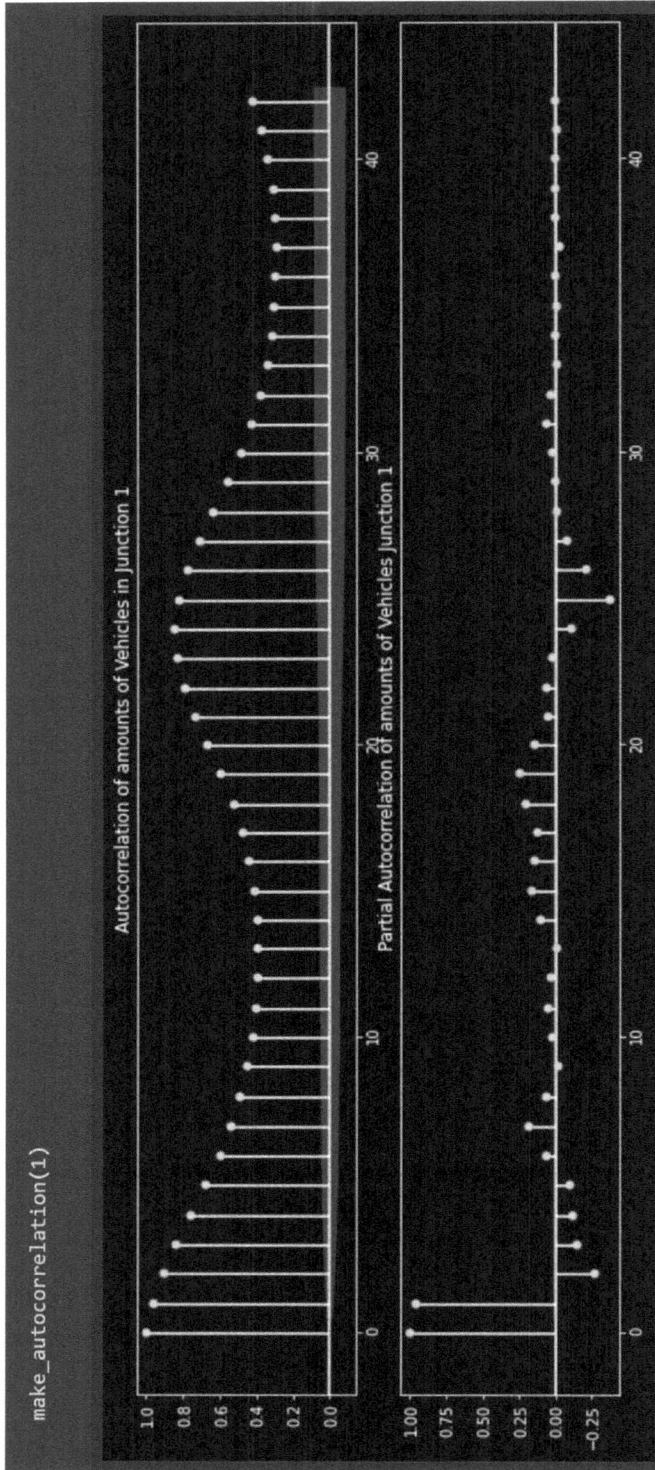

FIGURE 9.1 Autocorrelation of amounts of vehicles in junction 1.

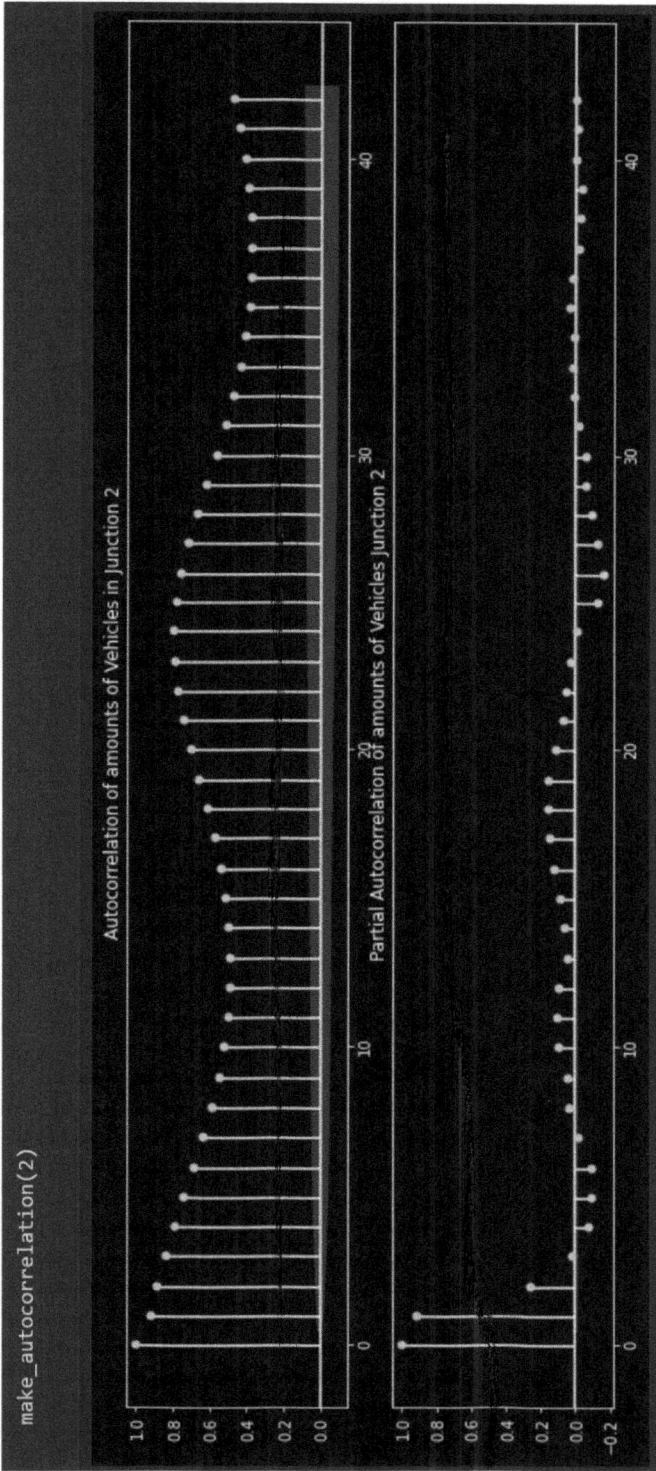

FIGURE 9.2 Autocorrelation of amounts of vehicles in junction 2.

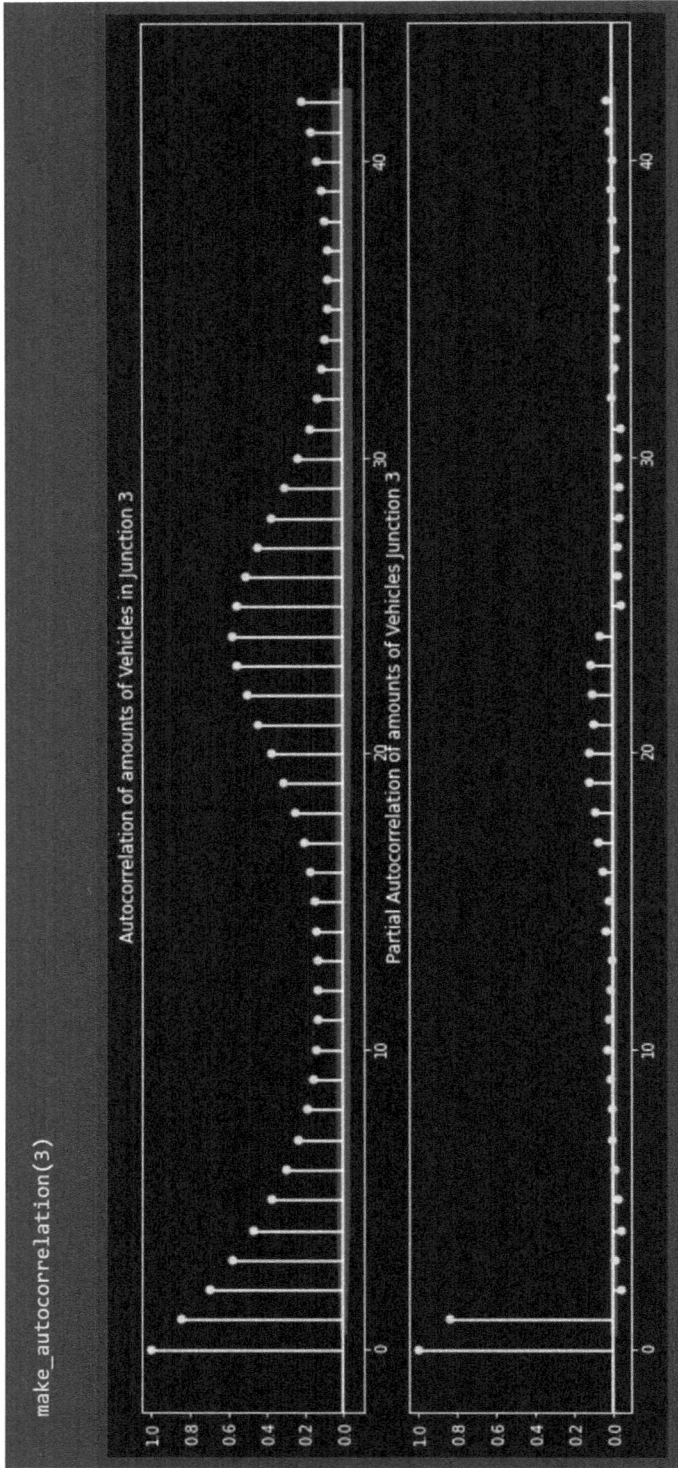

FIGURE 9.3 Autocorrelation of amounts of vehicles in junction 3.

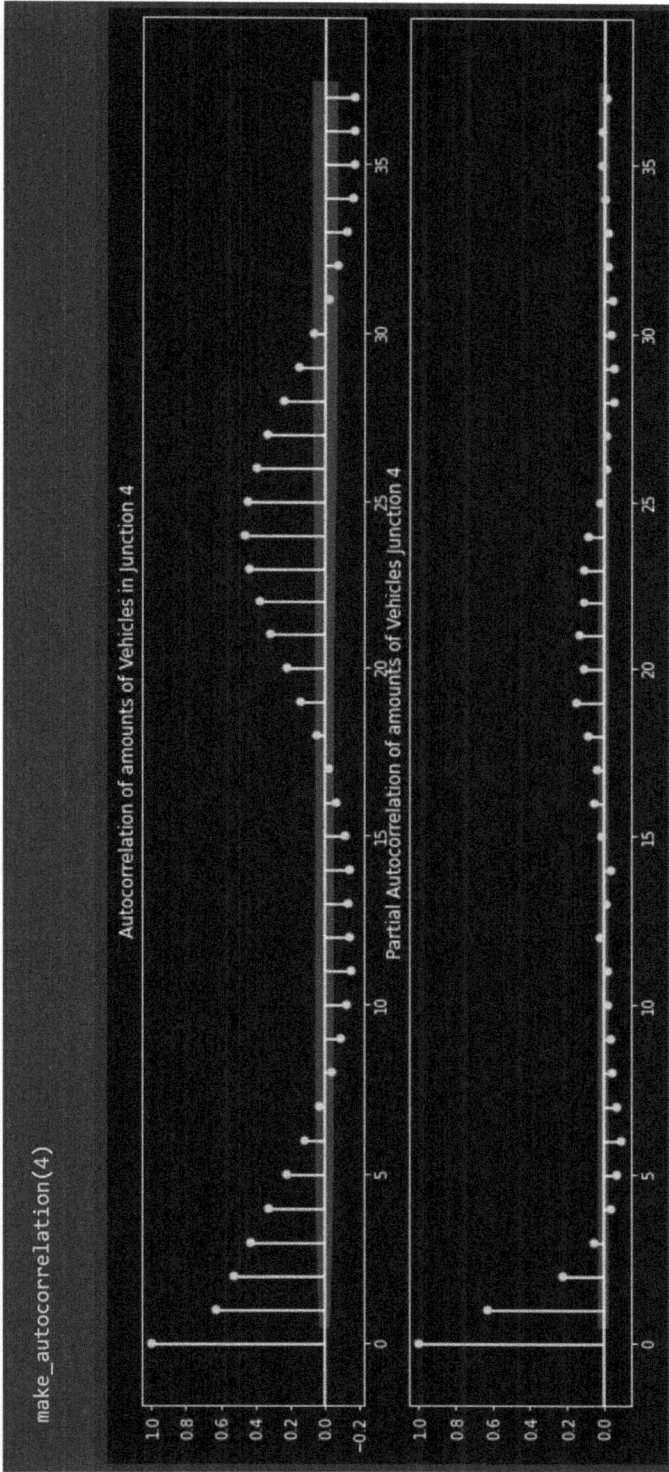

FIGURE 9.4 Autocorrelation of amounts of vehicles in junction 4.

it to suit the requirements or to maximize the precision of the machine learning models used. This is done by inputting the values that lie too far from the general concentration of data points, called outliers. It can also be done by removing some independent variables that have very high correlation, as it can result in overfitting. This feature engineering is only possible after visualizing the data. Data visualization is done using many types of plots. For example, bar charts are used to compare the data points, scatter plots are used to see the concentration of the data points, line plots are used to find the pattern in data points, box plots are used to see the spread of data points. One such plot used for data visualization is autocorrelation. Autocorrelation measures the relationship between a variable's current and past values. Autocorrelation represents the degree of similarity between a given time series and a lagged version of itself over successive time intervals. Autocorrelation helps us uncover hidden patterns in our data and helps us select the correct forecasting methods. It helps identify seasonality in our time series data. Analyzing the AutoCorrelation Function (ACF) and the Partial AutoCorrelation Function (PACF) in conjunction is necessary for the selection of the appropriate model for any time series prediction. The following plots show the autocorrelation in the dataset used in the chapter.

9.5 CONCLUSIONS

In conclusion, the data acquisition phase encompasses essential steps such as data loading, pattern visualization, feature engineering, and model performance evaluation. These processes establish a solid groundwork for leveraging machine learning in traffic flow prediction. By following this approach, accurate and dependable predictions can be achieved, leading to enhanced traffic management and effective measures to alleviate congestion.

REFERENCES

1. Bishop, C. M. (2006). *Pattern Recognition and Machine Learning*. Springer.
2. Chen, S., et al. (2018). Short-term traffic flow prediction with long short-term memory network. *IEEE Transactions on Intelligent Transportation Systems*, 19(2), 570–579.
3. Chen, X., Xu, H., Liu, Y., Cao, D., & Zhao, H. (2021). Adaptive traffic flow prediction based on self-attention mechanism and local trend. *IET Intelligent Transport Systems*, 15(5), 590–600.
4. Hastie, T., Tibshirani, R., & Friedman, J. (2009). *The Elements of Statistical Learning*. Springer.
5. He, J., Zhang, M., Ding, Z., & Zhao, D. (2019). ST-MetaNet: Learning to predict spatiotemporal properties of traffic flow with deep meta-learning. *IEEE Transactions on Intelligent Transportation Systems*, 21(10), 4428–4438.
6. Hu, X., et al. (2019). Traffic flow prediction using a hybrid clustering and recurrent neural network approach. *IEEE Transactions on Intelligent Transportation Systems*, 20(6), 2132–2142.
7. Jain, A. K., et al. (1999). Data clustering: A review. *ACM Computing Surveys (CSUR)*, 31(3), 264–323.
8. Jain, A. K., et al. (1999). Data clustering: A Review. *ACM Computing Surveys (CSUR)*, 31(3), 264–323.
9. James, G., et al. (2013). *An Introduction to Statistical Learning*. Springer.
10. Kaelbling, L. P., Littman, M. L., & Moore, A. W. (1996). Reinforcement Learning: A Survey. *Journal of Artificial Intelligence Research*, 4, 237–285.
11. Liu, X., Wang, L., & Lv, Y. (2020). Traffic flow prediction with multi-granularity temporal correlation learning. *IEEE Transactions on Intelligent Transportation Systems*, 22(4), 2194–2204.
12. Lv, Y., Duan, Y., Kang, W., & Li, L. (2014). Traffic flow prediction with spatiotemporal correlations using big data. *IEEE Journal of Selected Topics in Applied Earth Observations and Remote Sensing*, 7(6), 1814–1823.
13. Lv, Y., Wang, L., Li, L., & Lv, Q. (2020). Traffic flow prediction with multi-channel graph convolutional networks. *IEEE Transactions on Intelligent Transportation Systems*, 22(7), 4389–4399.
14. Ma, W., et al. (2015). Traffic flow prediction with big data: A deep learning approach. *IEEE Transactions on Intelligent Transportation Systems*, 16(2), 865–873.

15. Ma, Y., & Hu, S. (2015). A review on the prediction of traffic flow pattern using machine learning techniques. *Journal of Traffic and Transportation Engineering (English Edition)*, 2(3), 157–164.
16. Mitchell, T. (1997). *Machine Learning*. McGraw-Hill.
17. Mnih, V., et al. (2015). Human-level control through deep reinforcement learning. *Nature*, 518(7540), 529–533.
18. Qi, Y., et al. (2017). Traffic flow prediction with spatial-temporal correlation in big data environment. *IEEE Transactions on Intelligent Transportation Systems*, 18(11), 2965–2974.
19. Russell, S. J., & Norvig, P. (2016). *Artificial Intelligence: A Modern Approach (3rd ed.)*. Pearson.
20. Silver, D., et al. (2016). Mastering the game of go with deep neural networks and tree search. *Nature*, 529(7587), 484–489.
21. Sutton, R. S., & Barto, A. G. (2018). *Reinforcement Learning: An Introduction (2nd ed.)*. MIT Press.
22. Tang, J., et al. (2019). Spatiotemporal traffic flow prediction based on long short-term memory neural networks with a novel attention mechanism. *Transportation Research Part C: Emerging Technologies*, 107, 129–142.
23. Wang, D., Liu, Y., Xie, K., Zhang, Z., & Yan, X. (2019). Urban traffic flow prediction using a spatio-temporal graph convolutional neural network. *Sensors*, 19(16), 3561.
24. Wang, Z., et al. (2020). Traffic flow prediction via hybrid deep learning: A survey. *IEEE Transactions on Intelligent Transportation Systems*, 21(1), 427–439.
25. Xu, Y., Wu, S., & Zhang, S. (2020). Short-term traffic flow prediction based on long short-term memory neural network with historical data augmentation. *IEEE Access*, 8, 12404–12415.
26. Yin, H., et al. (2018). Short-term traffic flow prediction with a hybrid model combining LSTM and ARIMA. *Transportation Research Part C: Emerging Technologies,* 89, 110–124.
27. Yu, H., Wang, X., & Yin, B. (2017). Traffic flow prediction using LSTM recurrent neural networks. *IEEE Access,* 5, 20551–20559.
28. Zhang, X., et al. (2016). Short-term traffic flow prediction: A spatiotemporal graph convolutional neural network approach. *Transportation Research Part C: Emerging Technologies*, 72, 74–93.
29. Zheng, S., Lv, Y., & Yang, Q. (2019). Traffic flow prediction with graph convolutional neural networks. *IEEE Transactions on Intelligent Transportation Systems*, 20(3), 1018–1030.
30. Zheng, Y., et al. (2017). Traffic flow prediction with big data: A deep learning approach. *IEEE Transactions on Intelligent Transportation Systems*, 16(2), 865–873.

10 Enhancing Autistic Spectrum Disorder Diagnosis Using ML Techniques

A Study on Deep Neural Network and Drop-out Deep Neural Network

Sanat Kumar Sahu and Sushil Kumar Sahu

10.1 INTRODUCTION

Autistic Spectrum Disorder (ASD) is a complex neurological situation that profoundly affects the way individuals interact, communicate, and behave. It typically manifests during early childhood and exhibits a wide range of severity across different individuals (Mohanty, Parida, and Patra 2021). People diagnosed with ASD often face challenges in understanding social cues, engaging in repetitive behaviors, and experiencing heightened sensitivity to sensory stimuli (Thabtah 2017). It is widely believed that an amalgamation of genetic tendencies and environmental aspects contribute to its development. Genetic studies have identified certain gene variations that may increase the risk of ASD, but the complete genetic landscape and its interaction with environmental influences are yet to be fully understood. Early detection and intervention are critical in improving outcomes for individuals with ASD (Thabtah and Peebles 2020; Sahu and Verma 2022). Timely identification allows for the implementation of targeted therapies and support services, which can help individuals develop crucial social and communication skills, manage behavioral challenges, and adapt to their sensory sensitivities. These interventions aim to enhance their overall quality of life and promote successful integration into society. Currently, scientists have been using a type of computer technology called ML to help diagnose influential factors in the disorder ASD more quickly (P. Verma, Awasthi, Sahu, et al. 2022). The solution to the screening of the ASD disorder is by using special computer programs called Deep Neural Networks (DNN) and Drop-out Deep Neural Networks (Dout-DNN) as tools to help classify and diagnose ASD. These DNNs use complicated step-by-step instructions to study any complex data sets and find patterns and characteristics that can be helpful in sorting and diagnosing ASD. By analyzing ASDS data, these models can pick out important details that can assist in classifying and identifying ASD. By utilizing DNN and Dout-DNN classifiers, our aim is to enhance the performance of ASDS classification models. These models are designed to accurately differentiate between individuals with ASD and those without the disorder. By leveraging the power of deep learning and neural networks, these classifiers can effectively learn from vast amounts of data, automatically extracting relevant features and making accurate predictions (P. Verma, Awasthi, Shrivas, et al. 2022; Sahu and Shrivas 2020). By analyzing an ASDS dataset we can identify the most influential factors contributing to the disorder. This knowledge can inform the development of more precise screening tools, allowing for earlier and more accurate diagnoses.

DOI: 10.1201/9781003433309-10

ASD is a complex neurological condition that significantly impacts social interaction, communication, and behavior. Early detection and intervention are crucial for individuals with ASD to achieve better outcomes (Pratibha Verma, Awasthi, and Sahu 2021a). ML techniques, such as DNN and Dout-DNN classifiers, hold promise in expediting the diagnosis of ASD and identifying influential factors. Continued research in this field has the potential to revolutionize ASD screening and contribute to improved understanding and support for individuals on the autistic spectrum.

10.2 REVIEW OF LITERATURE

Thabath et al. (2020) used various classification methods including AdaBoost, Bagging, Nnge, RIDOR, PRISM, CART, RML, C4.5 and RIPPER to classify three ASD datasets for children, adults and adolescents. RML outperformed other methods significantly (Thabtah and Peebles 2020). The main focus of the work of Wang et al. (2019) was on building and designing specific features and using techniques to convert information into a suitable format. They used a complicated type of computer program called a deep learning classifier to screen for ASD. Their efforts resulted in a high level of accuracy, with a sensitivity and specificity of 99% (Wang et al. 2019). Diabat (2019) explored classifiers such as PART, RIPPER, voted perceptron and C4.5, with the Ensemble Classification for Autism Screening (ECAS) achieving the highest accuracy of 100% contrast to other models (Diabat and Al-shanableh 2019). Akyol et al. (2018) in their study used FR and LR-FR, for the purpose of classifying ASDS problems. Among the two models, LR-FR exhibited superior performance compared to FR, attaining an impressive accuracy rate of 97.33%. The outcome indicates that the LR-FR model was more effective in accurately classifying and identifying ASD-related issues. The findings suggest that utilizing the LR-FR model can lead to enhanced accuracy and precision in the classification of ASD problems, thereby offering potential benefits in the diagnosis and understanding of this disorder (Akyol, Gultepe, and Karaci 2018). Vaishali et al. (2017) used a method called BFA wrapper based FST, along with various classifiers including J48 Decision Tree, Naive Bayes, SVM, K-NN, and MLP to classify ASD. The outcome demonstrates that the classifiers MLP and SVM achieved the highest accuracy rate of 99.66% when using FS by the BFA method. This exceptional accuracy surpassed the performance of other approaches used in the study (Vaishali and Sasikala 2017).

10.3 PROPOSED METHODOLOGY

Figure 10.1 demonstrates the workflow used to build up the Autism Spectrum Disorder Screening (ASDS) model, which includes the DNN and Dout-DNN. This outline serves as a helpful visual

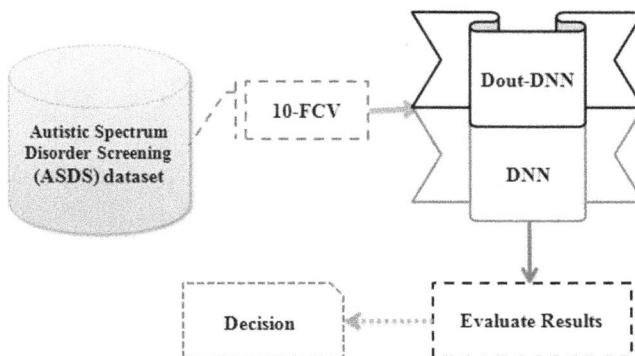

FIGURE 10.1 The proposed DNN architecture for ASD classification.

guide, making it easier to understand and imitate the study's process in ASD modelling. It gives a clear representation of the steps and processes involved in building a computational model suitable for ASD screening.

10.3.1 DATASET

Our research work utilizes the ASD children's scan data obtained from the UCI machine learning repository. This ASD dataset comprises valuable information that can be employed to identify autistic traits and has been specifically curated for future studies aiming to increase the classification performances of ASD cases. The ASD dataset covers a comprehensive collection of 20 characteristics derived from the screening data of children diagnosed with autism (Thabtah, 2017).

10.3.2 DEEP NEURAL NETWORK

The DNN starts by taking the input ASDS dataset and establishing an input pipeline (Tripathi et al. 2021; Pratibha Verma, Awasthi, and Sahu 2021b; Zhao et al. 2019; Misman et al. 2019). This pipeline involves defining iterators that enable scanning through the ASDS dataset. Additionally, the algorithm incorporates the functionality to shuffle the ASDS dataset, introducing randomness. Once the input pipeline is defined, the next step involves feeding the input ASDS data into the training model (Lang et al. 2019). To increase the training process and overall performance of the DNN model an optimization technique such as Stochastic Gradient Descent (SGD) is employed. The SGD is operated to optimize and refine the models by iteratively adjusting the parameters based on the observed errors during training. We utilize the bias updater. As SGD updates the biases, the DNN can fine-tune the model to better fit the training data and improve its overall performance. The model then proceeds with training, evaluation, and prediction tasks (Lecun, Bengio, and Hinton 2015). During training, arrays of hidden layers are defined, and pre-initialized weights are assigned to each layer. This process generates and saves the model within the processing system. Finally, an evaluation of the resulting DNN classifier is performed. The DNN consists of an initial input layer comprising 21 units, which is then followed by multiple hidden layers with varying numbers of neurons in each layer. Figure 10.2 shows the proposed Architecture of Deep Neural Network.

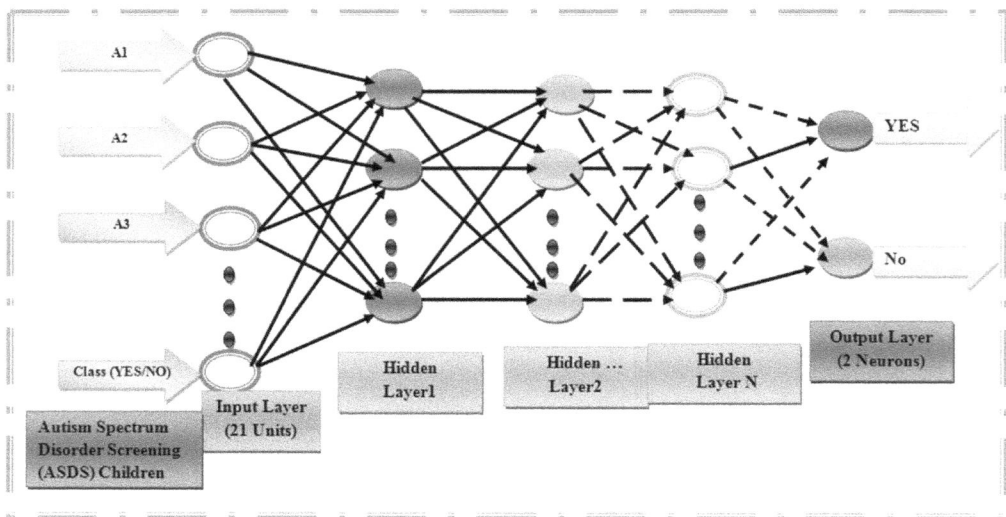

FIGURE 10.2 Architecture of Deep Neural Network.

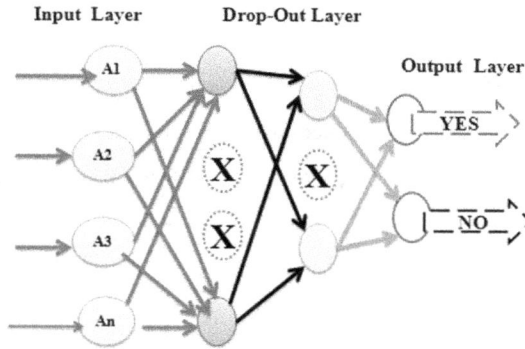

FIGURE 10.3 Architecture of Dropout Deep Neural Network.

10.3.3 DROPOUT DEEP NEURAL NETWORK

Dropout is a regularization technique used in Deep Neural Networks (DNNs) to mitigate the issue of overfitting when working with ASDS data (Wan et al. 2013). It involves the insertion of dropout layers between the network's layers. During the training process, a fraction of the input units or neurons are randomly deactivated or "dropped out" at each update. This random deactivation allows the network to learn from different subsets of the ASDS data during each update, promoting the acquisition of more resilient and generalized features. By incorporating dropout, the network becomes less sensitive to the specific weights of individual neurons, relying instead on the collective knowledge obtained from multiple overlapping sub networks. This reduction in dependency on specific neurons enhances the network's resilience and diminishes the risk of overfitting the training data specific to ASDS. During inference or testing, dropout is typically turned off, and the full network is utilized for making predictions on ASDS data (Srivastava et al. 2014). Dropout acts as a form of regularization, enhancing the model's ability to generalize and reducing overfitting, thereby improving its performance on unseen ASDS data. In summary, dropout is a widely adopted technique in deep learning that effectively addresses the issue of overfitting in DNNs, making it a valuable tool in developing more robust features for ASDS classification. Figure 10.3 shows the proposed Architecture of Dropout Deep Neural Network.

10.4 RESULT AND DISCUSSION

The developed models were applied to analyse the performance of the approaches in order to determine their efficacy. The classification performance was carefully appraised using a 10-fold Cross-Validation (CV) method. As the assessment criteria for this analysis we choose parameters like accuracy, sensitivity and specificity. The results of the categorization of ASD patients using the DNN and Dout-DNN algorithms in each cross-validation fold are shown in Table 10.1.

Accuracy plays a crucial role in ASDS classification as it represents the overall correctness of the model's predictions. A high accuracy value indicates that the model effectively distinguishes between individuals with ASD and non-ASD, enabling early detection and intervention for improved outcomes. The table represents the classification performances of the Dout-DNN and DNN models. In terms of classification accuracy, the Dout-DNN achieved 92.81%, while the DNN achieved 93.84% accuracy. Comparing the DNN and Dout-DNN, the DNN exhibited 1.03% higher accuracy.

In the context of other metrics like sensitivity, the Dout-DNN achieved 88.74%, while the DNN model achieved 92.72% sensitivity. Sensitivity holds significant importance in ASDS classification as it checks the model's talent to correctly recognize individuals with ASD. A high sensitivity value

TABLE 10.1
Performances of models

Sr. No.	Model	Accuracy	Sensitivity	Specificity
1	Dout-DNN	92.81	88.74	97.16
2	DNN	93.84	92.72	95.04

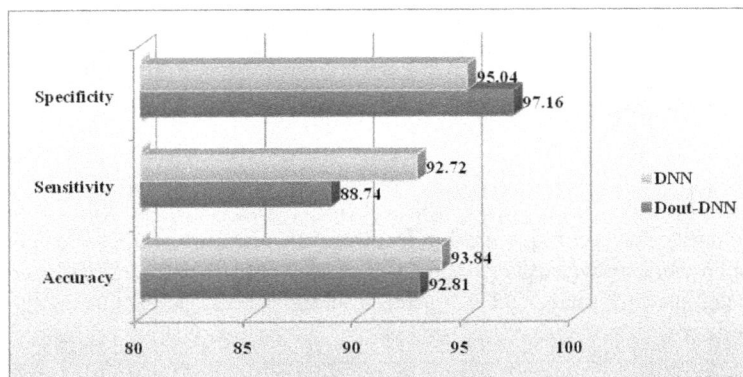

FIGURE 10.4   A comparative Analysis of the Dout-DNN and DNN in ASDS Classification.

indicates that the model is effective in identifying affected individuals, enabling early detection and intervention for improved outcomes.

Similarly in the classification of ASD, specificity holds significant importance as it measures the model's capacity to accurately identify individuals who do not have ASD. By minimizing false positives, specificity ensures accurate classification of negative instances, facilitating appropriate allocation of resources and interventions to those who truly require them. The Dout-DNN achieved 97.16% specificity, while the DNN model achieved 95.04% specificity.

The bar chart in Figure 10.4 represents the presentation comparison of two models, namely Dout-DNN and DNN, based on their accuracy, sensitivity, and specificity. The chart shows three key metrics: accuracy, sensitivity and specificity along with the corresponding model levels Dout-DNN and DNN. For the first model Dout-DNN, it achieved an accuracy of 92.81%, indicating the proportion of correct predictions overall. The sensitivity of this model is 88.74%, representing the percentage of exactly recognized positive cases (ASD cases), while the specificity is 97.16%, signifying the percentage of rightly identified negative cases (non-ASD cases). The next model DNN performed a little better with an accuracy of 93.84%, involving a higher proportion of correct predictions overall compared to Dout-DNN. The sensitivity of the DNN model is 92.72%, representing a higher percentage of correctly identified positive cases (ASD cases). The specificity of this model is 95.04%, demonstrating a faintly lower percentage of correctly identified negative cases (non-ASD cases) compared to Dout-DNN.

10.5   CONCLUSION

The main objective of the proposed methods is to identify ASD in children early on to enhance their quality of life. Early identification facilitates timely interventions and support, essential for enhanced development and reduced symptoms. Customized therapies, support and education services tackle specific needs, helping children obtain skills, progress communication, and promote positive social interactions, enhancing overall well-being. In this study, we have explored how well DNN and Dout-DNN classifiers work to improve ASD diagnosis. The outcomes demonstrated that both models performed very well to identifying ASD. Compared to the Dout-DNN model's accuracy of 92.81%, the DNN model had an almost higher accuracy of 93.84%. However, the DNN model's sensitivity and specificity were 95.04% and 88.74%, as well, compared to 92.72% and 95.04% for the Dout-DNN model. The final results indicate that both models have the ability to accurately identify individuals with ASD and differentiate them from those without the disorder. The outcome of the DNN and Dout-DNN models demonstrate their potential as effective tools in identifying individuals with ASD. These models can aid in early detection and intervention, leading to improved outcomes and quality of life for individuals on the autism spectrum. The performance of ASDS classification models can be further enhanced by investigating additional ML techniques, optimizing model architectures, and incorporating other relevant factors. Continued improvements in this field have the potential to revolutionize ASD screening, leading to earlier and more accurate diagnoses and improved support for individuals with ASD.

REFERENCES

Akyol, K., Gultepe, Y., and Karaci, A. 2018. "A Study on Autistic Spectrum Disorder for Children Based on Feature Selection and Fuzzy." *International Congress on Engineering and Life Science*, 804–7.

Diabat, M. Al, and Al-shanableh, N. 2019. "Ensemble Learning Model For Screening." *International Journal of Computer Science & Information Technology (IJCSIT)* 11 (2): 13–14.

Lang, S., Bravo-marquez, F., Beckham, C. and Hall, M. 2019. "WekaDeeplearning4j: A Deep Learning Package for Weka Based on DeepLearning4j." no. May.

Lecun, Y., Bengio, Y., and Hinton, G. 2015. "Deep Learning."

Misman, M. F., et al. 2019. "Classification of Adults with Autism Spectrum Disorder Using Deep Neural Network." University of Edinburgh. Downloaded on June 14, 2020, at 19:47:17 UTC from IEEE Xplore.

Mohanty, A. S., Parida, P., and Patra, K. C. 2021. "ASD Classification for Children Using Deep Neural Network." *Global Transitions Proceedings* 2 (2): 461–66.

Sahu, S. K., and Shrivas, A. K. 2020. "Comparative Study of Classification Models with Genetic Search Based Feature Selection Technique." *Cognitive Analytics*, 773–83.

Sahu, S. K., and Verma, P. 2022. "Classification of Autistic Spectrum Disorder Using Deep Neural Network with Particle Swarm Optimization." *International Journal of Computer Vision and Image Processing (IJCVIP)* 12 (1): 1–11.

Srivastava, N., Hinton, G., Krizhevsky, A., Sutskever, I., and Salakhutdinov, R. 2014. "Dropout: A Simple Way to Prevent Neural Networks from Overfitting." *Journal of Machine Learning Research* 15: 1929–58.

Thabtah, F. 2017. "Autism Spectrum Disorder Screening: Machine Learning Adaptation and DSM-5 Fulfillment." 1–6.

Thabtah, F, and Peebles, D. 2020. "A New Machine Learning Model Based on Induction of Rules for Autism Detection." *Health Informatics Journal* 26 (1): 265–86.

Tripathi, N., Goshisht, M. K., Sahu, S. K., and Arora, C. 2021. "Applications of Artificial Intelligence to Drug Design and Discovery in the Big Data Era: A Comprehensive Review." *Molecular Diversity* 25 (3): 1643–64.

Vaishali, R., and Sasikala, R. 2017. "A Machine Learning Based Approach to Classify Autism with Optimum Behaviour Sets." *International Journal of Engineering & Technology* 5 (x): 1–6.

Verma, P., Awasthi, V. K., Sahu, S. K., and Shrivas, A. K. 2022. "Coronary Artery Disease Classification Using Deep Neural Network and Ensemble Models Optimized by Particle Swarm Optimization." *International Journal of Applied Metaheuristic Computing (IJAMC)* 13: 1–25.

Verma, P., Awasthi, V. K., Shrivas, A. K., and Sahu, S. K. 2022. "Stacked Generalization Based Ensemble Model for Classification of Coronary Artery Disease." In *Internet of Things and Connected Technologies (Lecture Notes in Networks and Systems 340)*, edited by Rajiv Misra, Nishtha Kesswani, Muttukrishnan Rajarajan, Veeravalli Bharadwa, and Ashok Patel, Vol. 340, 1: 57–65. Springer International Publishing Cham.

Verma, P., Awasthi, V. K., and Sahu, S. K. 2021a. "A Novel Design of Classification of Coronary Artery Disease Using Deep Learning and Data Mining Algorithms." *Revue d ' Intelligence Artificielle* 35 (3): 209–15.

Verma, P., Awasthi, V. K., and Sahu, S. K. 2021b. "An Ensemble Model with Genetic Algorithm for Classification of Coronary Artery Disease." *International Journal of Computer Vision and Image Processing* 11 (3): 70–83.

Wan, L., Zeiler, M., Zhang, S., LeCun, Y. and Fergus, R. 2013. "Generalizing Pooling Functions in CNNs: Mixed, Gated, and Tree." In *30th International Conference on Machine Learning*, Atlanta, Georgia, USA, 2013, 40: 863–75.

Wang, H., Li, L. , Chi, L., and Zhao, Z. 2019. "Autism Screening Using Deep Embedding Representation." In *Lecture Notes in Computer Science*, Springer, Cham.

Zhao, R., Yan, R., Chen, Z., Mao, K., Wang, P., and Gao, R. X. 2019. "Deep Learning and Its Applications to Machine Health Monitoring." *Mechanical Systems and Signal Processing* 115: 213–37.

11 Deep Learning: A State-Of-The-Art Approach To Artificial Intelligence "AI"

Sangeet Vashishtha and Pooja Sharma

11.1 INTRODUCTION: AI – TRANSFORMING INDUSTRIES AND IMPROVING PEOPLE'S LIVES

Artificial Intelligence (AI) has emerged as a transformative force, revolutionizing various industries and enhancing the quality of human life. With its ability to analyze vast amounts of data, learn from patterns, and make intelligent decisions, AI has opened new horizons of possibilities. In healthcare, AI enables more accurate diagnoses, personalized treatments, and drug discovery. Industries such as finance and transportation benefit from AI's ability to automate processes, optimize operations, and detect anomalies. AI-powered virtual assistants and smart home devices have simplified daily tasks, making homes more efficient and convenient. Moreover, AI has made significant strides in areas like computer vision, natural language processing, and robotics, enabling breakthroughs in facial recognition, language translation, and autonomous systems. As AI continues to advance, it holds the potential to reshape industries, drive innovation, and empower individuals, transforming the way we work, live, and interact with technology [1] [2].

11.1.1 THE RISE OF DEEP LEARNING: A STATE-OF-THE-ART APPROACH

Deep learning has experienced a meteoric rise as a state-of-the-art approach within the realm of artificial intelligence. Figure 11.1 shows the various concepts of deep learning and its ascent can be attributed to a convergence of factors, including the availability of vast amounts of data, significant advancements in computational power, and breakthroughs in algorithmic techniques. Deep learning has proven itself to be a powerful tool for tackling complex problems and has demonstrated remarkable capabilities in various domains. By utilizing neural networks with multiple layers, deep learning algorithms can automatically learn intricate patterns and high-level representations from raw data [3]. Deep learning has achieved new levels of accuracy and performance in domains such as "computer vision, natural language processing, and robotics" due to its ability to grasp hierarchical connections. The scalability and parallel processing capabilities of modern hardware, such as GPUs and TPUs, have further accelerated the training and inference processes, making deep learning a practical and efficient approach. With its state-of-the-art capabilities, deep learning is revolutionizing industries, empowering innovation, and driving the next generation of artificial intelligence applications. [4].

DOI: 10.1201/9781003433309-11

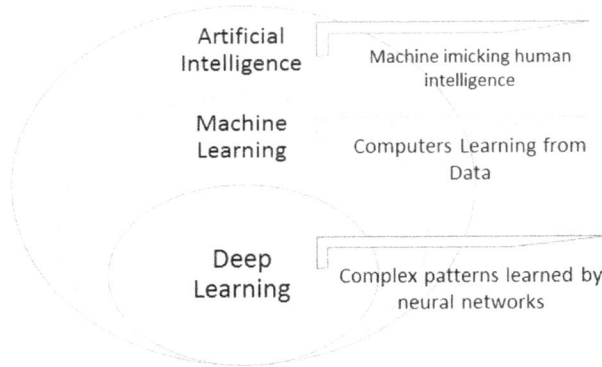

FIGURE 11.1   concepts of Deep Learning.

11.2 DEEP LEARNING FUNDAMENTALS

Deep learning fundamentals refer to the foundational concepts, principles, and techniques that form the basis of deep learning, a subfield of machine learning. These fundamentals provide an understanding of how deep learning works, its underlying principles, and the key components involved. By grasping these fundamentals, individuals can build a solid knowledge base to further explore and apply deep learning techniques effectively.

11.2.1 UNDERSTANDING PRINCIPLES AND CONCEPTS: DEEP LEARNING

Deep learning is a subset of Artificial Intelligence (AI) that has gained significant attention due to its exceptional ability to learn from vast amounts of data and solve complex problems. To understand deep learning, it is essential to grasp the underlying principles and concepts that form its foundation. [5]

At the core of deep learning are artificial neural networks, which are inspired by the structure and functionality of the human brain. Neural networks consist of interconnected nodes called neurons, each with its activation function. These neurons are organized into layers, with each layer responsible for extracting specific features from the input data. The connections between neurons, along with the weights assigned to those connections, determine the neural network's ability to learn and make predictions. [6]

Deep neural networks as shown in Figure 11.2, also known as deep feed forward networks or multilayer perceptions', have multiple hidden layers between the input and output layers. This depth allows them to capture increasingly complex patterns and representations as information flows through the network. The feed forward process involves passing data through the network from input to output, with each layer transforming the data to make increasingly abstract representations. [7]

Training deep neural networks is a crucial aspect of deep learning. The back propagation algorithm plays a central role in this process, as it enables the network to learn from its mistakes by adjusting the weights based on the error signal propagated backward from the output layer to the input layer. By iteratively updating the weights using optimization algorithms like stochastic gradient descent, the network gradually improves its performance.

Convolutional Neural Networks (CNNs) are a specialized type of deep neural network that excel in computer vision tasks. CNNs leverage the concept of convolution, which involves sliding a filter or kernel over the input data to extract relevant features. This spatial hierarchy allows CNNs to

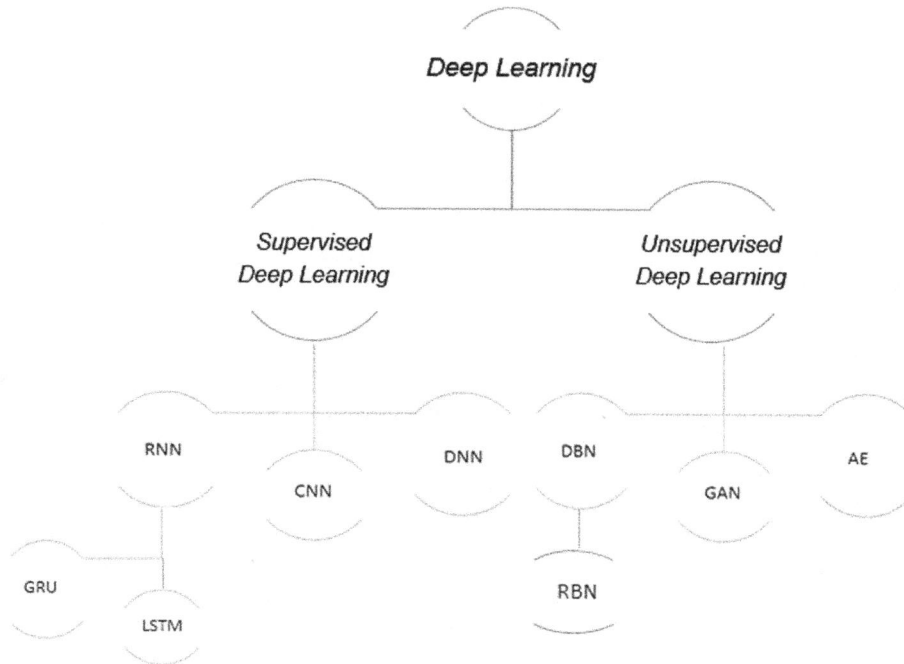

FIGURE 11.2 Deep Neural Networks.

detect patterns at different levels of abstraction, making them highly effective for tasks such as image classification, object detection, and image segmentation.

In contrast, Recurrent Neural Networks (RNNs) are designed to process sequential data by utilizing feedback connections that allow information to persist across time steps. This capability makes RNNs suitable for tasks involving sequences, such as natural language processing, speech recognition, and time series analysis. However, traditional RNNs suffer from the vanishing gradient problem, which hampers their ability to capture long-term dependencies. To address this, Long Short-Term Memory (LSTM) networks were introduced. LSTMs have specialized memory cells that selectively retain and update information, enabling them to overcome the vanishing gradient problem and model long-term dependencies more effectively. [8]

Generative models, including Variation Auto Encoders (VAEs) and Generative Adversarial Networks (GANs), are also part of the deep learning landscape. VAEs are used to generate new data samples by learning the underlying distribution of the training data. GANs, on the other hand, consist of a generator and a discriminator that compete with each other. The generator aims to produce realistic samples, while the discriminator's task is to distinguish between real and generated samples. This adversarial training process leads to the generation of highly realistic data samples. [9]

Transfer learning is another key concept in deep learning, where pre-trained models are leveraged to tackle new tasks or datasets with limited labelled data. By transferring knowledge from models trained on large-scale datasets, the performance and efficiency of deep learning algorithms can be significantly improved.

Regularization techniques play a vital role in preventing overfitting in deep learning models. Dropout regularization randomly disables a fraction of neurons during training to encourage robustness and prevent reliance on specific neurons. Batch normalization is another technique that normalizes the inputs to each layer, making training more stable and accelerating convergence.

11.2.2 AI NEURAL NETWORKS: MIMICKING THE HUMAN BRAIN

Artificial Neural Networks (ANNs) form the foundation of deep learning, with their structure and functionality inspired by the intricate workings of the human brain. ANNs are composed of interconnected nodes, or artificial neurons, which work collectively to process and learn from data. The concept of artificial neurons stems from the biological neurons found in the human brain. Like their biological counterparts, artificial neurons receive inputs, perform computations, and produce outputs. Each artificial neuron computes a weighted sum of its inputs and applies an activation function to determine its output. This output serves as the input to other neurons, forming a network of interconnected processing units.

The connectivity pattern in artificial neural networks can be compared to the connections between neurons in the brain. Neurons are organized into layers, where each layer receives inputs from the previous layer and passes its outputs to the next layer. The input layer receives the raw data, such as images or text, while the output layer produces the final predictions or classifications. [10]

The strength of connections between neurons, known as weights, play a crucial role in the learning process of artificial neural networks. During training, the network adjusts these weights based on the observed errors, aiming to minimize the discrepancy between predicted outputs and the true values. This iterative optimization process enables the network to learn and improve its performance over time. The architecture of artificial neural networks can vary, depending on the complexity of the task at hand. Deep neural networks, also known as multilayer perceptrons, have multiple hidden layers between the input and output layers. These hidden layers allow the network to learn and capture increasingly abstract features and representations as information flows through the network. The depth of the network enables it to model intricate relationships in the data, making deep learning highly effective for complex tasks.

The ability of artificial neural networks to mimic the brain's functionality has contributed to their success in various domains. For instance, in computer vision tasks, such as image classification or object detection, Convolutional Neural Networks (CNNs) have shown remarkable performance. CNNs leverage the concept of convolution, where filters or kernels slide over the input data to detect relevant features. This spatial hierarchy of convolutional layers enables CNNs to automatically learn hierarchical representations of visual data. Furthermore, Recurrent Neural Networks (RNNs) have revolutionized the field of sequential data processing. RNNs use feedback connections to maintain information over time steps, thereby rendering them ideal for applications like time series analysis and natural language processing. RNNs can capture dependencies in sequential data by incorporating information from previous time steps into the current computation [11].

11.2.3 ARCHITECTURE AND STRUCTURE OF ARTIFICIAL NEURAL NETWORKS

Artificial Neural Networks (ANNs) are computational models inspired by the structure and functionality of the human brain. The architecture and structure of ANNs play a crucial role in their ability to learn and make predictions from data. The basic building block of an artificial neural network is the artificial neuron, also known as a node, shown in Figure 11.3. Neurons are organized into several layers, where one works on input data while others are on output data and produce the final predictions or classifications.

Each neuron in the network receives input from the layer above, adds the inputs together according on their weights, and then applies an activation function to the inputs to generate an output. As the network learns, the weights are modified to optimize the connections between neurons, thus improving its overall performance. The activation function introduces non-linearity, allowing the network to learn intricate relationships between inputs and outputs. The arrangement of neurons and connections in an ANN is determined by its architecture. Different architectures are suited

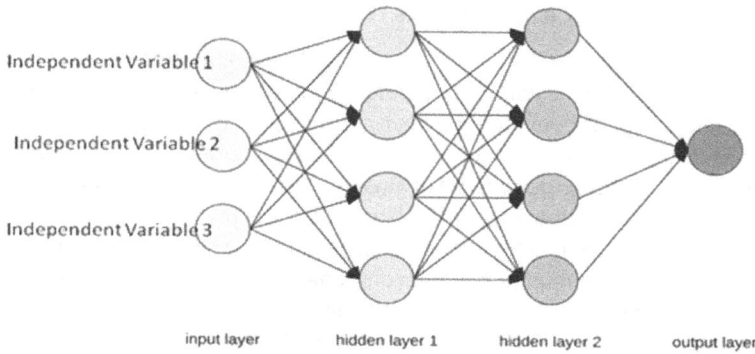

FIGURE 11.3 Artificial neural networks.

to different types of tasks. For example, feed forward neural networks, also known as multilayer perceptrons, consist of one or more hidden layers. The presence of hidden layers enables these networks to capture increasingly complex patterns and representations as data flows through the network [12].

Convolutional Neural Networks (CNNs) are another popular architecture, particularly effective in computer vision tasks. CNNs leverage the concept of convolution, where filters or kernels slide over the input data to extract relevant features. Convolutional layers capture spatial hierarchies of visual information, enabling CNNs to automatically learn hierarchical representations of images.

The incorporation of feedback connections in RNNs empowers them to effectively process sequential data, enabling information to persist across different time steps. This inherent architectural feature renders RNNs highly suitable for a wide range of tasks that involve sequences, such as NLP and time series analysis. RNNs can capture dependencies in sequential data by incorporating information from previous time steps into the current computation [13].

The architecture and structure of an ANN can be further customized through the choice of activation functions, regularization techniques, and optimization algorithms used during training. These choices impact the network's ability to learn and generalize from data. In summary, the architecture and structure of artificial neural networks define how neurons are organized, interconnected, and compute outputs. Different architectures offer specific advantages in modeling different types of data. Understanding the architecture and structure of ANNs is essential for designing and deploying effective deep learning models across various domains.

11.2.4 WORKING MECHANISMS OF ARTIFICIAL NEURAL NETWORKS

Artificial Neural Networks (ANNs) operate through a series of working mechanisms that enable them to process information and make predictions. ANNs consist of interconnected artificial neurons that receive inputs, compute weighted sums, apply activation functions, and produce outputs. During the training process, the network adjusts the weights of the connections based on observed errors, using algorithms such as back propagation and gradient descent optimization. This iterative process allows ANNs to learn from data and improve their performance over time. The working mechanisms of ANNs, including feed forward propagation, weighted sum computation, activation functions, and weight adjustments, play a crucial role in their ability to model complex relationships and make accurate predictions. Understanding these mechanisms is essential for comprehending the functionality of ANNs and harnessing their power in various applications [14].

11.2.5 CONVOLUTIONAL NEURAL NETWORKS (CNNs)

Convolutional Neural Networks (CNNs) are a specialized type of artificial neural network that excel in processing and analyzing visual data, making them particularly effective in computer vision tasks. CNNs leverage the concept of convolution, where small filters or kernels slide over the input data to extract relevant features. This spatial hierarchy of convolutional layers allows CNNs to automatically learn hierarchical representations of images, capturing low-level features such as edges and textures, and gradually building up to higher-level concepts and objects. CNNs also incorporate pooling layers to downsample the feature maps and reduce spatial dimensions while retaining important information. The combination of convolutional and pooling layers in CNNs enables them to learn and recognize complex visual patterns, making them highly effective in tasks such as image classification, object detection, and image segmentation. The success of CNNs in computer vision has revolutionized fields such as autonomous driving, medical imaging, and facial recognition, amongst others.

11.2.6 RECURRENT NEURAL NETWORKS (RNNs)

Recurrent Neural Networks (RNNs) are a type of artificial neural network specifically designed to process sequential data, such as time series, natural language, and speech. Unlike feed forward neural networks, RNNs have feedback connections that allow information to persist across time steps, enabling them to capture temporal dependencies in the data. This recurrent nature makes RNNs well-suited for tasks that involve sequences and context. RNNs process sequential data by maintaining a hidden state that updates at each time step, incorporating information from previous steps into the current computation. This hidden state serves as the memory of the network, allowing it to learn long-term dependencies. RNNs can be stacked to form deep architectures, known as deep recurrent neural networks, to capture even more complex patterns and representations. Applications of RNNs include language modeling, sentiment analysis, machine translation, speech recognition, and music generation. However, RNNs suffer from the vanishing or exploding gradient problem, which hampers their ability to capture long-term dependencies. To overcome this, variants such as Long Short-Term Memory (LSTM) and Gated Recurrent Unit (GRU) were developed, which introduce specialized gates to regulate the flow of information and alleviate the gradient problem. RNNs have proven to be a powerful tool for modeling sequential data and have made significant contributions to various domains that involve temporal information [15].

11.3 TRAINING DEEP NEURAL NETWORKS

Deep Neural Networks (DNNs) have revolutionized various fields of artificial intelligence, however, effectively training these complex models requires careful consideration of various techniques, addressing challenges, and adopting best practices. This article aims to highlight the key aspects involved in training deep neural networks with the help of Figure 11.4 [16].

11.3.1 KEY ASPECTS

11.3.1.1 Architecture design

1. Network depth and width: understanding the impact of network depth and width training dynamics, model capacity, and computational requirements.
2. Activation functions: exploring popular activation functions and their suitability different network architectures.
3. Regularization techniques: introducing regularization methods such as dropout, L1/L2 regularization, and batch normalization to prevent overfitting and improve generalization.

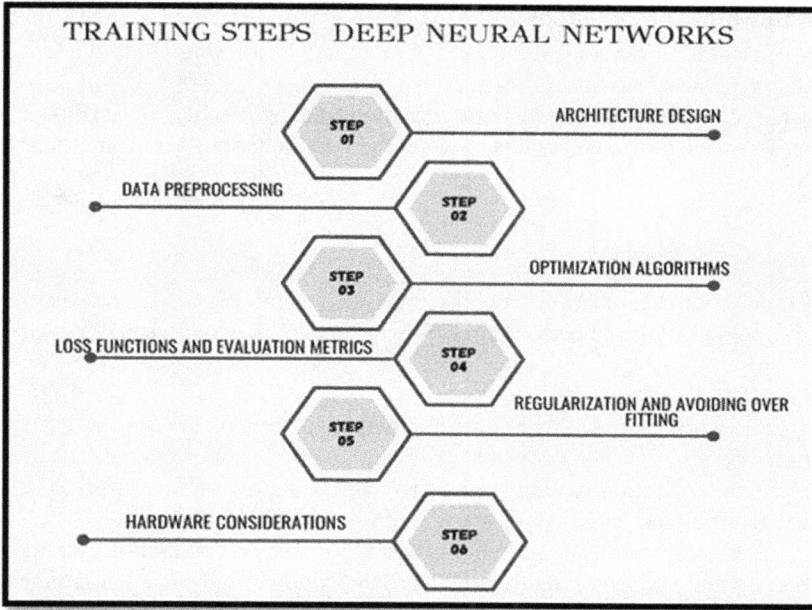

FIGURE 11.4 Training steps.

11.3.1.2 Data preprocessing

1. Data cleaning: techniques for handling missing data, outliers, and noise to ensure high-quality training data.
2. Data augmentation: data augmentation serves multiple purposes in machine learning and helps prevent overfitting by dropping the model's reliance on specific patterns or characteristics present in the original training data.
3. Data normalization: scaling and standardizing input features to a consistent range for efficient training and improved convergence.

11.3.1.3 Optimization algorithms

1. Gradient descent variants: overview of popular optimization algorithms such as Stochastic Gradient Descent (SGD), Adam, and RMSprop, highlighting their advantages and limitations.
2. Learning rate scheduling: strategies for adapting the learning rate during training, including step decay, exponential decay, and cyclic learning rates.
3. Momentum techniques: understanding the role of momentum and techniques like Nesterov momentum to accelerate convergence and escape local optima.

11.3.1.4 Regularization and avoiding overfitting

1. Dropout: implementing dropout layers to randomly deactivate a subset of neurons during training, preventing over-reliance on specific features and improving generalization.
2. Early stopping: monitoring validation loss to determine the optimal training epoch and prevent overfitting.
3. Model ensemble: utilizing ensemble techniques, such as bagging or boosting, to combine predictions from multiple models and enhance overall performance.

11.3.1.5   Hardware considerations

1. GPU acceleration: leveraging the power of Graphics Processing Units (GPUs) for efficient parallel computation and faster training times.
2. Distributed training: exploring dispersed training frameworks like TensorFlow's Horovod or PyTorch's dispersed data analogous to distributed computations across multiple machines or devices.

11.3.2   THE TRAINING PROCESS

The training process of deep neural networks involves several steps that collectively enable the model to learn improvisation and performance from data. Here is an overview of the typical training process [17]:

1. Define the network architecture: Set the number and arrangement of layers in the deep neural network architecture first. The types of layers (such as convolutional, recurrent, or fully connected), the activation functions, and the connections among the layers. The architecture design depends on the specific problem and the nature of the data.
2. Prepare the training data: The training data is a decisive component of training deep neural networks. It should be representative of the problem domain and contain a sufficient amount of labeled examples. The data needs to be preprocessed, which involves steps such as onslaught, address missing values, normalize or scaling features, and separating the data into training and validation sets are all part of data preparation.
3. Initialize the model parameters: At the beginning of the training process, the parameters "weights and biases" are randomly initialized. The initial values play a vital role in determining the model's starting point and can influence the training dynamics.
4. Forward propagation: Forward propagation involves sending input data down the network while doing calculations. Each layer applies a set of mathematical operations, such as matrix multiplications and activation functions, to produce output activations. The output of the last layer represents the model's predictions.
5. Calculate the Loss: This function trials the discrepancy between the predicted output and the ground truth labels. This function depends on the explicit task for regression or cross-entropy for classification. The goal of training is to diminish this loss.
6. Backpropagation: Backpropagation is the key algorithm for updating the model's parameters based on the loss. It calculates the gradients of the loss with respect to each parameter in the network using the chain rule of calculus. The gradients flow backward through the network, and the parameter updates are performed to minimize the loss. [18]
7. Optimization algorithm: An optimization algorithm, for example, "Stochastic Gradient Descent (SGD)," is used to update the model's parameters based on the computed gradients. These algorithms determine the direction and magnitude. Advanced optimization techniques like Adam or RMSprop can also be employed to accelerate convergence and improve training efficiency.
8. Iterative training: This is an iterative process, with each iteration referred to as an epoch. In each epoch, the training dataset is processed in its entirety. Using the estimated gradients as a guide, the network and parameters are modified. This course of action continues for numerous epochs until the model's performance converges or a predefined stopping criterion is met.
9. Validation and model selection: During training, it is important to evaluate the model's performance on a separate validation set. The validation set provides an unbiased estimate of the model's generalization ability. Metrics such as accuracy, precision, recall, or loss are calculated on the validation set to monitor the model's progress and guide the selection of the best-performing model.

10. Hyper parameter tuning: Deep neural networks have several hyper parameters, such as learning rate, batch size, regularization strength, and network architecture. These hyper parameters significantly influence the training process and the model's performance. Hyper parameter tuning involves systematically adjusting these values and evaluating the model's performance to find the optimal combination.

11. Test and deployment: After completing the training process and selecting the best model based on the validation set, the model's performance is assessed on a separate test set. This provides an unbiased evaluation of the model's capabilities. If the model meets the desired performance criteria, it can be deployed for real-world applications and inference tasks.

11.3.3 CHALLENGES IN TRAINING: DEEP NETWORKS

Training deep neural networks come with various challenges that practitioners often encounter. These challenges can impact the convergence, performance, and generalization of the models. Here are some common challenges in training deep networks:

Headings	Components of Challenges	Description
3.2.1	**Vanishing and exploding gradients**	Deep networks with many layers are prone to gradient-related issues. The gradients can diminish exponentially (vanishing gradients) or explode (exploding gradients) as they propagate backward through the network during backpropagation. This can hinder the convergence of the model or lead to unstable training.
3.2.2	**Overfitting**	A model fails to generalize when it becomes highly specialised in training data. Deep networks, with their high capacity and flexibility, are particularly susceptible to overfitting. This can happen when the model learns noise or spurious patterns present in the training data. Regularization techniques can help mitigate overfitting
3.2.3	**Computational complexity**	Training intricate neural networks can be a resource and time-intensive process, particularly when handling extensive models and datasets. The forward and backward pass calculations, as well as the optimization process, require significant computational resources. Utilizing GPUs and distributed computing frameworks can help alleviate this challenge.
3.2.4	**Hyper parameter tuning**	Deep networks include a variety of hyper parameters that must be properly tweaked for best performance. Determining the appropriate learning rate, batch size, regularization strength, or network architecture can be a challenging and iterative process. Hyper parameter search strategies, such as grid search or Bayesian optimization, can assist in finding suitable configurations.
3.2.5	**Dataset size and quality**	Deep networks classically require a large amount of labeled training data to learn complex patterns effectively. However, acquiring such datasets can be challenging, especially in domains with limited annotated data. Additionally, the quality of the training data, including biases, noise, or class imbalance, can affect the training process and model performance.
3.2.6	**Network architecture design**	Designing effective network architecture for a specific problem is a non-trivial task. Determining the appropriate depth, width, and layer types requires domain knowledge and experimentation. Selecting an architecture that balances model capacity and computational efficiency is crucial. Pre-trained models or transfer learning can be helpful in leveraging architectures trained on large datasets.

(continued)

Headings	Components of Challenges	Description
3.2.7	**Generalization to unseen data**	Ensuring that a trained deep network generalizes well to unseen data is a critical challenge. The model should be able to extract meaningful and relevant features from the training data that can be applied to new instances. Regularization techniques, data augmentation, and extensive validation testing can aid in improving generalization capabilities.
3.2.8	**Interpretability and explainability**	Deep networks often lack interpretability, making it difficult to understand and explain their decisions. The black-box nature of these models can pose challenges in certain domains where interpretability is essential, such as healthcare or finance. Techniques like layer visualization, attention mechanisms, or model-agnostic interpretability methods can provide insights into the model's behavior.

Addressing these challenges requires a combination of theoretical knowledge, practical experience, and experimentation. Staying up-to-date with the latest research and leveraging community resources can help navigate and overcome these hurdles in training deep networks. [19] [20]

11.3.4 SOLUTIONS TO TRAINING CHALLENGES

To address the challenges in training deep neural networks, practitioners have developed various solutions and techniques. Here are some commonly employed strategies to overcome these challenges:

1. Initialization techniques: To mitigate vanishing or exploding gradients, careful initialization of the network parameters is crucial. Techniques like Xavier or He initialization, which set the initial weights based on the size of the I/O dimensions, can help stabilize the training process and improve convergence.
2. Normalization techniques: The vanishing/exploding gradients issue can be resolved and convergence speeded up using normalisation approaches like batch normalisation or layer normalisation. These techniques normalize the activations or gradients within each layer, making the optimization process more stable.
3. Regularization methods: Regularization techniques play a vital role in preventing overfitting. Dropout, which randomly deactivates a subset of neurons during training, helps to reduce co-adaptation and promotes model generalization. L1/L2 regularization or weight decay can also be employed to add regularization pressure and control the complexity of the model.
4. Advanced optimization algorithms: Optimization algorithms like Adam, RMSprop, or AdaGrad offer adaptive learning rates and momentum-based updates. These algorithms help accelerate convergence, handle sparse gradients, and adaptability the learning rate to each parameter individually. Selecting the appropriate optimization algorithm may improve training efficiency and convergence speed.
5. Data augmentation: These techniques artificially expand the training dataset by applying random transformations or perturbations to the input data. Techniques like rotation, scaling, flipping, or adding noise can help create diverse samples and improve model generalization. Data augmentation is predominantly useful when the available labeled data is limited.
6. Transfer learning and pretrained Models: Transfer learning leverages pretrained models that are trained on all-encompassing datasets, such as ImageNet or BERT. By using these pretrained models as a starting point, practitioners can benefit from the learned representations and adapt the models to their specific tasks. Transfer learning reduces the need for large amounts of labeled data and can improve training efficiency.

7. Ensembling and model averaging: Ensembling combines the predictions of multiple independently trained models to ameliorate performance and generalization. Techniques like bagging, boosting, or Stacking can be used to group models. Model averaging, where the parameters of multiple checkpoints during training are averaged, can also help stabilize the training process and enhance performance.

8. Model regularization techniques: To enhance model generalization, techniques like early stopping, this stops training when the validation loss stops, improving, can be employed. Additionally, techniques like weight sharing, dropout, or regularization within specific layers or connections (for example, convolutional layers or attention mechanisms) can help prevent overfitting and improve model performance.

9. Transfer learning from similar domains: If labeled data is limited in the target domain, transfer learning from a related domain with more data can be beneficial. By leveraging knowledge from a similar domain, models can learn useful representations and adapt them to the target domain, reducing the need for extensive training on limited data.

10. Progressive training and layer-wise fine-tuning: Training deep networks progressively or layer-wise fine-tuning involves training the model in stages. Starting from shallow layers and gradually adding complexity can help the model converge more efficiently and avoid getting stuck in poor local optima. This approach can improve training stability and performance.

11. Debugging techniques and visualizations: During training, monitoring and analyzing the training process can be helpful in identifying issues or anomalies. Techniques like learning rate schedules, gradient or activation visualizations, or monitoring metrics like loss and accuracy can aid in debugging and understanding the behavior of the model during training.

By applying these solutions and techniques, practitioners can overcome the challenges associated with training deep neural networks and improve the performance, convergence, and generalization capabilities of their models [21] [22].

11.4 APPLICATIONS OF DEEP LEARNING

Deep learning, with its ability to model complex patterns and extract meaningful representations from large datasets, has found applications in various domains. Here are some prominent areas where deep learning has made significant contributions:

1. Convolutional Neural Networks (CNNs) are widely used for these tasks, enabling applications like autonomous vehicles, facial recognition, medical imaging analysis, and augmented reality.

2. Natural Language Processing (NLP): Language understanding, text generation, sentiment analysis, machine translation and question answering are all important advances in deep learning. These applications have been facilitated by the popular Bidirectional Encoder Representation from Recurrent Neural Networks (RNN) and Transformer models (BERT).

3. Speech recognition and synthesis: Deep learning techniques, particularly recurrent neural networks and convolutional neural networks, have significantly improved speech recognition accuracy. Applications include voice assistants, speech-to-text transcription, speaker identification, and text-to-speech synthesis.

4. Recommender systems: Collaborative filtering and deep neural networks have been instrumental in building advanced recommender systems and provide humanized recommendations for products, movies, music and content, enhancing the user experience and increasing business revenue.

5. Autonomous vehicles: In order for autonomous cars to detect and comprehend their environment, deep learning is essential. Convolutional neural networks are employed for object

detection, recognition, and tracking, while recurrent neural networks are used for tasks like behavior prediction and path planning.

6. Financial services: Deep learning techniques are widely employed in financial services for tasks like fraud detection, credit scoring, algorithmic trading, and risk assessment. Deep neural networks can process large volumes of financial data, extract relevant features, and make accurate predictions, aiding decision-making and risk management.

7. Robotics: Deep learning enables robots to perceive and interact with their environment. Applications include object recognition, grasping and manipulation, navigation, and human-robot interaction. Deep reinforcement learning, in particular, has been successful in training robots to perform complex tasks through trial and error.

8. Gaming and virtual reality: Deep learning has made significant contributions to the gaming industry, enhancing game graphics, character animation, and intelligent game agents. Deep neural networks are used to model game environments, learn from player behavior, and create realistic virtual worlds.

9. Energy and sustainability: Deep learning techniques are applied in energy management, demand forecasting, renewable energy optimization, and power grid maintenance. Models can analyze large-scale energy data, identify patterns, and make predictions to optimize energy consumption and improve sustainability.

11.5   RECENT ADVANCES IN DEEP LEARNING

Deep learning has been an active area of research, and several recent advances have contributed to furthering the capabilities of deep neural networks. Here are some notable recent advancements in deep learning:

1. Transformer models: Transformer models, introduced in the "Attention is All You Need" paper, have revolutionized natural language processing tasks. Transformers leverage self-attention mechanisms to capture contextual relationships in sequential data, enabling significant improvements in machine translation, language understanding, and text generation. Models like GPT-3 (Generative Pre-trained Transformer 3) have achieved remarkable performance on a range of language tasks.

2. Generative Adversarial Networks (GANs): GANs have gained popularity for their ability to generate realistic synthetic data. Recent advancements in GANs have led to improved image synthesis, such as StyleGAN and BigGAN, which produce high-quality and highly customizable images. GANs have also been extended to other domains, including text generation, video synthesis, and even 3D object generation.

3. Reinforcement learning: Deep reinforcement learning has witnessed significant progress in recent years. Algorithms like Proximal Policy Optimization (PPO) and Soft Actor-Critic (SAC) have demonstrated impressive performance in complex control tasks and game playing. Deep reinforcement learning has achieved breakthroughs in areas such as robotics, autonomous navigation, and game AI, with agents surpassing human-level performance in games like GoChess, and Dota 2.

4. Meta-learning: Meta-learning, or learning to learn, focuses on training models that can quickly adapt to new tasks or domains. Recent advances in meta-learning, such as Model-Agnostic Meta-Learning (MAML), have shown in few-shot learning scenarios, where models can learn from only a few examples and generalize to new tasks with minimal training.

5. Efficient deep learning: The efficiency and scalability of deep learning models have been a topic of research. Several techniques have been proposed to reduce model size, memory footprint, and inference time while maintaining performance. Examples include knowledge distillation, neural architecture search, quantization, and pruning methods. These

techniques enable deployment of deep learning models on resource-constrained devices and improve their practical usability.

6. Self-supervised learning: Self-supervised learning has gained attention as an approach to learn from unlabeled data. Instead of relying on labeled examples, models are trained to predict or reconstruct data patterns from unlabeled samples. Pretraining models with self-supervised learning, followed by fine-tuning on labeled data, has achieved remarkable results in various domains.

7. Explainable and Interpretable (E&I) Deep Learning: The E&I of deep learning models have become increasingly important. Researchers have been exploring techniques to understand and interpret the decisions made by deep neural networks. Methods like attention mechanisms, saliency maps, and concept attribution have been developed to provide insights into model behavior and enhance trust and transparency.

8. Federated learning: Federated learning enables training models across multiple decentralized devices or servers while maintaining data privacy. This approach allows models to be trained without centrally aggregating sensitive data, making it suitable for applications in healthcare, finance, and the IoT. Recent advancements in federated learning aim to address challenges such as communication efficiency, security, and scalability.

9. Multimodal learning: Multimodal learning involves processing and integrating information from different modalities such as text, images, and audio. Recent advancements have focused on developing deep learning architectures that can effectively model and fuse multimodal data. Multimodal models have originated applications in areas such as multi-media analysis, social media understanding, and autonomous systems.

10. Continual learning: Continual learning addresses the challenge of retaining knowledge from previous tasks while learning new ones. Recent research has explored methods like regularization techniques and episodic memory.

11.6 CHALLENGES AND FUTURE DIRECTIONS IN DEEP LEARNING

Although deep learning has had great success in a number of different fields, the technology still faces a number of obstacles. Additionally, there are promising directions that researchers and practitioners are exploring to overcome these challenges. Here are the key challenges and future directions in deep learning [23] [24]

11.6.1 Challenges

1. Data scarcity: Deep learning models frequently need a lot of labelled data, which might be difficult in fields with little labelled data. Addressing data scarcity through techniques like transfer learning, unsupervised learning, or active learning is essential to enable deep learning models to learn from limited labeled data.

2. Interpretability and explainability: Deep learning models are repeatedly seen in airplane black boxes, to understand their decisions and trust their outputs. Enhancing model interpretability and developing explainable AI techniques are crucial for building trust, ensuring accountability, and enabling adoption in critical domains.

3. Robustness and generalization: Deep learning models can be sensitive to adversarial attacks, noise, or out-of-distribution data, leading to a lack of robustness and limited generalization. Developing models that are more robust, resilient to perturbations, and capable of generalizing to unseen examples is a significant challenge.

4. Ethical considerations: Deep learning raises ethical concerns related to fairness, bias, privacy, and accountability. Addressing these ethical considerations is crucial to ensure that deep

learning models are fair, unbiased, and respectful of privacy, and to establish guidelines for their responsible development and deployment.

5. Computational resources and efficiency: Developing efficient algorithms, hardware optimizations, and alternative computing platforms can help make deep learning more accessible and sustainable.

11.6.2   Future directions

1. Continual learning: Enabling deep learning models to continuously learn and adapt to new information throughout their deployment is an important future direction. Continual learning techniques aim to overcome catastrophic forgetting and allow models to acquire new knowledge while retaining previously learned information [25].

2. Transfer and multitask learning: Transfer learning and multitask learning approaches leverage knowledge learned from one task to improve performance on related tasks or enable learning from limited data. Further advancements in these areas will enhance the ability of deep learning models to transfer knowledge across domains and tasks.

3. Hybrid models and interdisciplinary approaches: Combining deep learning with other AI techniques, such as symbolic reasoning, probabilistic modeling, or reinforcement learning, can lead to more powerful and interpretable models. Interdisciplinary collaborations with fields like cognitive science, neuroscience, and psychology can provide insights into human intelligence and inspire new directions in deep learning.

4. Domain-specific challenges: Different domains, such as healthcare, finance, robotics, or cybersecurity, have specific challenges and requirements. Future research will focus on developing domain-specific deep learning techniques to address these challenges effectively, including privacy-preserving methods, safety considerations, real-time processing, and interpretability in critical applications.

5. Active learning and human-in-the-loop: Incorporating human feedback and guidance in the learning process can improve the performance and efficiency of deep learning models. Active learning approaches that select informative samples for labeling and involve humans in the decision-making loop can reduce the reliance on large, labeled datasets and improve model performance.

6. Federated learning and privacy-preserving techniques: Advancements in federated learning and privacy-preserving techniques will facilitate collaborative learning across organizations, domains, or devices, opening up new possibilities for deep learning applications.

7. Meta-learning and adaptive systems: Meta-learning aims to build models that can quickly adapt to new tasks or environments, advancing meta-learning techniques and developing adaptive deep learning systems

11.7   SUMMARY OF DEEP LEARNING AS A STATE-OF-THE-ART APPROACH:

Deep learning has emerged as a state-of-the-art approach in the field of artificial intelligence and machine learning. It has revolutionized various domains by enabling machines to learn and make predictions from complex and large-scale data. Here is a summary of this research:

1. Hierarchical representations: These representations enable the model to extract high-level features and patterns automatically, allowing for better understanding and analysis of complex data.

2. Unsupervised and self-supervised learning: Deep learning techniques have made significant progress in unsupervised and self-supervised learning. This means that models can learn from unlabeled data, discovering meaningful structures and representations without explicit human annotations. Unsupervised learning has the potential to unlock knowledge from vast amounts of unannotated data.

3. Image and speech recognition: Deep learning has achieved remarkable success in image and speech recognition tasks. Convolutional Neural Networks (CNNs) have become the go-to model architecture for computer vision, accurately recognizing objects, faces, and scenes in images and videos. Similarly, deep learning models, such as Recurrent Neural Networks (RNNs) and transformers, have significantly advanced speech recognition and natural language processing tasks.

4. Natural language processing: Deep learning has propelled the field of Natural Language Processing (NLP) to new heights. Models like Recurrent Neural Networks (RNNs), Long Short-Term Memory (LSTM), and transformer architectures like BERT and GPT have enabled machines to understand and generate human-like text, improve machine translation, sentiment analysis, and chatbots.

5. Reinforcement learning: Deep learning contributions to reinforcement learning; a paradigm where an agent learns to make decisions based on feedback from its environment. Deep reinforcement learning techniques have achieved groundbreaking results in complex domains, such as game-playing (for example, AlphaGo and OpenAI Five), robotics, and autonomous systems.

6. Big data and scalability: Deep learning has demonstrated its capability to handle big data and scale effectively. With the advent of high-performance computing resources, deep learning models can efficiently process large datasets and train complex architectures, enabling the analysis of massive amounts of information in real-time.

7. Transfer learning and pretrained models: Deep learning models trained on massive datasets, such as ImageNet or large-scale language corpora, have become valuable resources for transfer learning. Pretrained models serve as a starting point for developing new models, allowing practitioners to leverage existing knowledge and accelerate training on specific tasks with limited data.

8. Interdisciplinary applications: Deep learning has found applications across diverse domains, including healthcare (for example, medical imaging, disease diagnosis), finance (for example, fraud detection, stock market prediction), autonomous driving, robotics, recommendation systems, and more. Its versatility and ability to learn complex patterns make it a powerful tool across various industries.

9. Deep learning has demonstrated its effectiveness in tackling challenging problems and approaching the precincts of what machines can achieve. it is expected to play an even more momentous role in shaping the future of artificial intelligence and impacting various aspects of our lives [26].

11.7.1 IMPLICATIONS FOR AI AND SOCIETY

The rapid development and widespread adoption of AI, including deep learning, have profound implications for society. Here are some key implications:

1. Automation and job displacement: AI technologies, including deep learning, have the potential to automate many tasks and jobs across various industries. Automation can boost production and efficiency, but it can also result in job loss and social problems. Efforts are needed to re-skill and up-skill the workforce to adapt to the changing job landscape.

2. Privacy and security: AI systems safeguarding privacy and ensuring data security are paramount to protect individuals from potential misuse or unauthorized access to their personal information. Strong data protection measures and responsible data handling practices are necessary to address privacy concerns.

3. Trust and explainability: Deep learning models are often considered "black boxes" due to their complex and opaque nature. Enhancing the explainability and interpretability of AI systems,

especially in critical applications such as healthcare or finance, is crucial to build trust among users, regulators, and society as a whole.

4. Economic impact: The widespread adoption of AI, including deep learning, can have a significant economic impact. It can drive innovation, create new job opportunities, and lead to economic growth. However, it may also exacerbate wealth inequality if the benefits of AI are not distributed equitably across society. Ensuring inclusive economic growth and mitigating socioeconomic disparities are important considerations.

5. Human-AI collaboration: AI, including deep learning, has the capacity to expand human capabilities and improve decision-making processes. Emphasizing human-AI collaboration, where AI systems work alongside humans rather than replacing them, can lead to more effective and responsible use of AI technology.

6. Legal and regulatory frameworks: The rapid advancements in AI, including deep learning, raise questions about legal and regulatory frameworks. Addressing concerns related to liability, accountability, intellectual property rights, and data governance is crucial to ensure that AI systems operate within legal and ethical boundaries.

7. Social impact and bias: AI systems if not carefully designed and trained, can perpetuate biases and amplify existing societal inequalities. Awareness of the social impact of AI and addressing bias through inclusive data collection, diverse model development teams, and rigorous evaluation processes is essential to avoid discriminatory outcomes.

8. Healthcare and well-being: Deep learning has the potential to revolutionize healthcare, from medical imaging analysis to drug discovery. AI-enabled technologies can enhance diagnostics, treatment planning, and patient care. However, careful consideration is necessary to ensure patient privacy, regulatory compliance, and the ethical use of AI in healthcare.

9. Environmental sustainability: AI contributes to environmental sustainability efforts by energy consumption optimization, improving resource management, and facilitating atmosphere change modeling. Leveraging AI for sustainable development can lead to positive environmental impacts.

11.7.2 INSPIRING FUTURE RESEARCH AND INNOVATION

Deep learning has already achieved remarkable advancements in various domains [27], but there are still numerous avenues for future research and innovation. Here are some areas that can inspire future exploration:

1. Explainable AI: Enhancing the explainability of deep learning models is a crucial area of research. Developing techniques and methodologies that provide insights into the decision-making processes of AI systems can help build trust, ensure transparency, and enable users to understand and interpret the outcomes of deep learning models.

2. Robustness and adversarial defense: Deep learning models are vulnerable to adversarial attacks, where maliciously crafted inputs can lead to incorrect predictions. Future research can focus on developing robust models that are resistant to such attacks and improving the overall security and reliability of deep learning systems.

3. Lifelong and continual learning: Enabling deep learning models to learn continuously from new data and adapt to changing environments is an important research direction. Lifelong and continual learning techniques can help mitigate catastrophic forgetting, improve model flexibility, and enable efficient adaptation to new tasks without requiring retraining from scratch.

4. Interdisciplinary applications: Exploring the application of deep learning techniques in interdisciplinary domains such as healthcare, climate science, sustainability, social sciences, and finance can unlock new possibilities for addressing complex challenges and making significant societal impact.

5. Reinforcement learning: Advancing reinforcement learning techniques can lead to more efficient and effective learning algorithms for training autonomous agents. Research can focus on improving sample efficiency, exploration strategies, handling high-dimensional state spaces, and addressing safety and ethical concerns in reinforcement learning scenarios.
6. Federated learning and privacy-preserving techniques: Federated learning enables collaborative training of models on decentralized data without compromising data privacy. Research in this area can focus on developing robust federated learning algorithms, privacy-preserving techniques, and secure aggregation protocols to enable large-scale distributed training while protecting sensitive user data.
7. Multimodal learning: Integrating multiple modalities, such as vision, language, and audio, can enhance the capabilities of deep learning models. Future research can explore techniques for multimodal representation learning, fusion mechanisms, and developing models that can effectively understand and generate information from different modalities.
8. Meta-learning: Meta-learning involves training models to learn how to learn. By leveraging prior knowledge and experience, meta-learning algorithms can enable fast adaptation to new tasks with limited data. Further research can focus on improving meta-learning techniques, transfer learning capabilities, and generalization to diverse domains.
9. Efficient deep learning: Deep learning models are resource-intensive and often require substantial computational power and energy. Developing efficient deep learning algorithms, model compression techniques, hardware accelerators, and low-power implementations can enable the deployment of deep learning on edge devices and resource-constrained environments.
10. Ethical and fair AI: Continuing research on ethical considerations, fairness, and bias in deep learning is crucial. Developing methods for mitigating biases in data, ensuring fairness in algorithmic decision-making, and addressing ethical dilemmas in AI system development and deployment are important areas of exploration.

11.8 CONCLUSION

In conclusion, deep learning has emerged as a state-of-the-art approach in the field of artificial intelligence, revolutionizing various domains and pushing the boundaries of what machines can achieve. With its ability to learn hierarchical representations from complex data, deep learning has enabled significant advancements in image and speech recognition, natural language is processing, and reinforcement learning, among other areas [28] [29].

Deep learning fundamentals, such as neural networks, training algorithms, activation functions, and optimization techniques, form the backbone of this approach. Challenges in training deep networks, including data scarcity, overfitting, and computational complexity, have been addressed through solutions like transfer learning, regularization, and improved optimization algorithms.

The applications of deep learning are vast and impactful, spanning areas like healthcare, finance, robotics, recommendation systems, and more. It has demonstrated its ability to handle big data, scale efficiently, and learn from unlabeled or limited data through unsupervised and self-supervised learning. While deep learning has achieved remarkable success, it also brings forth important implications for AI and society. These include considerations related to job displacement, ethical concerns, privacy and security, trust and explainability, economic impact, and social implications. Addressing these implications requires responsible development, strong regulatory frameworks, and ensuring equitable distribution of AI benefits [30].

Looking ahead, inspiring future research and innovation in deep learning include areas such as explainable AI, robustness and adversarial defense, lifelong and continual learning, interdisciplinary applications, reinforcement learning, federated learning, multimodal learning, efficient deep learning, meta-learning, and ethical and fair AI. Overall, deep learning has transformed the field of artificial intelligence, offering powerful tools to analyze complex data, make accurate predictions, and drive advancements across various domains. As research and development in deep learning continue to evolve, it is crucial to ensure its responsible and ethical use to maximize its potential in positively impacting society.

REFERENCES

1. Goodfellow, I., Bengio, Y., & Courville, A. (2016). *Deep Learning*. MIT Press.
2. He, K., Zhang, X., Ren, S., & Sun, J. (2015). Delving deep into rectifiers: Surpassing human-level performance on ImageNet classification. In *Proceedings of the IEEE international conference on computer vision* (pp. 1026–1034).
3. Huang, G., Liu, Z., Van Der Maaten, L., & Weinberger, K. Q. (2017). Densely connected convolutional networks. In *Proceedings of the IEEE conference on computer vision and pattern recognition* (pp. 4700–4708).
4. LeCun, Y., Bengio, Y., & Hinton, G. (2015). Deep learning. *Nature*, 521(7553), 436–444.
5. Schmidhuber, J. (2015). Deep learning in neural networks: An overview. *Neural Networks*, 61, 85–117.
6. Dai, J., Li, Y., He, K., & Sun, J. (2016). R-FCN: Object detection via region-based fully convolutional networks. In *Advances in neural information processing systems* (pp. 379–387).
7. Krizhevsky, A., Sutskever, I., & Hinton, G. E. (2012). ImageNet classification with deep convolutional neural networks. In *Advances in neural information processing systems* (pp. 1097–1105).
8. He, K., Zhang, X., Ren, S., & Sun, J. (2016). Deep residual learning for image recognition. In *Proceedings of the IEEE conference on computer vision and pattern recognition* (pp. 770–778).
9. Siddiqui, S. A., Yadav, A. K., & Anwar, S. (2019). Deep learning-based prediction models for stock market forecasting: A survey. *Expert Systems with Applications*, 115, 419–435.
10. Szegedy, C., et al. (2015). Going deeper with convolutions. In *Proceedings of the IEEE conference on computer vision and pattern recognition* (pp. 1–9).
11. Simonyan, K., & Zisserman, A. (2014). Very deep convolutional networks for large-scale image recognition. arXiv preprint arXiv:1409.1556
12. Vaswani, A. et al. (2017). Attention is all you need. In *Advances in neural information processing systems* (pp. 5998–6008).
13. Bahdanau, D., Cho, K., & Bengio, Y. (2014). Neural machine translation by jointly learning to align and translate. arXiv preprint arXiv:1409.0473
14. Silver, D., et al. (2016). Mastering the game of Go with deep neural networks and tree search. *Nature*, 529(7587), 484–489.
15. Mnih, V., et al. (2015). Human-level control through deep reinforcement learning. *Nature*, 518(7540), 529–533.
16. Arjovsky, M., Chintala, S., & Bottou, L. (2017). Wasserstein generative adversarial networks. In *International conference on machine learning* (pp. 214–223).
17. Radford, A., Metz, L., & Chintala, S. (2015). Unsupervised representation learning with deep convolutional generative adversarial networks. arXiv preprint arXiv:1511.06434
18. Goodfellow, I., et al. (2014). Generative adversarial nets.
19. Singh, D., & Ghosh, P. (2019). Deep learning techniques for sentiment analysis: A review. *International Journal of Information Management,* 49, 427–436.
20. Kaiming, H., Zhang, X., Ren, S., & Sun, J. (2016). Identity mappings in deep residual networks. In *European conference on computer vision* (pp. 630–645).
21. Kingma, D. P., & Ba, J. (2014). Adam: A method for stochastic optimization. arXiv preprint arXiv:1412.6980
22. Kingma, D. P., & Ba, J. (2014). Adam: A method for stochastic optimization. arXiv preprint arXiv:1412.6980
23. Abadi, M., et al. TensorFlow: A system for large-scale machine learning. In *12th {USENIX} Symposium on Operating Systems Design and Implementation ({OSDI} 16)* (pp. 265–283).

24. Redmon, J., Divvala, S., Girshick, R., & Farhadi, A. (2016). You only look once: Unified, real-time object detection. In *Proceedings of the IEEE conference on computer vision and pattern recognition* (pp. 779–788).

25. Liu, W., Anguelov, D., Erhan, D., Szegedy, C., Reed, S., Fu, C. Y., & Berg, A. C. (2016). SSD: Single shot multibox detector. In *European conference on computer vision* (pp. 21–37).

26. Szegedy, C., Ioffe, S., Vanhoucke, V., & Alemi, A. A. (2017). Inception-v4, Inception-ResNet and the impact of residual connections on learning. In *AAAI conference on artificial intelligence* (pp. 4278–4284).

27. Sutskever, I., Vinyals, O., & Le, Q. V. (2014). Sequence to sequence learning with neural networks. In *Advances in neural information processing systems* (pp. 3104–3112).

28. Hochreiter, S., & Schmidhuber, J. (1997). Long short-term memory. *Neural Computation*, 9(8), 1735–1780.

29. Zhang, X., Zhou, X., Lin, M., & Sun, J. (2017). Shufflenet: An extremely efficient convolutional neural network for mobile devices. In *Proceedings of the IEEE conference on computer vision and pattern recognition* (pp. 6848–6856).

30. Kingma, D. P., & Welling, M. (2013). Auto-encoding variational Bayes. arXiv preprint arXiv:1312.6114

12 An Approach through Different Mathematical Models to Enhance the Utility in Different Areas of Machine Learning

Pooja Swaroop Saxena

12.1 INTRODUCTION

Machine learning is an algorithm which uses computational methods to learn directly from data without being reliant upon a predetermined equation. It allows computers to do what humans and animals do naturally. With greater samples available, the algorithm adapts to improve performance. This type of machine learning is known as deep learning.

A data analytics technique in which computers help to learn to do what comes naturally to humans and animals, is machine learning. It is an algorithm in which we directly learn from data by using computational methods without preparing a model with the help of predetermined equations. The algorithms improve their performance for large sample sizes.

12.2 CLASSIFICATION OF MACHINE LEARNING

Depending on the nature of the learning "signal" or "response" machine learning is classified into four major categories, which are as follows:

12.2.1 SUPERVISED LEARNING

Supervised learning is a type of machine learning approach where an algorithm learns from labeled training data to make predictions or decisions. In supervised learning, we have a dataset consisting of input variables (features) and their corresponding output variables (labels or target values). The goal is to train a model that can accurately predict the labels for new, unseen data.

12.2.2 UNSUPERVISED LEARNING

Unsupervised education is a type of machine intelligence used to draw conclusions from datasets including data outside expectations. In alone education algorithms, categorization or classification is not predetermined. It looks for patterns to determine whether objects or occurrences are from similar classes, decided by observation.

DOI: 10.1201/9781003433309-12

12.2.3 Reinforcement learning

Reinforcement Learning (RL) is a type of machine learning paradigm where an agent learns to make decisions by interacting with an environment. The agent's objective is to maximize its cumulative reward over time. It is inspired by behavioral psychology, where learning is achieved through trial and error.

In RL, the agent operates in an environment and takes actions to achieve specific goals. Each action leads to a new state, and the agent receives feedback in the form of a reward signal from the environment based on the action taken.

Reinforcement learning has been successfully applied in various domains, including robotics, game playing, recommendation systems, autonomous vehicles, and resource management, amongst others. It has the advantage of learning directly from interactions with the environment without requiring labeled training data, making it suitable for scenarios where explicit supervision is difficult or expensive.

12.2.4 Semi-supervised learning

Semi-supervised learning is a type of machine learning paradigm that falls between supervised learning and unsupervised learning. In this approach, the model is trained on a combination of labeled data (data with corresponding target labels) and unlabeled data (data without target labels). The goal of semi-supervised learning is to leverage the additional information present in the unlabeled data to improve the model's performance compared to using only labeled data.

In many real-world scenarios, obtaining labeled data can be expensive, time-consuming, or even infeasible. However, unlabeled data is often more abundant and easier to collect. Semi-supervised learning aims to make use of this abundance of unlabeled data to augment the learning process and obtain better generalization on unseen data.

12.3 LEARNING MODELS

Learning models are computational algorithms or systems designed to improve their performance on a specific task through experience, often by processing data and adjusting their internal parameters accordingly. Machine learning is concerned with using the right features to build the right models that achieve the right tasks. Some of the Learning models are:

12.3.1 Logical models

In machine learning, logical models are approaches that use logical rules or constraints to model relationships between input features and target outputs. These models are typically rule-based and can be useful in various applications, especially when the data has a clear and interpretable structure.

Here are some examples of logical models in machine learning:

1. Decision trees: In this model data splits into subsets based on the values of input features. Each internal node of the tree represents a decision based on a feature, and each leaf node represents a prediction or a class label. Decision trees are easy to interpret and can capture non-linear relationships in the data.
2. Rule-based classifiers: A set of IF-THEN rules to make predictions has been used in this type of classifier. Each rule consists of a condition based on input features and a corresponding predicted class label. These models are interpretable and can be useful in domains where human experts often use rule-based decision-making.

3. Logical rule induction: Logical rule induction algorithms aim to discover logical rules that accurately describe patterns in the data. These rules can be expressed in terms of IF-THEN statements and are often used for classification tasks.

4. Fuzzy logic: Fuzzy logic extends traditional binary logic by allowing degrees of truth. Instead of crisp binary values (TRUE/FALSE), fuzzy logic assigns a degree of membership between 0 and 1 to each proposition. Fuzzy logic can be useful in situations where there is uncertainty or ambiguity in the data.

5. Constraint Satisfaction Problems (CSP): CSP is a type of logical model used in various AI applications, where the goal is to find solutions that satisfy a set of constraints. It can be used for tasks like scheduling, planning, and optimization.

6. Inductive Logic Programming (ILP): This is the combination of logic programming and machine learning. It has been used to induce logical rules or programs from data, allowing the model to generalize from specific examples.

7. Logic-based ensemble methods: Ensemble methods that combine the predictions of multiple models can also be formulated using logical rules. For example, logical rules can be used to combine the outputs of individual classifiers to make a final decision.

Logical models have several advantages in machine learning, including interpretability, human-understandable representations, and the ability to handle noisy or incomplete data. However, they may have limitations in capturing complex relationships present in large and high-dimensional datasets, which are better addressed by other machine learning approaches like deep learning. As such, logical models are often used in combination with other techniques to improve model performance and interpretability.

12.3.2 Geometric models

In geometric models, looks may be detailed as points in two ranges (x- and y- arbor) or a three spatial room (x, y and z). Even when faces are not essentially lines, they may be posed in a lines form (for example, hotness as a momentary function may be represented on two axes).

These are a few examples of geometric models in machine learning:

1. Support Vector Machines (SVM): For classification and regression tasks, the SVM geometric model has been used.

2. Convolutional Neural Networks (CNN): For image and video analysis CNNs are widely used. They are inspired by the structure and organization of the visual cortex in humans. CNNs exploit the local spatial correlations and hierarchical representations in images by using convolutional layers that apply filters to capture local patterns and pooling layers that aggregate information across spatial regions.

3. Graph Neural Networks (GNN): GNNs are designed to process structured data represented as graphs. They leverage the graph structure to propagate and exchange information between nodes, capturing relational dependencies. GNNs enable learning on graph-structured data, such as social networks, molecule structures, or recommendation systems, by incorporating geometric principles.

4. Manifold learning: Manifold learning techniques aim to uncover the underlying geometric structure of high-dimensional data. They transform the data into a lower-dimensional representation while preserving its intrinsic structure. Methods like t-SNE (t-distributed Stochastic Neighbor Embedding) and Isomap use geometric principles to visualize and explore high-dimensional data in a lower-dimensional space.

5. Geometric deep learning: This is a field that focuses on extending deep learning techniques to process and analyze structured and geometric data. It combines deep learning architectures

with geometric representations, such as graphs, point clouds, or meshes, to enable learning and reasoning on structured and spatially connected data.

Geometric models in machine learning offer advantages in capturing spatial relationships, handling structured data, and interpreting patterns in a geometric context. However, they may face challenges with scalability, generalization to unseen data, and sensitivity to data transformations.

12.3.3 Linear models

Linear models are nearly natural. In this case, the function is presented as a uninterrupted blend of allure inputs. Thus, if x_1 and x_2 are two scalars or headings of the alike measure and a and b are dictatorial scalars, then $ax_1 + bx_2$ shows an uninterrupted consolidation of x_1 and x_2. In the most natural case place f(x) shows a direct route, and we have an equation of the form:

f(x) = mx + c, where "c" shows the interrupt and "m" shows the slope. Linear models are constant, that is, narrow variations in the data have only a small effect on the developing model. In contrast, considering models of processed wood, these are likely to change when reflecting on the details of how the wood is prepared, for example, the choice of a particular split at the root of the timber usually influences the rest of the wood. As a result of having comparatively few complication, linear models tend to reduce difference and extreme bias. This indicates that linear models are less inclined to overfit the initial data than other models. However, they are more inclined to underfit.

12.3.4 Distance-based models

Distance-based models are inferior to geometric models. Like linear models, distance based models are based on the arithmetic of the data. Distance-based models reflect the concept of distance. From the perspective of machine learning the idea of absolute distance is not established only the relative positions of the mid-points of sets of two points. They reflect the median of two points, considering the ways of moving between two points.

12.3.5 Probabilistic models

The triennial classification of machine intelligence algorithms is the probabilistic model. We have visualised before that the k-most familiar neighbor uses the distance to categorize individuals and reasonable models use a reasonable verbalization to partition the range of points. In this section, we visualize in what way or manner the probabilistic models use the plan of contingency to categorize new systems.

12.4 PERSPECTIVES AND ISSUES IN MACHINE LEARNING

This includes probing a huge field of attainable theories to decide individually the ones that best fit the data with some pre-planning by the researcher. For example, consider the range of theories that might work for one particular researcher. This range of theories exists for all of the judgement functions that may be represented by a few choices for the weights w_0 through w_6. The LMS treasure for fitting weights achieves this aim by iteratively bringing into harmony the weights, making an adjustment for each weight. At each opportunity the speculated judgement function anticipates a profit that varies according to the previous adjustments.

Machine learning, as a rapidly evolving field, encompasses various perspectives and raises numerous issues. Some key perspectives and issues in machine learning:

1. Ethical considerations: One of the most critical perspectives in machine learning revolves around ethics. As algorithms increasingly impact people's lives, questions arise concerning bias, fairness, transparency, and accountability. Ensuring that machine learning models do not discriminate against individuals or perpetuate existing social biases is a significant concern.

2. Bias and fairness: Biases present in training data can be inadvertently learned and perpetuated by algorithms, leading to discriminatory outcomes. Addressing bias and ensuring fairness in algorithmic decision-making is an ongoing challenge in machine learning.

3. Interpretability and explainability: The lack of interpretability and explainability can reduce trust in the technology, especially in critical domains like healthcare or legal systems. Efforts are underway to develop techniques for model interpretability and explainable AI.

4. Privacy and data security: The models require large amounts of data for training and data security. The collection, storage, and use of personal data must be handled responsibly to safeguard individuals' privacy rights and prevent misuse or unauthorized access to sensitive information.

5. Scalability and efficiency: As machine learning models grow more complex, scalability and efficiency become critical issues. Training and deploying large-scale models require substantial computational resources, energy consumption, and time. Developing efficient algorithms and hardware architectures to support the scalability of machine learning is an ongoing area of research.

12.5 CONCLUSION

Machine learning is a field of machine intelligence that handles the design and implementation of algorithms that can get or give an advantage and form forecasts on datasets. The aim of machine intelligence is to seek to examine model construction and allow calculations that might gain an advantage. Machine learning is a strong way of producing indicators from data. However, it is important to remember that machine intelligence is only as good as the data used to train the algorithms.

REFERENCES

1. Churkin, A., et al. (2022). Machine learning for mathematical models of HCV kinetics during antiviral therapy. *Mathematical Biosciences, 343*, 108756.
2. Wilmott, P. (2022). Machine learning: An applied mathematics introduction. *Machine Learning and the City: Applications in Architecture and Urban Design*, 217–248.
3. Masum, M., Masud, M. A., Adnan, M. I., Shahriar, H., & Kim, S. (2022). Comparative study of a mathematical epidemic model, statistical modeling, and deep learning for COVID-19 forecasting and management. *Socio-Economic Planning Sciences*, 80, 101249.
4. Pavlyutin, M., et al. (2022). COVID-19 spread forecasting, mathematical methods versus machine learning, Moscow case. *Mathematics*, 10(2), 195.
5. Shanmugam, B. K., et al. (2022). Comparison of the prediction performance of separating coal in separation equipment using machine learning based cubic regression modelling and cascade neural network modelling. *International Journal of Coal Preparation and Utilization*, 1–16.
6. Varshneya, M., Mei, X., & Sobie, E. A. (2021). Prediction of arrhythmia susceptibility through mathematical modeling and machine learning. *Proceedings of the National Academy of Sciences*, 118(37), e2104019118.
7. Chaudhary, K., Alam, M., Al-Rakhami, M. S., & Gumaei, A. (2021). Machine learning-based mathematical modelling for prediction of social media consumer behavior using big data analytics. *Journal of Big Data*, 8(1), 1–20.
8. Iantovics, L. B. (2021). Black-box-based mathematical modelling of machine intelligence measuring. *Mathematics*, 9(6), 681.

9. Gupta, R. K., Sahu, Y., Kunhare, N., Gupta, A., & Prakash, D. (2021). Deep learning based mathematical models for feature extraction to detect Coronavirus disease using chest X-ray images. *International Journal of Uncertainty, Fuzziness and Knowledge-Based Systems*, 29(06), 921–947.

10. Clement, J. C., Ponnusamy, V., Sriharipriya, K. C., & Nandakumar, R. (2021). A survey on mathematical, machine learning and deep learning models for COVID-19 transmission and diagnosis. *IEEE Reviews in Biomedical Engineering*, 15, 325–340.

11. Bolte, J., & Pauwels, E. (2020). A mathematical model for automatic differentiation in machine learning. *Advances in Neural Information Processing Systems*, 33, 10809–10819.

12. Dutta, N., Subramaniam, U., & Padmanaban, S. (2020, January). Mathematical models of classification algorithm of Machine learning. In *International Meeting on Advanced Technologies in Energy and Electrical Engineering* (Vol. 2020, No. 1, pp. 3). Hamad bin Khalifa University Press (HBKU Press).

13. Boso, D. P., Di Mascolo, D., Santagiuliana, R., Decuzzi, P., & Schrefler, B. A. (2020). Drug delivery: Experiments, mathematical modelling and machine learning. *Computers in Biology and Medicine*, 123, 103820.

14. Khalafalla, M., Elmohr, M. A., & Gebotys, C. (2020, December). Going deep: Using deep learning techniques with simplified mathematical models against XOR BR and TBR PUFs (attacks and countermeasures). In *2020 IEEE International Symposium on Hardware Oriented Security and Trust (HOST)* (pp. 80–90). IEEE.

15. Fokas, A. S., Dikaios, N., & Kastis, G. A. (2020). Mathematical models and deep learning for predicting the number of individuals reported to be infected with SARS-CoV-2. *Journal of the Royal Society Interface*, 17(169), 20200494.

13 Study of Different Regression Methods, Models and Application in Deep Learning Paradigm

Arpita Shome, Gunjan Mukherjee, Arpitam Chatterjee, and Bipan Tudu

13.1 INTRODUCTION

The deep learning [7] regression specialization of machine learning focuses on foretelling discrete numerical values. Deep learning is utilised in artificial neural networks to execute intricate computations on enormous datasets and is structured like the human brain. A deep neural network prototype has been generated to discover patterns and correlations in a dataset, which may subsequently be employed to provide predictions on brand-new, untainted input. The aim of conventional regression is to develop a model based on mathematics that can forecast a continuous outcome variable from a set of characteristics or input variables. Deep neural networks, which comprise numerous layers of linked artificial neurons, are used in deep learning regression to develop this idea. Deep learning regression's main benefit is its capacity to automatically extract intricate patterns and attributes from provided data without the requirement for feature engineering by hand. Through a process known as training, the network acquires these properties by modifying the weights and prejudices of the neurons in response to the input-output interactions contained in the data used for training.

In recent years, we have seen a great deal of interest in the field of deep learning, primarily due to its significant function in numerous application fields. By expanding the variety of streams and the range of factors, it acts as decision-support software in the technological sector by using high-performance computing power in the stages of big data to expose the most fundamental abstractions contained in the original dataset.

In general, AI adds behavioral insights and cognitive ability to technology or systems [8], whereas Machine Learning (ML) is an approach that facilitates the construction of analytical frameworks by acquiring knowledge from input or experience [9]. Deep learning also implies data-driven approaches to learning that use multi-layer neural systems to process information.

Figure 13.1 illustrates the essential distinctions between deep learning, machine learning and artificial intelligence. In the context of the deep learning technique, the "deep" conveys the theory of multiple phases of processing information before creating an information based technique. Therefore, Deep Learning (DL) may be seen as a breakthrough for artificial intelligence that could be used to create smart and automated systems. Precisely, it enhances the technology to new heights known as "Smarter AI." due to DL's ability to acquire knowledge from information. Furthermore, there is an intricate relationship between DL and "Data Science" [10]. Data science typically implies the whole process of uncovering insights or meaning in input data in a certain

DOI: 10.1201/9781003433309-13

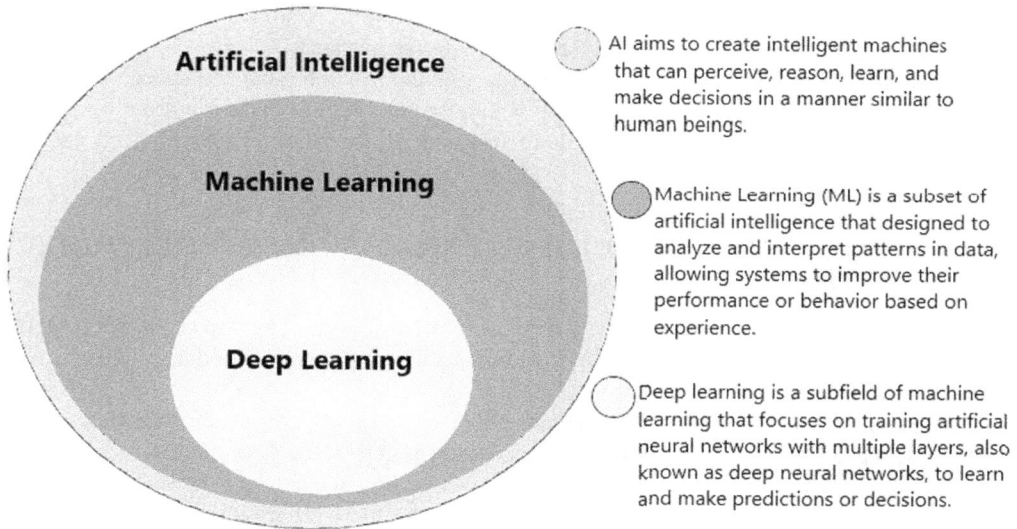

AI aims to create intelligent machines that can perceive, reason, learn, and make decisions in a manner similar to human beings.

Machine Learning (ML) is a subset of artificial intelligence that designed to analyze and interpret patterns in data, allowing systems to improve their performance or behavior based on experience.

Deep learning is a subfield of machine learning that focuses on training artificial neural networks with multiple layers, also known as deep neural networks, to learn and make predictions or decisions.

FIGURE 13.1 Essential distinctions among Artificial Intelligence, Machine Learning & Deep Learning.

area of interest, and is often used for the betterment of critical analytics and precise prediction [11, 12]. In broad terms, we are able to demonstrate that DL technology, particularly as regards aspects of a robust computing processor and impact, has the ability to change the world as we understand it.

Deep learning is a particular kind of prediction model that has gained a lot of popularity recently thanks to the latest innovations in the area of machine learning (LeCun et al., 2015) [1]. The basic idea simply pertains to deep neural networks and is not particular to machine learning or data analytics methods from operations research. The size of the networks, which may today easily encompass as many as hundreds of stages, millions of neural networks, and intricate architectures of connections between them, has altered from early tests with neural networks (for example, He et al., 2016) [2]. With the ability to represent even very complicated, non-linear connections between predictor and result variables, deep neural networks are now more flexible than ever before and surpass models from standard machine learning in a number of tasks.

A sizable, labelled dataset is often needed for developing a deep-learning regression model. Trained input data is used to fine-tune the framework's parameters, and a set of assessments is employed to evaluate how well the model performed during preparation and make any required changes. In order to decrease the inconsistency among its probable outputs and the real results, the model iteratively updates its weights and biases.

Deep Learning (DL) approaches in particular are a desirable choice since they increase computation and accuracy by arranging many layers of networks [3]. For instance, scientists in [4] looked at sequence-to-sequence neural networks to forecast the long-term energy usage of buildings.

Feedforward neural networks [80], convolutional neural networks, and recurrent neural networks are typically prepared and applied in deep learning regression. For jobs where the input data has a constant size and no temporal relationships, feedforward neural networks are appropriate. In regression problems involving signal or image data, CNNs are frequently employed because they are efficient at capturing spatial or temporal patterns. For applications involving data that is consecutive, such as time-series analysis or speech processing, RNNs are appropriate [81]. By passing input across the network and retrieving the output, the deep learning regression approach may be used to generate predictions on fresh, unexplored data once it has been trained.

Many industries, including health care, banking, the field of robotics, and speech recognition, have found use for deep learning regression. It is an effective tool for tasks requiring accurate and exact continuous value forecasts as a result of its capacity for handling complicated and high-dimensional data.

13.2 LITERATURE REVIEW OF DIFFERENT REGRESSION METHODS, MODELS AND APPLICATIONS

Due to being able to acquire knowledge from the information given to them, the deep learning approaches have recently emerged as one of the trendiest areas in the realm of data science, machine learning and artificial intelligence.

Although many uses of neural networks are effective, curiosity concerning this field of study subsequently waned. (Hinton et al., 2006) [5] first demonstrated "Deep Learning", which is founded on the idea of an artificial neural network. They saw a resurgence as deep networks evolved into a popular subject of conversation. For this reason, the term "new generation neural networks" has occasionally been used. This is due to the fact that efficiently trained deep networks have shown remarkable success in cases of regression and classification issues [6].

Market movements, financial indicators, and stock prices have all been forecast via deep learning regression algorithms. These models can help in the selection of investments and risk-management tactics by examining historical data and considering various financial parameters. Convolutional neural networks [16], deep neural networks with beliefs [13], and laying autoencoders [14] are the three basic deep learning algorithms that are frequently employed in research. The first two strategies have been incorporated into analysis conducted by the pertinent work on deep neural networks applied to accounting. For instance, Ding et al. [15] use a mixture of deep convolutional neural networks and neural tensor architecture for predicting both the short- and long-term effects of changes in stock prices. Deep belief networks are also used in several studies to anticipate the financial markets, such as those by Shen et al. [17], and Kuremoto et al. [18]. There haven't been many attempts to research whether the stacked autoencoder techniques can be used to forecast the financial markets.

Medical image analysis, illness diagnosis, and prognosis have all employed deep learning regression. For instance, deep learning models in medical image analysis can find and categorise anomalies in X-ray, MRI, and CT scans. Depending on clinical data, genetic data, and electronic health records, they can also forecast outcomes for patients. Continuously struggling with disruption, several nations have had healthcare catastrophes brought on by the severe acute respiratory syndrome Coronavirus 2 [1]. Effective outcomes for patient prediction could alleviate strain on medical facilities by aiding clinical staff in giving appropriate levels of care and allocating scarce healthcare facilities. Additionally, it would make things simpler for directors of services and legislators to react swiftly to any future outbreaks, whose scales of magnitude could be difficult to anticipate [2].

Energy demand and consumption have been predicted [28] using deep learning regression models. These models can assist utilities and electricity grid operators in making precise load estimates and maximizing energy distribution by looking at previous energy consumption trends, weather information, and other pertinent aspects. Deep learning regression techniques have been used to analyse environmental data and make predictions about things like air quality [29], water contamination [30], and weather patterns. These models may be used to manage natural resources, detect possible problems, and support environmental policy choices. As seen in Table 13.1 below there are some deep learning regression applications in a range of domains and industries. The following are a few notable instances:

TABLE 13.1

A overview of deep learning applications in widely recognised practical domains

Research Field	Research	Methodologies used	Result Accuracy
Financial Forecasting	Fischer et al., 2018 [23]	On S&P 500, the study applies an extensive explanation to a significant financial market forecasting task utilising LSTM networks, random forests and the logistic regression model.	Based on a short-term reversal approach, before transaction expenses, yields 0.23%. In comparison, this reveals moderate susceptibility of typical sources for deliberate threats.
Healthcare and Medical Diagnosis	Abdulaal & Patel, 2020 [21]	Focused on creating and contrasting two different forecasting models: ANN and a Cox regression model for SARS-CoV-2-related fatalities during admission. The evaluation of the model included calibration, discrimination, and validation.	High accuracy was attained by the Cox regression model and ANN algorithms 83.8%, 95% confidence interval (CI) 73.8–91.1 and 90.0%, 95% CI 81.2–95.6, accordingly.
Natural Language Processing (NLP)	Yang et al., 2020 [27]	The pre-processed corpus Google opensource database, which comprises 30 speech command words) is trained using DNN, RNN, and CNN algorithms to provide the model with the greatest accuracy.	Contrasting the trial outcomes of CNN, RNN, and DNN that employ CNN estimates accurate prediction.
Energy Load Forecasting	Heghedus et al., 2018 [26]	The study assesses the digital metre energy consumption dataset, and utilise autocorrelation to determine significant parameters using GRU (Gated Recurrent Unit) and RNN-based recurrent Deep Learning.	Approaches containing one and three hidden layers fared worse than those with two GRUs concealed, each with twenty nodes.
Environmental Analysis	Li et al., 2017 [22]	The study runs experiments on Acoustic Scenes and Events (DCASE) data collected in 2016 applying the Deep Neural Network, Gaussian Mixture, Recurrent Neural Network, Convolutional Neural Network, and i-vector.	A mean CV predictive value of 84.2% and test performance of 84.1% are attained by the model (DNN employing the Smile6k features) with the highest accuracy. and the 4-fold Cross Validation average is 72.5% by 15.6%.
Customer Behaviour and Market Analysis	Generosi et al., 2018 [24]	A convolutional neural network is embedded in the looks and emotions recognition module. The FER+ and EmotioNet crowd-sourced datasets were used to train a convolutional neural network for this research project. The CNN used for both the age and gender parts was likewise trained using the IMDB-Wiki dataset.	The algorithm succeeded in identifying faces in 96% and 94% of the video slices analysed by the Logitech Quickcam and webcam, respectively. The algorithm can accurately anticipate a customer's sex (the overall accuracy is 0.92; for the age range, the prediction accuracy is 0.83).
Autonomous Vehicles	Miglani et al., 2019 [19]	This study compares MLP, LSTM, combinations of LSTM, and CNN for predicting traffic flow in autonomous cars and explores their usefulness in contemporary smart transportation systems.	MLP, LSTM, a combination of LSTM, and CNN models have been extensively researched. But Hybrid structures are shown to be far more precise than separate models.

13.3 ESSENTIAL ELEMENTS OF DEEP LEARNING REGRESSION METHODS

Neural networks: These are primarily the brains of deep learning regression, and are made up of neurons, or linked nodes, arranged into layers. Each layer's neurons take in inputs, modify those inputs with weights and biases, and then transmit the modified data to the following layer. Deep neural networks may learn complex links involving input and output since they contain many hidden layers.

Activation function: The neural network may learn complicated associations in the data because activation functions provide it with irregularity. ReLU (Rectified Linear Unit), sigmoid geometry tanh, and other popular activation functions are frequently employed in deep learning regression.

Loss function: Several measures, including Mean Squared Error (MSE) and Mean Absolute Error (MAE), which measure the disparity between the values that were predicted and the actual values, penalize greater prediction mistakes more severely [78]. The kind of regression issue determines the loss function to be used.

Optimization algorithm: The optimisation algorithm is in charge of minimizing the loss function throughout the training procedure. Deep learning regression frequently employs gradient descent and its derivatives, including Stochastic Gradient Descent (SGD), Adam, and RMSprop.

Training: Using backpropagation, the neural network progressively modifies its weights and biases throughout the training phase. The loss function's gradients are calculated according to the model's assumptions and updated in a way that minimises loss.

Overfitting: It is a prevalent issue in terms of deep learning regression analysis, where the model works fine on the trained input data yet fails on new, untrained data. Overfitting is reduced using strategies including dropout, regularization, and early termination. It is an effective tool for tackling actual regression issues since it is able to identify complex structures and correlations in the data. However, in order to train properly, a lot of processing power and data are also needed.

13.4 TYPES OF DEEP LEARNING APPROACHES

Like machine learning, there are primarily three variations of deep learning paradigms: supervised, unsupervised, and a hierarchy of features or ideas is used in DL applications, where high-level characteristics could be established from low-level characteristics and vice versa. A detailed taxonomy of deep learning methods is shown in Figure 13.2.

The deep learning approach is further broken down into other variations like MLP, CNN, GAN, AE and its variations and also other combinations of many hybrid algorithms.

Further, there is a feature learning algorithm that relies on automatic extraction of characteristics. Deep learning uses artificial neural networks to execute complex calculations on huge volumes of data. It is an aspect that depends on how the human mind operates and functions.

13.4.1 SUPERVISED LEARNING

The most popular and commonly applied technique in deep learning is supervised learning, makes use of labelled input data [51]. This contains a number of matching inputs and outputs. As an illustration, an iterative change in the network's parameters can improve the outputs' approximations. After effective training, the researchers will be capable of extracting accurate information from their surroundings. Recurrent Neural Networks (RNN), Convolutional Neural Networks (CNN), Long Short Term Memory (LSTM), and Gated Recurrent Units (GRU) are some of the supervised learning methods used in deep learning. In supervised learning, inputs (features) and target outputs (labels) are coupled, and the neural network undergoes training using labelled data. Learning a function for

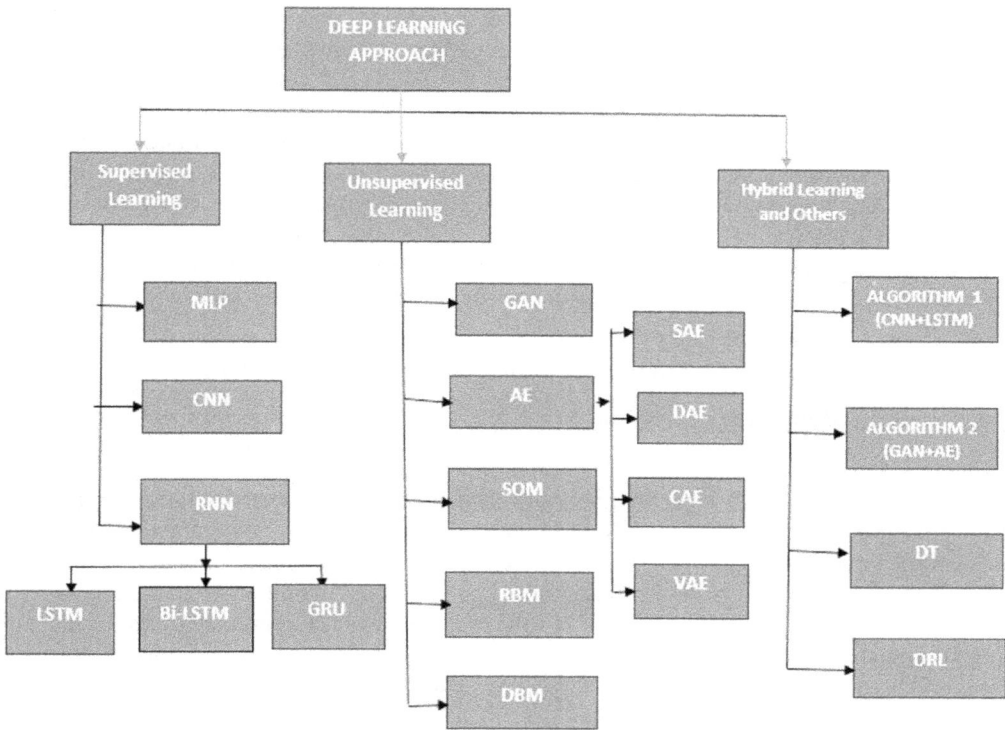

FIGURE 13.2   A Taxonomy of Deep Learning Approaches.

mapping that accurately predicts the labels for novel, unseen inputs is the objective. During training, the network adjusts its parameters using an optimisation approach such as backpropagation and gradient descent depending on the disparity between its anticipated output and the true labels. For tasks like speech recognition, recognition of objects, sentiment analysis, and picture categorization, supervised learning is applied [79].

13.4.2   UNSUPERVISED LEARNING

Unsupervised learning is primarily evaluating a neural network without a clear target output using unlabeled input. The network seeks to identify and understand the fundamental patterns, structures, and representations contained in the data. For applications including clustering, decreasing dimensionality, and generative modelling, this kind of learning is frequently employed. Autoencoders, which seek to reassemble the input data from a compressed representation such as Variational AutoEncoders (VAEs) and Generative Adversarial Networks (GANs), which lead to processing new samples similar to the training data, are examples of popular unsupervised learning algorithms in deep learning.

13.4.3   HYBRID LEARNING

Hybrid networks are boosted by the possibility of simultaneously developing deep generative and discriminative models within the same environment and reaping their benefits. Generative models may learn from both labelled and unlabeled data, making them flexible. Conversely, discriminative models surpass their generative equivalents in supervised tasks while becoming incapable of acquiring knowledge from unlabeled input. Hybrid deep learning algorithms frequently consist of multiple basic deep learning paradigms, where the fundamental architecture is one of the aforementioned selective or procedural deep learning models.

A successful hybridization can surpass others when it comes to performance and uncertainty in high-risk scenarios. Hybridization, such as using combinations of different parameters of DL, may prove to be an important function in the domain as an assortment of neural networks are developed with independent training databases. Designing efficient integrated supervised and unsupervised models in this way, compared to employing single techniques, might be an important scientific opportunity to address a number of real-world obstacles, such as model unpredictability and semi-supervised learning assignments. A detailed taxonomy of deep learning methods has been elaborated.

13.5 DEEP LEARNING IN SUPERVISED LEARNING METHODS

13.5.1 MULTI-LAYER PERCEPTRON (MLP)

This is a supervised learning method [43] also regarded as the deep learning basic architecture. It is a sometimes referred to as a perceptron or artificial neuron that is composed of several layers of linked nodes. MLPs have been widely utilised for a variety of projects, involving regression, detection, classification, and pattern recognition. These are an essential component of deep learning.

The raw data, which might be a sequence of vectors or a collection of features, is delivered to the input layer. A characteristic or an aspect of the provided data is represented by each input node. Amongst all the layers of an MLP, there may be one or more hidden layers [41]. Each of the hidden layers is an assortment of nodes that use weighted connections to process and change the incoming data. Each link between nodes in neighboring levels has a weight attached to it. Based on the information gathered from the hidden layers that have been modified, the output layer generates the final conclusions or outputs. The specific job will determine how many nodes are in the output layer. The modification of weights is used to measure the link strength. Furthermore, every link in the hidden and outcome layers has a corresponding bias term, enhancing control and adaptability over the predictions made by the model. The output of every node in the hidden and resultant layers is subjected to an activation function. The sigmoid, tanh, and ReLU activation functions are mainly used in MLPs [40]. A multi-class classification job might have numerous output nodes that reflect the probabilities of various classes, whereas a binary classification work could only have one output node that represents the likelihood of belonging to one class. MLPs move the provided data from the layer of input to the layer of output by propagating it across the layers. The weighted total of the inputs is calculated for each layer, then transmitted to the following layer by passing via the activation function. The backpropagation technique, which determines the gradients of the network's design parameters (weights and biases) compared to a selected loss function, is used to train MLPs. The parameters are then repeatedly updated using the gradients, often using a stochastic gradient descent optimisation process.

MLPs are adaptable models that can discover intricate patterns and asymmetrical connections in the data. Although they can handle tasks requiring sequential or spatial data, Recurrent Neural Networks (RNNs) and Convolutional Neural Networks (CNNs) are better suited for these tasks.

13.5.2 CONVOLUTIONAL NEURAL NETWORK

This is referred to as CNN, which represents a deep learning framework made for handling organised, grid-like data, especially visual data like photographs and movies. In several vision related applications, including categorization of an image, object identification, advance segmentation of photographs, and others, CNNs have achieved outstanding results [83] [82]. CNN's main principle is to use layers with various capabilities to develop hierarchical ways to represent visual information. CNN's primary ideas and elements are listed below:

Convolutional layer: Firstly, the input image is subjected to convolutional filters, sometimes known as kernels, by the convolutional layer. Each filter gathers spatial information by performing element-wise multiplication and summing across a limited receptive field. The filters are trained to recognise many visual elements, including edges, textures, and more complex patterns. A kernel uses a feature map to accomplish convolution. The component-wise summation of the component-wise multiplication between the input's matching receptive field and the filter yields the resultant feature map. The following is a representation of the convolution operation's formula:

$$\text{Output Feature Map} = \text{ReLU}\sum\left(Input\,image \times Filtration\right) + Bias \tag{13.1}$$

Activation Function: By introducing non-linearity into the network, an activation function [36, 37] enables the network to learn intricate mappings among inputs and outputs. Nonlinear activation functions are supplied with feature maps from a convolutional layer. New tensors, which are sometimes applied as feature maps, are generated by processing the feature maps via an activation function. The Rectified Linear Unit (ReLU) [38] is a commonly used activation function in CNNs because it quickly completes training and effectively solves the vanishing gradient problem. The architecture of CNNs is shown in Figure 13.3.

$$(\text{ReLU})\,f(y) = \max(0, y) \tag{13.2}$$

Pooling Layer: The feature map's spatial scopes in the convolution layer are lessened by this layer. It does this by taking the largest amount (max-pooling) or the average amount (average pooling) inside each pooling window while down sampling the input. Figure 13.3 shows the basic architecture of CNN where pooling aids in preserving the most crucial traits while lowering computing costs and controlling overfitting.

$$\text{Output} = \text{Max (Values of pooling window)} \tag{13.3}$$

CNN's convolution and pooling are specifically affected by the ideas of simple cells and complicated cells that are well-established in visual neuroscience.

Fully Connected Layers: The attribute values are streamlined into a 1D vector via fully connected layers after numerous convolutional and pooling layers. Comparison of these layers with feedforward neural networks is possible, as every neuron in the layer before it and the layer after it is linked to every other neuron. For multi-class classification problems, fully linked layers are

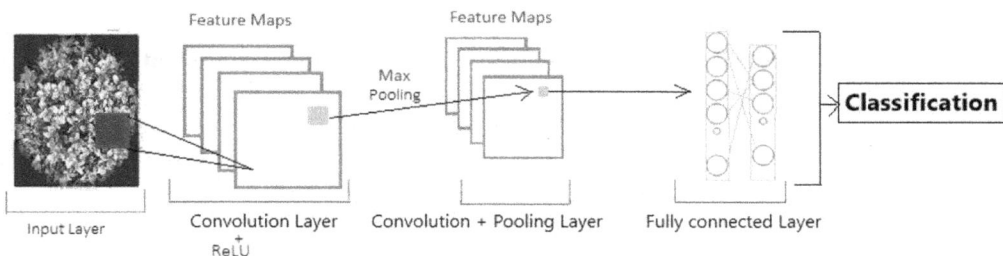

FIGURE 13.3 Architecture of CNN.

frequently accompanied by a softmax activation [66]. These layers aid in making estimations based on the learned characteristics.

Dropout: During training, a portion of the resultant activations are randomly set to zero using the regularization approach. Additionally, CNN employs a "dropout" [42] which can address the excess-adaptation problem that could occur in a typical network infrastructure. By making the network learn alternate interpretations and become more resilient, it helps prevent overfitting.

Backpropagation: Backpropagation [32] is a method for seeking the loss function of a network's parameter gradients. The technique of backpropagation may be used to compute gradients as long as the modules' internal weights and inputs are generally smooth functions of those inputs. The completely connected layers update in the backward propagation via the CNNs in accordance with the basic methodology of fully connected neural networks. By executing the convoluted procedure on the feature maps among the layers of the CNN architecture, the filters of the layers are modified.

13.5.3 RECURRENT NEURAL NETWORKS

Recurrent Neural Network, or RNN, is a specific kind of architecture for neural networks made to handle a sequence of input data, including time duration, text, audio, and more [33]. RNNs are able of capturing temporal relationships because they feature recurrent connections that enable information to survive and be transmitted from one phase to the next, in contrast to feedforward neural networks that analyse each input individually. The ideas of hidden states and time steps are central to RNN theory. The primary elements and equations that RNNs employ are listed below:

Hidden state: An RNN's hidden state is thought to be its internal memory or representation of the inputs that came before it in the sequence. In accordance with the present input and the prior hidden state, the hidden state gets modified at each time step. The hidden state may be thought of as an overview of the data that has already been processed. Using the input, all the parameters involve calculation at every single time step.

Figure 13.4 is a representation of the fundamental updating equation for a straightforward RNN:

$$h_t = \text{Activation}(Y * x_{t+} Z * h_{t-1} + b) \tag{13.4}$$

where the hidden state at each time stage; t is denoted by h_t; the input for each time stage t is x_t; Y,V, Z, and b are the weight matrices and bias vectors, respectively which are all learned during training.

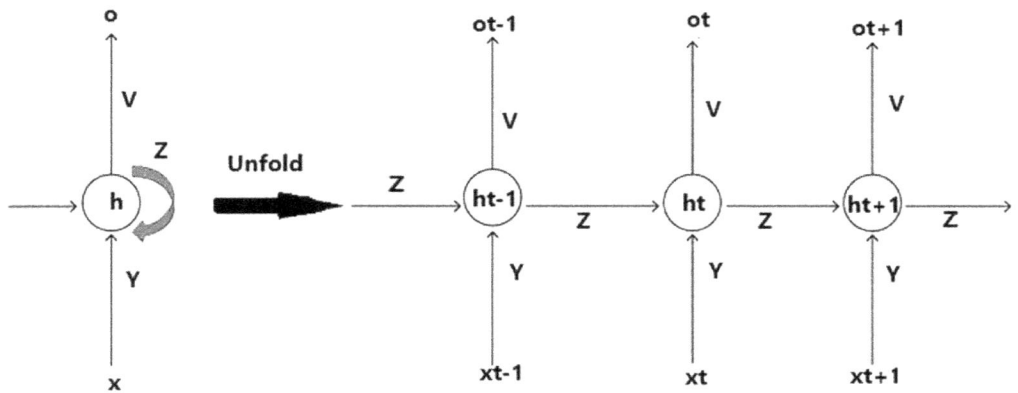

FIGURE 13.4   Basic structure of RNN.

Basic Propagation through time (BPTT): The backpropagation method, which has been modified for sequences, is used to train RNNs, with regard to the loss function, and determines the gradients of the network's characteristics over all time steps. The parameters are then updated using optimisation methods such as stochastic gradient descent, which utilizes the gradients. Whenever backpropagation was initially developed, learning Recurrent Neural Networks (RNNs) was its most intriguing use. RNNs are frequently preferable for tasks involving consecutive inputs, such as voice and language.

13.5.3.1 Long Short-Term Memory (LSTM)

Due to the concealed gradient issue, RNNs may have trouble capturing long-term dependencies. All of the layers in RNNs have the same weight. Although learning long-term dependency management is their prime objective, observation and mapping based research illustrates that learning to retain the relevant data for an extended duration is difficult. Addition of explicit RAM to the system can overcome it. Long-Short-Term Memory (LSTM) [34] methods, especially with hidden parts, whose natural nature is retaining inputs for a long period, are the very initial approach of this sort. Allowing it to copy its own real time data, A distinct element called the memory cell has a link to itself at the subsequent phase with a single element's weight. Since it contains more layers than typical RNNs, LSTM networks outperform them. Self-connection is monitored by an entity that performs storage maintenance. Specialised RNN designs like LSTM and GRU take this problem into account by including neurons for storage and filtering systems. These architectural designs provide better information flow regulation and preservation over longer sequences. This makes it feasible to develop a system for recognising speech that include transcribing order and acoustics. It's crucial to remember that this present hidden state, which in return depends on the prior hidden states, affects the result at each time step. RNNs can describe consecutive connections and collect temporal information in the data because of this recursive connection.

13.5.3.2 Bi-LSTM

By analyzing the sequence of input data in both upward and downward directions, Bi-LSTM [39] is a modification of the LSTM architecture, a system that considers both previous and future information. This enables a more thorough comprehension of the order of inputs by allowing the model to incorporate both past and prospective dependence. A Bi-LSTM uses two independent LSTM layers, one of which processes the sequence of inputs forward and the other of which processes it backward. Concatenating the concealed states from both sides results in a composite representation that includes input data from both the perspective times. In the processing of natural language jobs, bidirectional LSTM is a preferred option.

13.5.3.3 GRU (Gated Recurrent Unit)

In order to dynamically update and reset data in the hidden state, gating methods are incorporated into the RNN architecture known as GRU [35]. It was developed as a less complicated substitute for the Long Short-Term Memory (LSTM) architecture that produced comparable performance. Update and reset gates are the two new gates introduced by GRU. The reset gate regulates the removed hidden state measurement, whereas the update gate governs how much of the prior hidden state is maintained and integrated with the new information. These gates are used by the GRU to control information flow and solve the vanishing gradient issue. In GRU, the hidden state is modified according to the input and the update and reset gates, which are obtained using sigmoid activation functions. GRU has been extensively employed in domains like AI translation, voice recognition, and article synthesis because it can efficiently capture dependencies across lengthy sequences. Equations for computing the output in more sophisticated RNN designs such as the Gated Recurrent

Unit (GRU) might include extra components and gating methods. The core concept, however, stays the same: creating the output by merging the hidden state with the proper weights and biases.

RNNs have been extensively employed in various areas of applications using sequences of input data that employ the detection of voice, time series analysis, and natural language processing (including language modelling, machine translation, and sentiment analysis). They work best when the job at hand requires that the sequence and temporal connections between the data items be important. Recurrent connections in RNNs enable the transmission of the hidden state from one phase to another. The network can recognise cross-temporal relationships and discover sequences in sequential information thanks to the recurrent connection.

13.6   DEEP LEARNING IN UNSUPERVISED LEARNING METHOD

13.6.1   GENERATIVE ADVERSARIAL NETWORK

Ian Goodfellow [44] developed the Generative Adversarial Network (GAN) for generative modelling that provides new logical instances on request. A generator and a discriminator neural network make up this type of deep learning model. In generative modelling, when the objective is to create fresh samples that mimic the training data, GANs are employed.

The fundamental principle of GANs is to train the discriminator network to discriminate between genuine and false samples while simultaneously training the generator network to create real examples. The generator creates more real samples to deceive the discriminator, while it attempts to improve its potential to differentiate between genuine and fake information. The generator network creates synthetic samples from a vector or input of random noise. It generally comprises an outcome layer that produces the produced samples and a couple of hidden layers. Figure 13.5 shows the basic work of a generative adversarial network.

A binary classifier, the discriminator network gets both genuine samples from the trained input data and fictitious patterns via the generator. The purpose of this test is to establish if a sample is genuine or not. In order to maximise its ability to discriminate between genuine samples and fraudulent ones, the discriminator is trained. The generator and discriminator layers are trained in an order at the time of training. The discriminator seeks to maximise its capacity to discriminate between original and false patterns, while the generator seeks to minimise the discriminator's capability to identify the created samples as fake. The generator gradually becomes better at producing more realistic samples as a result of this adversarial training method.

The loss function directs the GAN's training methods. The discriminator seeks to maximise the loss by distinguishing between genuine and false inputs, while the generator's goal is to minimise the loss by demonstrating its ability to trick the discriminator. Binary cross-entropy and Wasserstein distance are two common loss functions in GANs. Training with GANs might be difficult and unstable. Mode collapse, in which the generator only generates a small variety of samples, and a

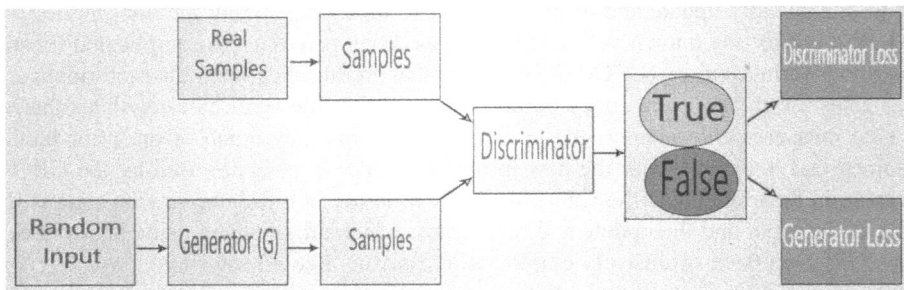

FIGURE 13.5   General structure of Generative Adversarial Network.

disparity between the generator and the discriminator are possible problems. Many strategies have been put out to address these issues, including mini-batch discrimination, comparing features, and regularization methods.

Modern transfer learning research also makes use of GANs to impose the symmetry of the implicit feature space [45]. Identical to the conventional GAN algorithm, a Bidirectional GAN (BiGAN) constructs a mapping from an implicit state to the information distribution. Inverse models, for instance, [46], may also map information to the concealed state. In general, GANs have proven to be a complete field of data-independent growth and a solution to issues needing generative approaches. Several fields, notably generative imaging, image-to-image interpretation, text-to-image development, and video production have successfully used GANs. They have proven that they can provide high-quality, varied samples that closely mirror the distribution of training data. With continued attempts to enhance training strength, sample effectiveness, and control over produced outputs, GANs remain an important research topic.

13.6.2 AUTOENCODER

Deep learning algorithms that use unsupervised learning fall under the category of "AE," or autoencoder. Data input is encoded into a lower-dimensional latent space, translated, and then assembled back into the original data. An autoencoder [52] is a neural network that seeks to learn effective depictions of the input data. The model is encouraged to capture the most crucial aspects of the data since the neural network has been configured to minimise the reconstruction error.

Encoder: The encoder converts input data into a latent space with a lower dimension. Usually, it is made up of a number of secret layers that gradually diminish the dimension of the data.

Latent space: The encoder's input data is compressed and represented in the latent space. In comparison to the original input data, it has less dimensions. The bottleneck layer is another name for the latent space.

Decoder: The decoder reassembles the provided data from the latent state. The framework of the encoder is replicated, but in reverse, and the input space's original dimensions are eventually reached.

Numerous unsupervised learning applications, including dimensionality reduction, extracting features, beneficial coding, constructive modelling, denoising, asymmetry or detection of outliers, and the like, make extensive use of the auto-encoder. For learning representations for upcoming classification problems, regularized autoencoders such as the sparse, the denoising, and the contractive are helpful, although variational autoencoders can be applied as generative models, as will be covered below.

13.6.2.1 Sparse AutoEncoder (SAE)

In order to motivate the latent model to be sparse, sparse autoencoders [47] impose a sparsity restriction during training. Fewer latent components are triggered at a time due to sparsity, which encourages more concise and meaningful representations. Sparse autoencoders can be used to choose essential features and find correlations in the data. Even though SAEs are allowed to contain more hidden states, only a handful of obscured phases can be active one at a time. A sparse autoencoder containing a number of active units has a schematically depicted hidden layer. Therefore, in accordance with its limitations, this model needs to react to the special statistical characteristics of the training database.

13.6.2.2 Denoising AutoEncoder (DAE)

A denoising autoencoder is a variation on the fundamental autoencoder that lowers the likelihood of acquiring the identity function by changing the reconstruction criterion in an effort to enhance representation (extract meaningful features) [48]. Using a faulty or noisy form of the input, denoising autoencoders are taught to recreate the original input. Noise-reducing autoencoders learn to identify pertinent features and represent the fundamental framework of the input by recreating accurate information from noisy data. They work well for both noise reduction and feature extraction. To put it another way, it is taught to reform the basic, unaltered input as its resultant by minimizing the standard reconstruction error throughout the trained information, purifying the contaminated input, or denoising. DAEs may therefore be used for automatic pre-processing in the realm of computing. For example, an image might be automatically pre-processed to improve its potential for accurate recognition using a denoising autoencoder.

13.6.2.3 Contractive AutoEncoder (CAE)

Contractive autoencoders were suggested by Rifai et al. [49] with the goal of making the autoencoders resilient to minor modifications in the trained data. An explicit regularization is included in the goal function of a CAE, to drive the approach to evaluate an encoding that is resilient to even modest variations in input data. The adaptability of the acquired depiction on the training data is thus diminished. While CAEs promote the validity of representation, DAEs promote the robustness of reconstruction, as previously described. By including penalties to the loss equation that reduces susceptibility to minor input data perturbations, contractive autoencoders are made to learn resilient representations. The autoencoder can better generalize the data and capture the most important characteristics thanks to this regularization.

13.6.2.4 Variational AutoEncoders (VAEs)

These generative models acquire a latent space with a particular probabilistic distribution, usually a Gaussian distribution. VAEs [50] understand the variables of the distribution of possibilities rather than just one latent representation, enabling filtering from the latent domain to offer fresh data samples. For data synthesis, latent space interpolation, and picture production, VAEs are often utilised. Their fundamentally different natures to the traditional autoencoder discussed above makes variational autoencoders particularly useful for generative modelling. In a VAE, original data is assumed by an underlying probability distribution, and the parameters of the distribution are subsequently sought. Although this method was primarily intended for unsupervised learning, it has been used successfully in other contexts, including supervised and semi-supervised learning.

Data visualization, feature extraction, and the lowering of computational complexity are all possible uses for autoencoders. Autoencoders may be used to discover irregularities or outliers in fresh, unseen data that depart from the taught representations by learning the usual characteristics of the input. This can be helpful for activities involving information and picture compression since it allows for the storage and reconstruction of the compressed representation. These are but a few instances of the uses and varieties of autoencoders seen in deep learning. Autoencoders offer effective methods for unsupervised learning representations and identifying the fundamental structure of data.

13.6.3 Self-Organizing Map

The Self-Organizing Map (SOM) is a kind of unsupervised learning method used for data grouping, dimensionality reduction, and visualization. In honour of Teuvo Kohonen, who created SOMs, they are often referred to as Kohonen maps [53]. The basic goal of SOMs is to preserve the topological links between the input samples while representing high-dimensional information provided into a lower-dimensional structure or lattice structure. To do this, SOMs learn a collection of prototype vectors, often known as neurons, that each symbolize a distinct area of the input space. A weight vector with the same dimensions as the input data is linked to every cell in the SOM.

A collection of neurons, often structured within a 1D, 2D, or 3D grid structure, make up the SOM [77]. Each neuron in the network of neurons is linked to the neurons next to it. A weight vector that corresponds to a particular location in the input space is attached to each cell in the SOM. These weighted vectors are first set up at random. The nodes, along with the weight vector, closely resemble the input sample and are considered to be the best match. The winning neuron then works with its neighboring neurons to modify their weight vectors.

In order to shift the selected nodes and their weight vectors into the input of the specimen, adaptation takes place. This aids the SOM in organising itself and capturing the fundamental structure of the incoming data. The neighbourhood function dictates the impact of nearby neurons on the weight update process, whereas it keeps updating the size and weight. The rate of learning is initially high, enabling more significant changes, and it steadily declines with time. In a similar manner, the neighbourhood function initially encompasses a broader area of nearby neurons and gradually narrows as training goes on. Using SOMs to evaluate high-dimensional input onto a lower-dimensional grid structure allows for data visualization and clustering [54]. It is possible to cluster and organise related samples because similar neurons located within the SOM frequently reflect related or comparable data points.

SOMs have found use in many domains, such as finding anomalies, data representation in high dimensions, picture and text interpretation, customer segmentation, and emotion analysis [76]. They offer a potent tool for unsupervised examination and comprehension of large datasets while maintaining the fundamental makeup of the provided data.

13.6.4 RESTRICTED BOLTZMANN MACHINE

This is referred to as RBM [57]. RBMs are mainly employed in unsupervised learning tasks, including collaborative filtering, dimensionality reduction, and feature learning. There are two levels in RBMs: exposed units and concealed units. The input data is represented by the visible units, while the latent characteristics or representations that the RBM has learned are recorded by the hidden units. An RBM's exposed and concealed units are both binary, which means they can have either a binary value of 0 or 1. Each exposed and concealed unit in RBMs has a set of bias words associated with it. These biases affect the units' chances of activation and regulate how much energy they provide to the RBM as a whole.

Boltzmann machines generally have visible and hidden linked nodes. The study can better discover anomalies by knowing how the system finally functions. There is a restriction on the value of interconnections between the exposed and unexposed layers in RBMs, a subset of Boltzmann machines [55]. This limitation enables techniques for Boltzmann machines in general to be more valuable [56]. An example of such an approach is the gradient-based opposite variance algorithm. Reduction of dimensionality, regression, categorization, collective filtering, extraction of characteristics, topic modelling, and many more tasks have all found use for RBMs.

For applications including interactive collective filtering, decreasing dimensionality, learning characteristics, and initial training deep neural networks, RBMs have been extensively employed. They offer a successful strategy towards unsupervised learning and may identify complex connections and patterns in the input data.

13.6.5 DEEP BELIEF NETWORK

The deep belief network is referred to as the DBN. A deep generative model is created using multiple levels of Restricted Boltzmann Machines (RBMs) in this kind of deep learning architecture. DBNs are being deployed to applications which involve learning features, preliminary training neural networks, and generative modelling. They are effective models for unsupervised learning. A DBN is built by layered RBMs. The hidden component of one RBM acts as the layer that is visible for the following RBM. As a result, a generative approach is built layer by layer, with each layer representing higher-level properties depending on the lower-level information learned by the one before it.

A method known as greedy layer-wise pre-training is frequently used to pre-train DBNs in an unsupervised way, one layer at a time. The DBN trains each RBM to reconstruct the triggers of the preceding hidden layer as the input (visible) layer [58]. This pre-training aids in the acquisition of practical representations and the detection of intricate patterns in the data. When training successive layers, the weights that were learned in each layer are used as fixed parameters. This layer-wise training method avoids the optimisation problems involved with building deep networks from the beginning and aids in accurately capturing hierarchical representations

A last fine-tuning phase is carried out using supervised learning with backpropagation after the DBN has been pre-trained. The DBN serves to establish a feedforward neural network, and labelled input is utilised to fine-tune the parameters. The DBN may pick up discriminatory traits and adjust to the particular job in hand thanks to this fine-tuning stage. DBNs offer a technique for launching complex structures and extracting accurate representations by preliminary training the structure layer by layer. DBNs have greatly aided in the understanding and creation of deep learning models, even if more recent developments have switched towards end-to-end training methods.

13.7   DEEP LEARNING: HYBRID STRATEGIES

It's vital to remember that there are hybrid strategies as well, which integrate different learning paradigms. For instance, using both labelled and unlabeled data, a deep neural network is trained in semi-supervised learning and the performance of the network is enhanced by using the unlabeled data. Another method is transfer learning, a technique that enables a neural network that was initially trained on a big dataset to be improved upon using a smaller, task-specific dataset. This allows the network to use the learned illustrations from the pre-training.

13.7.1   HYBRID STRATEGY NO 1: CNN + LSTM:

Trained convolutional neural networks, can be used in hybrid learning for extracting features and transfer learning in image recognition applications. A CNN equipped with LSTM used to identify brain tumors [59], and LSTMs can enhance CNN's feature extraction capabilities. The layered LSTM-CNN architecture surpasses traditional CNN classification when utilised for picture classification using the Kaggle data set.

13.7.2   HYBRID STRATEGY NO 2: GAN + AE

Unsupervised learning methods can be used to apply hybrid learning to problems involving anomaly detection. To train autoencoder or Generative Adversarial Networks (GANs) and comprehend a representation of the normal data distribution, normal data samples can be used. Then, to find abnormalities in fresh or unexplored data, hybrid learning can combine unsupervised learning with aberration detection techniques. Zixu et al. [60] developed a GAN and AE-based unsupervised hierarchical technique for anomaly identification. To solve the issue of privacy preservation a group of generators are applied from the various IoT platforms and a sample pool is generated at a centralised controller. After that, for identifying anomalies, a centralised universal AE is trained and provided to specific local networks.

These are but a handful of instances in which hybrid learning may be used in applications involving deep learning. The specific job, the data that is available, and the targeted performance gains all influence the hybrid learning approaches that are used. Hybrid learning enables models to take advantage of the benefits of numerous learning methodologies, improving performance, resilience, and generalisation across a range of domains.

13.7.3 Deep transfer learning

Deep Transfer Learning is referred to as DTL. This phrase alludes to the utilization of transfer learning techniques in algorithms for deep learning. Using information acquired from a single assignment or area to enhance a model's performance on an entirely distinct but comparable task or domain is known as transfer learning [61]. Transfer learning is particularly useful in the framework of deep learning when the target job contains little labelled data or when building an algorithm from the beginning is substantially costly. It is particularly well-liked in deep learning. Now as it enables deep neural networks to be trained using minimal information [73].

Transfer learning is a procedure with two phases for developing DL models, which include pre-training and fine-tuning using training data from the target task. Deep networks of neurons have been popular in a variety of disciplines, leading to the development of several DTL approaches, making it essential to classify and summarise them. An already trained and sizable dataset in a deep neural network model like ImageNet for image-related tasks, serves as the foundation for advanced transfer learning. These trained models have acquired representations of basic visual characteristics, effectively capturing different picture patterns and ideas. The subsequent layers of the pre-trained model, including fully linked layers, can be rebuilt or tuned with job-specific labelled data after the extraction of features. This enables the model to pick up aspects particular to the job at hand or modify its previous representations to suit the new one. DTL is frequently employed when the source area differs from the destination area yet contains a large scale of labelled data [62]. Using less labelled data, the pre-trained model that was developed for the source domain might be adjusted or transformed for the target domain. This strategy is beneficial when the original realm and the target realm have some overlap.

Deep transfer learning's current applications includes detection of speech, machine vision, and processing of natural languages. It speeds up the learning process and enables models to take advantage of the information stored by previously trained models, enhancing results on objective activities even with a dearth of labelled data.

13.7.4 Deep reinforcement learning

DRL, also known as Deep Reinforcement Learning, [63] a learning paradigm known as Reinforcement Learning (RL) teaches an agent to interact with its environment and make choices that will maximise its cumulative rewards. As shown in Figure 13.6 below, the agent acts in the

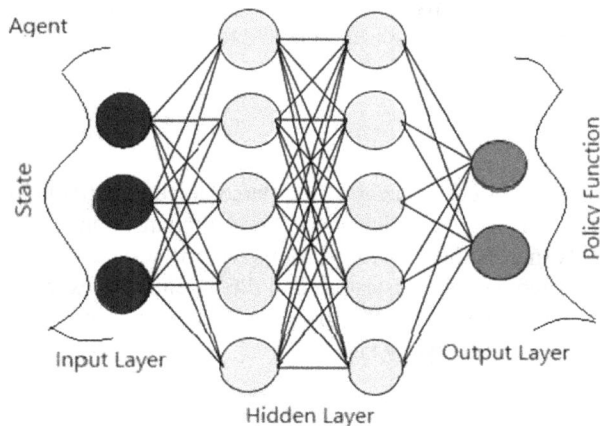

FIGURE 13.6 Schematic Architecture of Deep Reinforcement Learning.

world, gets feedback in the sense of benefits or costs, and modifies its behaviour in response to the results it observes. The agent is a sentient being that communicates with its surroundings. Based on its observations, it gets responses based on its performance in the form of benefits or punishments. The agent uses the state to reflect the environment's current state while making decisions. DRL frequently uses high-dimensional data, such as pictures or sensor inputs, to express states. An agent's decision to act is dependent on the situation at hand. To increase the anticipated cumulative benefit, the agent chooses its activities. The agent selects actions on the basis of conditions, which are strategies or regulations. It usually requires training to learn and links states to actions.

In order to figure out a parameter function or policy function, DRL methods like Deep Q-Networks (DQN) and Proximal Policy Optimisation (PPO) employ deep neural networks [75]. This allows the agent to learn complicated behaviours while executing actions in high-dimensional state spaces. DRL has made impressive progress in lots of fields, including robots, autonomous driving, and gaming [64]. Combination of deep neural networks and RL algorithms, and deep reinforcement learning study different characteristics directly from unprocessed sensory inputs. Deep RL has had success in a variety of fields, including autonomous driving, robotics, and game playing (such as AlphaGo) [74].

13.8   FEATURE LEARNING

The manner in which characteristics are retrieved in classical ML and DL differs significantly. Deep learning avoids this laborious and tedious phase by having computers automatically discover meaningful representations and features from the raw data. There are other deep learning models outside the many variations of artificial neural networks. Deep learning techniques rely on feature learning [31], which involves automatically noticing data representations. There are several uses for feature extraction algorithms, such as Scale Invariant Feature Transforms (SIFT) [65], Local Binary Patterns (LBP), GIST, RANSAC, steganography on visions using the Histogram Oriented Gradient (HOG) [20], Empirical Mode Decomposition (EMD) for analysis of voice, and many others, are common practices in traditional machine learning methods. Using alternative boosting techniques, a choice is made based on the many outcomes from the various algorithms, which apply lots of learning algorithms to the characteristics of a specific dataset.

13.9   FUTURE RESEARCH DIRECTION

Deep learning has an extremely promising future and is anticipated to have a big influence on many different fields and sectors. Deep learning is anticipated to succeed in the following important areas in the future:

13.9.1   HEALTHCARE

By allowing more precise and effective diagnostics, personalized treatment, and drug development, deep learning holds the capacity to revolutionize the healthcare industry. It can support medical decision-making by analyzing medical pictures, identifying illnesses, predicting patient outcomes, and more. A single integrated patient illustration and the potential to expand to encompass billions of pieces of information about patients will allow the forthcoming wave of anticipatory healthcare systems to assist doctors in their everyday tasks [72].

13.9.2   FUTURE-ORIENTED SMART DEVICES

Lightweight deep learning modelling has advantages for future-oriented smart devices and applications, such as faster inference, less memory utilization, and less electrical consumption. This broadens the variety of apps that can profit from AI capabilities through allowing the installation of deep learning techniques on devices with restricted resources. The growth of self-driving vehicles depends heavily on deep learning. It may facilitate autonomy and control mechanisms to ensure safe and dependable autonomous vehicle operation as well as perceptual tasks including object identification, recognition, and tracking. Numerous uses for these compact models include real-time object identification on cutting-edge, sophisticated surveillance systems, picture and footage processing on mobile devices, and personalized solutions on wearable technology. By facilitating on-device interpretation and minimizing the need to send confidential information to the cloud, they also help safeguard privacy.

13.9.3 Natural Language Processing (NLP)

The use of deep learning has made major strides in NLP. Its prospective applications include sentiment evaluation, chatbots, voice-controlled devices, and intricate language interpretation. It can improve interactivity and communication between people and machines. Software packages for Natural Language Processing (NLP) have previously prioritised effective models. By offering simple-to-use APIs that implement better options compared to brute-force grid searches for parameter tuning, such as random or Bayesian hyperparameter search approaches, machine learning and natural language processing developers can help reduce the amount of energy associated with model adjusting [67, 68]. Although there are software programs that implement these methods, modifying NLP models seldom uses them in practice.

13.9.4 Robotics

Deep learning improves the perception and interactivity of robots with their surroundings [70]. It can improve the capacity of robots to see, manipulate objects, grab objects, and navigate, making robotic systems more competent and intelligent. The overall trend is towards higher degrees of mental abilities, and some experts are even anticipating that deep learning will eventually achieve human-level capabilities [71].

13.9.5 Internet of Things (IoT)

By permitting IoT devices to analyse and interpret information locally, deep learning may enhance IoT devices' functionality and provide instantaneous knowledge and more effective decision-making processes. The Internet of Things (IoT), which consists of huge knowledgeable and interactive objects, and wireless communications, have been gaining popularity to recognise and obtain environmental and human information (such as geo-information, climate data, bio-data, behavioral patterns of people, and more) for an assortment of intelligent applications and services. These pervasive smart items or gadgets produce significant volumes of data every day, necessitating quick data processing on various smart mobile devices [69]. Automated homes, medical surveillance, automation in factories, and other areas can all benefit from this.

13.9.6 Scientific research

Deep learning has been utilised to analyse complex data, generate forecasts, and accelerate research and discoveries in a variety of scientific domains, including astronomy, physics, genetics, and drug development.

Deep learning has demonstrated potential in risk assessment, identifying fraudulent transactions, algorithmic trading, and financial forecasting. By analyzing vast amounts of financial data, it can aid in forecasting investments more precisely. The features of contemporary smart devices and apps are being improved by scientists and programmers as they seek to investigate and create approaches for compact deep learning modelling, permitting sophisticated and productive AI solutions in a variety of contexts. Improved interpretation speed, lessened memory utilization, and lower usage of energy are all advantages of a compact deep learning architecture for future-oriented smart devices and applications. Although deep learning has made great progress, there are still concerns to be resolved, like the need for more data, comprehension issues, and ethical issues. Deep learning, however, is positioned to continue spurring innovation and reshaping multiple sectors in the future thanks to continuous research and breakthroughs.

13.10   CONCLUSION

This chapter has presented an organised and detailed review of the deep learning paradigm, which has been recognised as a fundamental component of both artificial intelligence and neural networks. It started by reviewing the development of deep neural networks, eventually expanding to a variety of modern deep learning methods employed in various contexts. Deep learning approaches in various dimensions have been explored. We have also provided a categorization that considers the various deep learning regression techniques and their many applications. In our comprehensive examination, we have explored supervised and unsupervised learning using deep networks, hybrid learning and feature learning, which are potentially applicable to a range of real-world circumstances depending on the nature of the challenges.

In conclusion, by using the strengths of deep neural networks, deep learning regression approaches convey robust ways for overcoming regression issues. These techniques have been shown to be highly effective in a number of fields, such as time series analysis, audio and picture recognition, and language processing. Regression models using the deep learning perform outstandingly with complicated and high-dimensional information, rapidly acquiring meaningful representations and identifying nuanced patterns. Their capacity to cope with non-linear connections between source data and target variables permits them to operate more effectively and to make predictions that are more precise. Deep learning regression algorithm development entails minimizing the loss function by continuously modifying the framework's network measure using gradient descent and reverse propagation techniques. The model is better equipped to generalize to new cases by learning from and adapting to the training data. Deep learning regression techniques provide the additional advantage of being able to automatically separate specific characteristics from the data, eliminating the features for future forecasting. This streamlines the modelling process and conserves time.

Deep learning regression techniques nevertheless additionally present certain challenges. To perform at their best, algorithms frequently require a lot of training information that is labelled, which might be an obstacle in some fields. Deep neural networks are also computationally demanding, indicating that they might demand a lot of energy and time during training. Deep learning regression models' ability to generalize while preventing overfitting is frequently improved by normalisation approaches like weight decay and dropout. The most suitable model is determined, and overfitting can be prevented by using cross-validation and prompt stopping.

In conclusion, deep learning approaches to regression present an effective strategy for resolving regression issues by automatically discovering intricate patterns and representations in the information. Deep learning methods are supposed to keep improving and gaining use in a wide range of sectors as an outcome of the accessibility of bigger databases and ongoing advancements in these approaches.

REFERENCES

1.   LeCun, Y., Bengio, Y. and Hinton, G. "Deep learning." *Nature* 521, no. 7553 (2015): 436–444.
2.   He, K., Zhang, X., Ren, S., and Sun, J. "Deep residual learning for image recognition." In *Proceedings of the IEEE Conference on Computer Vision and Pattern Recognition*, pp. 770–778. 2016.

3. Li, K., Huang, W., Hu, G., and Li, J. "Ultra-short term power load forecasting based on CEEMDAN-SE and LSTM neural network." *Energy and Buildings* 279 (2023): 112666.

4. Marino, D. L., Amarasinghe, K., and Manic, M. "Building energy load forecasting using deep neural networks." In *IECON 2016–42nd Annual Conference of the IEEE Industrial Electronics Society*, pp. 7046–7051. IEEE, 2016.

5. Hinton, G. E., Osindero, S., and Teh, Y-W. "A fast learning algorithm for deep belief nets." *Neural Computation* 18, no. 7 (2006): 1527–1554.

6. Karhunen, J., Raiko, T., and Cho, K. H. "Unsupervised deep learning: A short review." *Advances in Independent Component Analysis and Learning Machines* (2015): 125–142.

7. Goodfellow, I., Bengio, Y., and Courville, A. *Deep Learning*. MIT Press, 2016.

8. Sarker, I. H., Furhad, Md H., and Nowrozy, R. "AI-driven cybersecurity: An overview, security intelligence modeling and research directions." *SN Computer Science* 2 (2021): 1–18.

9. Sarker, I. H. "Machine learning: Algorithms, real-world applications and research directions." *SN Computer Science* 2, no. 3 (2021): 160.

10. Sarker, I. H. "Data science and analytics: An overview from data-driven smart computing, decision-making and applications perspective." *SN Computer Science* 2, no. 5 (2021): 377.

11. Sarker, I. H., Moshiul Hoque, M., Uddin, Md K., and Alsanoosy, T. "Mobile data science and intelligent apps: Concepts, AI-based modeling and research directions." *Mobile Networks and Applications* 26 (2021): 285–303.

12. Sarker, I. H., Kayes, A. S. M., Badsha, S., Alqahtani, H., Watters, P., and Ng, A. "Cybersecurity data science: An overview from machine learning perspective." *Journal of Big Data* 7 (2020): 1–29.

13. Hinton, G. E., Osindero, S., and Teh, Y-W. "A fast learning algorithm for deep belief nets." *Neural Computation* 18, no. 7 (2006): 1527–1554.

14. Bengio Y., Lamblin, P., Popovici, D., and Larochelle, H. "Greedy layer-wise training of deep networks." *Advances in Neural Information Processing Systems* 19, no. 1 (2007).

15. Ding, X., Zhang, Y., Liu, T., and Duan, J. "Deep learning for event-driven stock prediction." In *Twenty-Fourth International Joint Conference on Artificial Intelligence*. 2015.

16. Krizhevsky, A., Sutskever, I., and Hinton, G. E. "Imagenet classification with deep convolutional neural networks." *Advances in Neural Information Processing Systems* 25 (2012).

17. Shen, F., Chao, J., and Zhao, J. "Forecasting exchange rate using deep belief networks and conjugate gradient method." *Neurocomputing* 167 (2015): 243–253.

18. Kuremoto, T., Kimura, S., Kobayashi, K., and Obayashi, M. "Time series forecasting using a deep belief network with restricted Boltzmann machines." *Neurocomputing* 137 (2014): 47–56.

19. Miglani, A., and Kumar, N. "Deep learning models for traffic flow prediction in autonomous vehicles: A review, solutions, and challenges." *Vehicular Communications* 20 (2019): 100184.

20. Hameed, M. A., Hassaballah, M., Aly, S., and Awad, A. I. "An adaptive image steganography method based on histogram of oriented gradient and PVD-LSB techniques." *IEEE Access* 7 (2019): 185189–185204.

21. Abdulaal, A., Patel, A., Charani, E., Denny, S., Alqahtani S. A., Davies, G. W., Mughal, N. and Moore, L, S. P. "Comparison of deep learning with regression analysis in creating predictive models for SARS-CoV-2 outcomes." *BMC Medical Informatics and Decision Making* 20 (2020): 1–11.

22. Li, J., Dai, W., Metze, F., Qu, S., and Das, S. "A comparison of deep learning methods for environmental sound detection." In *2017 IEEE International Conference on Acoustics, Speech and Signal Processing (ICASSP)*, pp. 126–130. IEEE, 2017.

23. Fischer, T., and Krauss, C. "Deep learning with long short-term memory networks for financial market predictions." *European Journal of Operational Research* 270, no. 2 (2018): 654–669.

24. Generosi, A., Ceccacci, S. and Mengoni, M. "A deep learning-based system to track and analyze customer behavior in retail store." In *2018 IEEE 8th International Conference on Consumer Electronics-Berlin (ICCE-Berlin)*, pp. 1–6. IEEE, 2018.

25. Mehtab, S., and Sen, J. "Stock price prediction using CNN and LSTM-based deep learning models." In *2020 International Conference on Decision Aid Sciences and Application (DASA)*, pp. 447–453. IEEE, 2020.

26. Heghedus, C., Chakravorty, A., and Rong, C. "Energy load forecasting using deep learning." In *2018 IEEE International Conference on Energy Internet (ICEI)*, pp. 146–151. IEEE, 2018.

27. Yang, X., Yu, H., and Jia, L. "Speech recognition of command words based on convolutional neural network." In *2020 International Conference on Computer Information and Big Data Applications (CIBDA)*, pp. 465–469. IEEE, 2020.

28. Olu-Ajayi, R., Alaka, H., Sulaimon, I., Sunmola, F., and Ajayi, S. "Building energy consumption prediction for residential buildings using deep learning and other machine learning techniques." *Journal of Building Engineering* 45 (2022): 103406.

29. Navares, R., and Aznarte, J. L. "Predicting air quality with deep learning LSTM: Towards comprehensive models." *Ecological Informatics* 55 (2020): 101019.

30. Barzegar, R., Aalami, M. T., and Adamowski, J. "Short-term water quality variable prediction using a hybrid CNN–LSTM deep learning model." *Stochastic Environmental Research and Risk Assessment* 34, no. 2 (2020): 415–433.

31. Goodfellow. I., Bengio,Y., and Courville, A. "Deep learning." (2016): 167–227.

32. LeCun, Y., Boser, B., Denker, J., Henderson, D., Howard, R., Hubbard, W., and Jackel, L. "Handwritten digit recognition with a back-propagation network." *Advances in Neural Information Processing Systems* 2 (1989).

33. Sherstinsky, A. "Fundamentals of recurrent neural network (RNN) and long short-term memory (LSTM) network." *Physica D: Nonlinear Phenomena* 404 (2020): 132306.

34. Schmidhuber, J., and Hochreiter, S. "Long short-term memory." *Neural Computer* 9, no. 8 (1997): 1735–1780.

35. Bengio, Y., Simard, P., and Frasconi, P. "Learning long-term dependencies with gradient descent is difficult." *IEEE Transactions on Neural Networks* 5, no. 2 (1994): 157–166.

36. Sonoda, S., and Murata, N. "Neural network with unbounded activation functions is universal approximator." *Applied and Computational Harmonic Analysis* 43, no. 2 (2017): 233–268.

37. Leshno, M., Lin, V. Y., Pinkus, A., and Schocken, S. "Multilayer feedforward networks with a nonpolynomial activation function can approximate any function." *Neural Networks* 6, no. 6 (1993): 861–867.

38. Clevert, D.-A., Unterthiner, T., and Hochreiter, S. "Fast and accurate deep network learning by exponential linear units (elus)." *arXiv preprint arXiv:1511.07289* (2015).

39. Siami-Namini, S., Tavakoli, N., and Namin, A. S. "The performance of LSTM and BiLSTM in forecasting time series." In *2019 IEEE International Conference on Big Data (Big Data)*, pp. 3285–3292. IEEE, 2019.

40. Sarker, I. H. "Deep cybersecurity: A comprehensive overview from neural network and deep learning perspective." *SN Computer Science* 2, no. 3 (2021): 154.

41. Sarker, I. H., Furhad, M. H., and Nowrozy, R. "AI-driven cybersecurity: An overview, security intelligence modeling and research directions." *SN Computer Science* 2 (2021): 1–18.

42. Géron, A. *Hands-on Machine Learning with Scikit-Learn, Keras, and TensorFlow*. O'Reilly Media, Inc. (2022).

43. Pedregosa, F., Varoquaux, G., Gramfort, A., Michel, V., Thirion, B., Grisel, O., Blondel, M., et al. "Scikit-learn: Machine learning in Python." *Journal of Machine Learning Research* 12 (2011): 2825–2830.

44. Goodfellow, I., Pouget-Abadie, J., Mirza, M., Xu, B., Warde-Farley, D., Ozair, S., Courville, A., and Bengio, Y. "Generative adversarial nets." *Advances in Neural Information Processing Systems* 27 (2014).

45. Li, B., François-Lavet, V., Doan, T., and Pineau, J. "Domain adversarial reinforcement learning." *arXiv preprint arXiv:2102.07097* (2021).

46. Donahue, J., Krähenbühl, P., and Darrell, T. "Adversarial feature learning." *arXiv preprint arXiv:1605.09782* (2016).

47. Makhzani, A., and Frey, B. "K-sparse autoencoders." *arXiv preprint arXiv:1312.5663* (2013).

48. Vincent, P., Larochelle, H., Lajoie, I., Bengio, Y., Manzagol, P-A., and Bottou, L. "Stacked denoising autoencoders: Learning useful representations in a deep network with a local denoising criterion." *Journal of Machine Learning Research* 11, no. 12 (2010).

49. Rifai, S., Vincent, P., Muller, X., Glorot, X., and Bengio, Y. "Contractive auto-encoders: Explicit invariance during feature extraction." In *Proceedings of the 28th International Conference on International Conference on Machine Learning*, pp. 833–840. 2011.

50. Kingma, D. P., and Welling, M. "Auto-encoding variational bayes." *arXiv preprint arXiv:1312.6114* (2013).

51. Kameoka, H., Li, S. I., and Makino, S. "Supervised determined source separation with multichannel variational autoencoder." *Neural Computation* 31, no. 9 (2019): 1891–1914.

52. Goodfellow, I., Pouget-Abadie, J., Mirza, M., Xu, B., Warde-Farley, D., Ozair, S., Courville, A., and Bengio, Y. "Generative adversarial nets." *Advances in Neural Information Processing Systems* 27 (2014).

53. Kohonen, T. "The self-organizing map." *Proceedings of the IEEE* 78, no. 9 (1990): 1464–1480.

54. Vesanto, J., and Alhoniemi, E. "Clustering of the self-organizing map." *IEEE Transactions on Neural Networks* 11, no. 3 (2000): 586–600.

55. Memisevic, R., and Hinton, G. E. "Learning to represent spatial transformations with factored higher-order Boltzmann machines." *Neural Computation* 22, no. 6 (2010): 1473–1492.

56. Wason, R. "Deep learning: Evolution and expansion." *Cognitive Systems Research* 52 (2018): 701–708.

57. Marlin, B., Swersky, K., Chen, B., and Freitas, N. "Inductive principles for restricted Boltzmann machine learning." In *Proceedings of the Thirteenth International Conference on Artificial Intelligence and Statistics*, pp. 509–516. JMLR Workshop and Conference Proceedings, 2010.

58. Hinton, G. E. "Deep belief networks." *Scholarpedia* 4, no. 5 (2009): 5947.

59. Vankdothu, R., Hameed, M. A., and Fatima, H. "A brain tumor identification and classification using deep learning based on CNN-LSTM method." *Computers and Electrical Engineering* 101 (2022): 107960.

60. Zixu, T., Liyanage, K. S. K., and Gurusamy, M. "Generative adversarial network and auto encoder based anomaly detection in distributed IoT networks." In *GLOBECOM 2020–2020 IEEE Global Communications Conference*, pp. 1–7. IEEE, 2020.

61. Weiss, K., Khoshgoftaar, T. M., and Wang, D. D. "A survey of transfer learning." *Journal of Big Data* 3, no. 1 (2016): 1–40.

62. Pan, S. J., and Yang, Q. "A survey on transfer learning." *IEEE Transactions on Knowledge and Data Engineering* 22, no. 10 (2009): 1345–1359.

63. Arulkumaran, K., Deisenroth, M. P., Brundage, M., and Bharath, A. A. "Deep reinforcement learning: A brief survey." *IEEE Signal Processing Magazine* 34, no. 6 (2017): 26–38.

64. Kaelbling, L. P., Littman, M. L., and Moore, A. W. "Reinforcement learning: A survey." *Journal of Artificial Intelligence Research* 4 (1996): 237–285.

65. Lindeberg, T. "Scale invariant feature transform." (2012): 10491.

66. Sharma, S., Sharma, S., and Athaiya, A. "Activation functions in neural networks." *Towards Data Science* 6, no. 12 (2017): 310–316.

67. Bergstra, J., Bardenet, R., Bengio, Y., and Kégl, B. "Algorithms for hyper-parameter optimization." *Advances in Neural Information Processing Systems* 24 (2011).

68. Bergstra, J., and Bengio, Y. "Random search for hyper-parameter optimization." *Journal of Machine Learning Research* 13, no. 2 (2012).

69. Ma, X., Yao, T., Hu, M., Dong, Y., Liu, W., Wang, F., and Liu, J. "A survey on deep learning empowered IoT applications." *IEEE Access* 7 (2019): 181721–181732.

70. Miyajima, R. "Deep learning triggers a new era in industrial robotics." *IEEE Multimedia* 24, no. 4 (2017): 91–96.

71. Pratt, G. A. "Is a Cambrian explosion coming for robotics?." *Journal of Economic Perspectives* 29, no. 3 (2015): 51–60.

72. Miotto, R., Wang, F., Wang, S., Jiang, X., and Dudley, J. T. "Deep learning for healthcare: Review, opportunities and challenges." *Briefings in Bioinformatics* 19, no. 6 (2018): 1236–1246.

73. Miikkulainen, R., et al. "Evolving deep neural networks." In *Artificial Intelligence in the Age of Neural Networks and Brain Computing*, pp. 293–312. Academic Press, 2019.

74. Granter, S. R., Beck, A. H., and Papke Jr, D. J. "AlphaGo, deep learning, and the future of the human microscopist." *Archives of Pathology & Laboratory Medicine* 141, no. 5 (2017): 619–621.

75. Schulman, J., Wolski, F., Dhariwal, P., Radford, A., and Klimov, O. "Proximal policy optimization algorithms." *arXiv preprint arXiv:1707.06347* (2017).

76. Ali, M. N. Y., et al. "Adam deep learning with SOM for human sentiment classification." *International Journal of Ambient Computing and Intelligence (IJACI)* 10, no. 3 (2019): 92–116.

77. Aly, S., and Almotairi, S. "Deep convolutional self-organizing map network for robust handwritten digit recognition." *IEEE Access* 8 (2020): 107035–107045.

78. Braun, S., and Tashev, I. "A consolidated view of loss functions for supervised deep learning-based speech enhancement." In *2021 44th International Conference on Telecommunications and Signal Processing (TSP)*, pp. 72–76. IEEE, 2021.

79. Asokan, R., and Vijayakumar, T. "Design of WhatsApp image folder categorization using CNN method in the android domain." *Journal of Ubiquitous Computing and Communication Technologies (UCCT)* 3, no. 03 (2021): 180–195.

80. Ozanich, E., Gerstoft, P., and Niu, H. "A feedforward neural network for direction-of-arrival estimation." *The Journal of the Acoustical Society of America* 147, no. 3 (2020): 2035–2048.

81. Hewamalage, H., Bergmeir, C., and Bandara, K. "Recurrent neural networks for time series forecasting: Current status and future directions." *International Journal of Forecasting* 37, no. 1 (2021): 388–427.

82. Ajmal, H., Rehman, S., Farooq, U., Ain, Q. U., Riaz, F. and Hassan, A. "Convolutional neural network based image segmentation: A review." *Pattern Recognition and Tracking XXIX* 10649 (2018): 191–203.

83. Verma, N. K., Sharma, T., Rajurkar, S. D., and Salour, A. "Object identification for inventory management using convolutional neural network." In *2016 IEEE Applied Imagery Pattern Recognition Workshop (AIPR)*, pp. 1–6. IEEE, 2016.

14 Deep Learning Impacts in the Field of Artificial Intelligence

Reshma Gulwani and Minal Aggarwal

14.1 INTRODUCTION: BACKGROUND AND DRIVING FORCES

In recent years, there have been remarkable advancements in the field of Artificial Intelligence (AI), and one of the key driving forces behind its progress is deep learning. It falls under the domain of AI that primarily centers on training artificial neural networks to perform complex tasks by analyzing and learning from vast amounts of data. This approach has revolutionized various domains, leading to significant breakthroughs and applications in areas spanning computer vision, natural language processing, speech recognition, and robotics.

Deep learning takes inspiration from the intricate design and functionality of the human brain particularly its interconnected neural networks, to develop Artificial Neural Networks (ANNs) for machine learning tasks. The concept of neural networks dates back several decades, but it is the recent advancements in computational power, the availability of large datasets, and breakthroughs in algorithmic techniques that have fueled the rapid growth and impact of deep learning.

The automatic extraction of meaningful features from raw data is a fundamental driver behind the success of deep learning. Traditionally, in machine learning, feature engineering required human experts to manually design and select relevant features from the data, which could be a time-consuming and error-prone process as shown in Figure 14.1. In deep learning, neural networks automatically learn hierarchical representations of data, uncovering intricate patterns and features without human intervention. This feature learning capability has enabled deep learning models to excel in tasks like image classification, object detection, and language translation.

Another driving force behind the impact of deep learning is the availability of large-scale labeled datasets. These algorithms thrive on vast amounts of data to effectively train complex models. In recent years, there has been a surge in the availability of labeled datasets, thanks to advancements in data collection and annotation techniques, as well as the proliferation of internet-connected devices generating massive amounts of data. These datasets, such as ImageNet for computer vision or the Common Crawl for natural language processing, have played a crucial role in training deep learning models and pushing the boundaries of AI performance.

14.2 DEEP LEARNING IN COMPUTER VISION

Deep learning has had a significant impact on computer vision, revolutionizing the way we analyze and understand visual data. These techniques have made remarkable advancements in areas such as image classification, object detection, segmentation, and image generation. Here are several fundamental aspects of deep learning within the domain of computer vision.

DOI: 10.1201/9781003433309-14

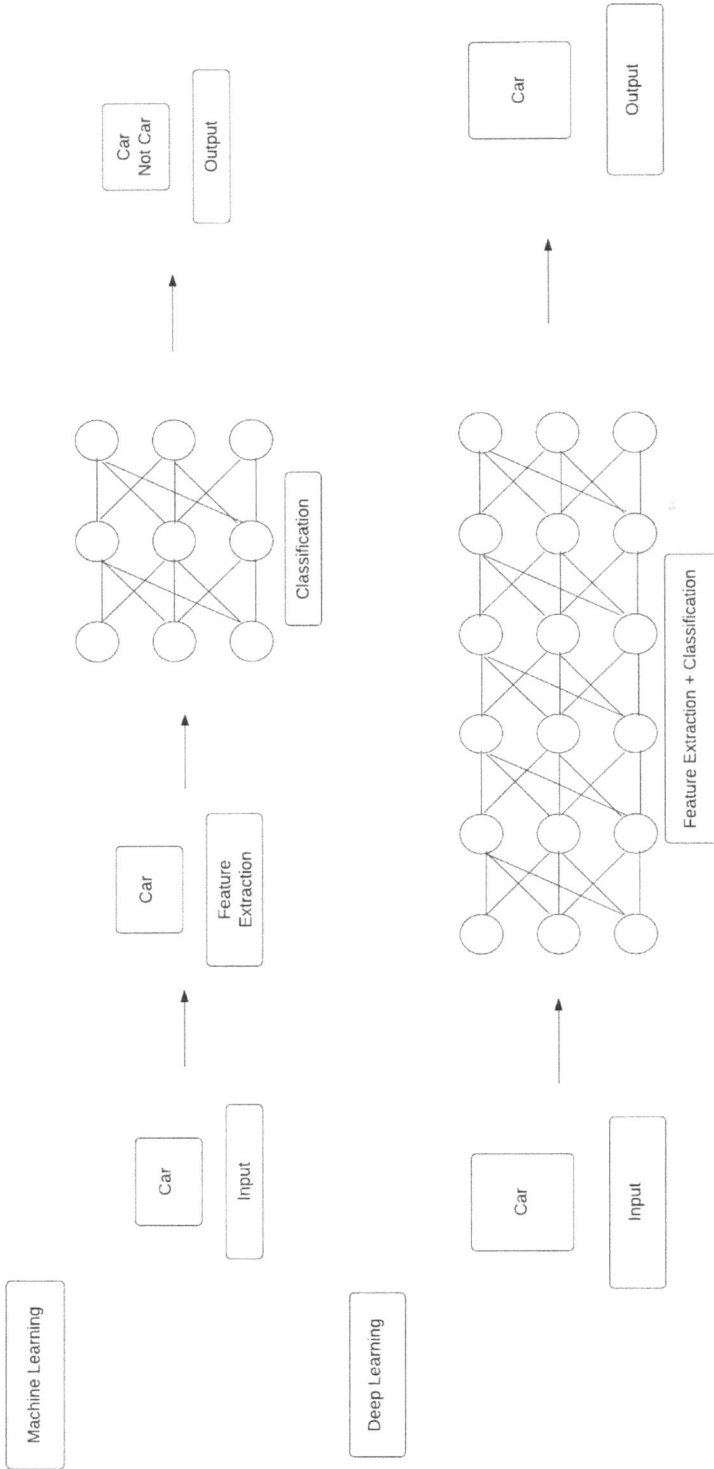

FIGURE 14.1 Deep Learning versus Machine Learning [1].

14.2.1 Image classification

Deep learning models have accomplished remarkable achievements in the domain of image classification tasks. Convolutional Neural Networks (CNNs) are purpose-built to analyze visual data, making them exceptionally potent in tasks like object identification, scene recognition, and detecting specific features within images. Models like AlexNet, VGGNet, GoogLeNet, and ResNet have achieved top performance on benchmark datasets like ImageNet, surpassing human-level accuracy as shown in Figure 14.2.

14.2.2 Object detection

The field of deep learning has witnessed substantial advancements in object detection techniques. It involves not only recognizing objects within an image but also precisely localizing them by drawing bounding boxes around them. Deep learning-driven object detection techniques, including region-based convolutional neural networks, Fast R-CNN, and Faster R-CNN, have attained state-of-the-art performance by combining object proposals and deep feature extraction.

14.2.2.1 Region-based Convolutional Neural Networks (R-CNN)

Region-CNN is an object detection framework that integrates selective search, a pre-trained CNN, Support Vector Machines (SVMs), and linear regression to identify and precisely locate objects within images. Ross Girshick et al. put forward R-CNN in 2014 as their pioneering work and have significant influence in the field of computer vision. It shows a step-by-step overview of the R-CNN framework:

1. Selective Search
   The selective search algorithm is utilized for generating region proposals within an image. It explores different combinations of image regions at multiple scales and sizes to find potential object boundaries. The algorithm is based on the intuition that objects can be composed of contiguous regions with similar properties, such as color, texture, or intensity. By combining these regions, the algorithm produces a set of region proposals.

2. Pretrained CNN
   Image recognition tasks commonly rely on the utilization of Convolutional Neural Networks (CNNs). The R-CNN approach relies on a pretrained CNN to extract features from every

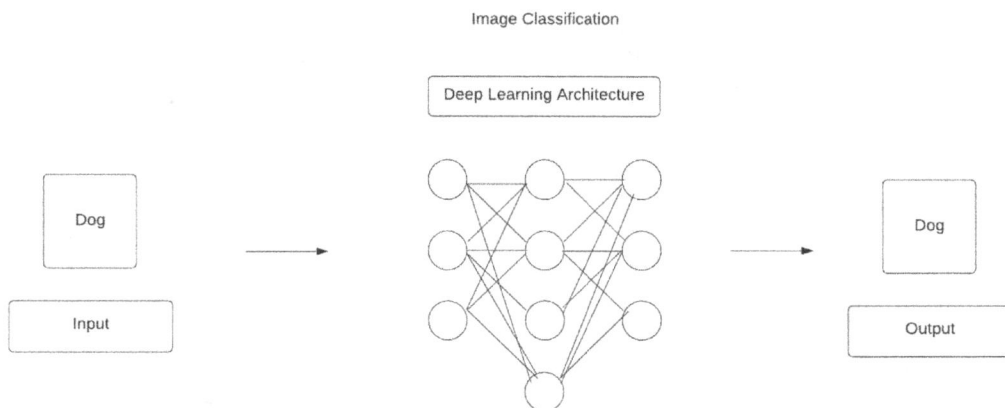

FIGURE 14.2 Image Classification using Deep Learning.

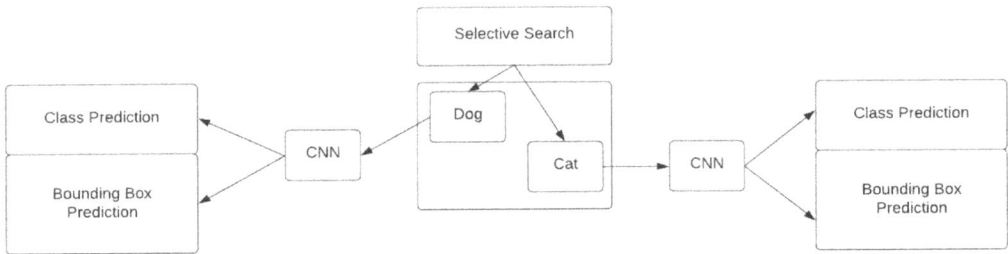

FIGURE 14.3   R-CNN Model with CNN features.

region proposal. The CNN is typically extensive datasets like ImageNet, to learn generic image representations. The CNN's before the final classification layer are used as a feature representation for each region proposal.

3. Classification with Support Vector Machines (SVM)
   For each region proposal, the features extracted from the pretrained CNN are passed through a support vector machine classifier. The SVM is trained to determine whether the region proposal contains a specific class of object. Multiple SVMs can be trained, one for each class that needs to be detected. Each SVM individually decides whether an example belongs to a specific class.

4. Bounding box regression
   In addition to classifying objects, R-CNN also aims to accurately localize the objects by predicting their bounding boxes. The features extracted from the CNN, along with the ground-truth bounding box labels for each region proposal, are used to train a linear regression model. This model learns to predict the coordinates of the ground-truth bounding box for a given region proposal.

14.2.2.2   Fast R-CNN

In Fast R-CNN, rather than performing independent CNN forward propagation for each region proposal, the CNN forward propagation is performed only once on the entire image. This is accomplished by incorporating a region of interest pooling layer. This layer extracts feature maps of a consistent size for every proposed region. These regions are derived from the convolutional feature maps that are produced by the CNN. The RoI pooling layer takes the region proposals and aligns them to a fixed spatial size, allowing for efficient computation and avoiding redundant calculations. These fixed-size feature maps are are passed into fully connected layers for subsequent processing and object classification. Figure 14.4 shows a step-by-step overview of the Fast R-CNN framework.

By sharing the CNN forward propagation across all region proposals, Fast R-CNN significantly reduces the computation time compared to the original R-CNN. This improvement helps make the object detection process faster and more efficient.

14.2.2.3   Faster R-CNN

This technique involves the substitution of the selective search algorithm with a region proposal network. The Faster R-CNN model as shown in Figure 14.5 integrates the region proposal generation directly into the network architecture. The Region Proposal Network (RPN) is a separate module that shares the convolutional layers of the object detection network. It takes feature maps from these

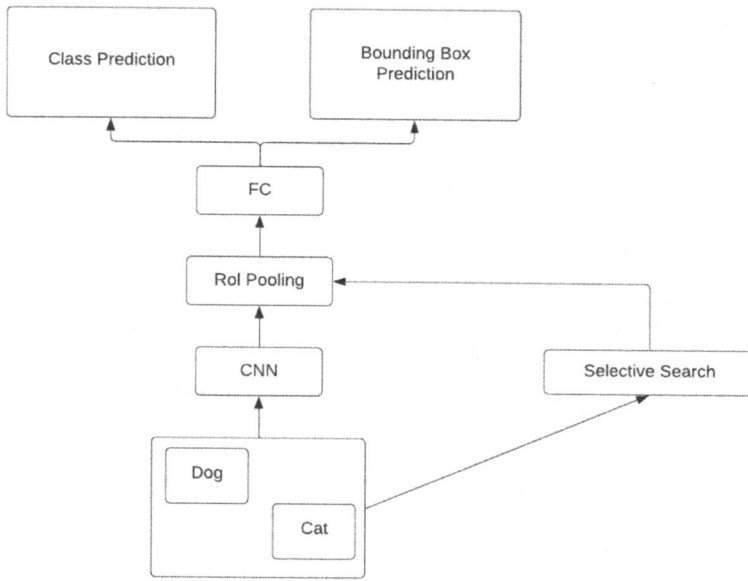

FIGURE 14.4 Fast R-CNN.

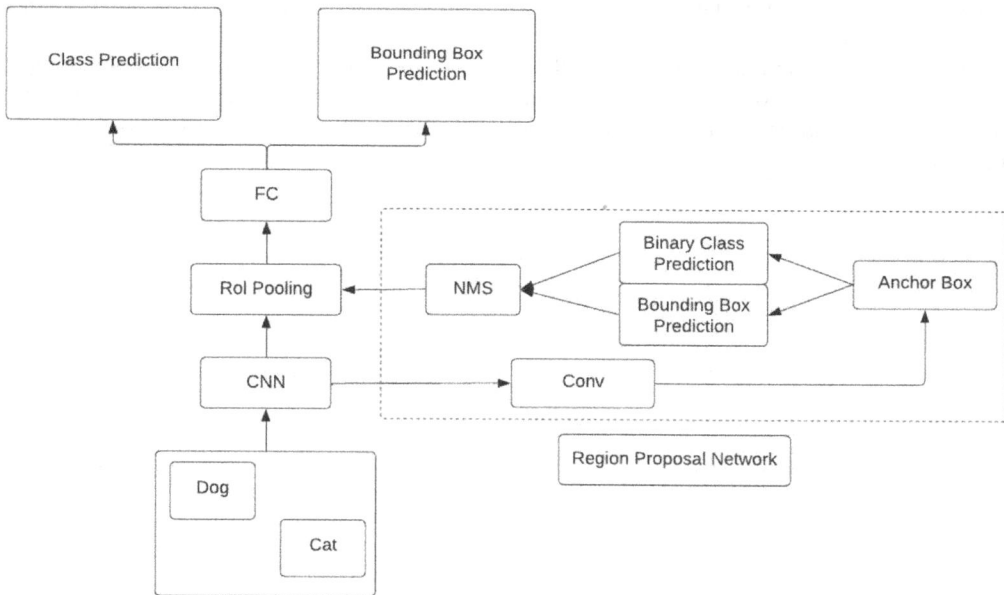

FIGURE 14.5 Faster R CNN.

shared layers and generates region proposals based on anchor boxes. Anchor boxes are predefined bounding boxes of different scales and aspect ratios.

The RPN performs convolutional operations on the shared feature maps and predicts two types of information for each anchor box: objectness scores (whether the anchor box contains an object or not) and bounding box regression offsets (adjustments required to match the ground truth bounding boxes). These predictions are used to filter out irrelevant proposals and refine the remaining proposals.

By incorporating the RPN into the network, the Faster R-CNN model effectively eliminates the need for selective search, reducing the computational overhead associated with generating region proposals. This approach leads to faster and more efficient object detection while maintaining or even improving accuracy compared to the previous methods.

14.2.3 SEMANTIC SEGMENTATION

It is a computer vision task where individual pixels within an image are allocated specific class labels, enabling a comprehensive understanding of the scene's structure and content. Deep learning models, such as Fully Convolutional Networks (FCNs), have significantly improved semantic segmentation accuracy by leveraging the spatial information encoded in convolutional layers. These advancements have facilitated various applications, such as image segmentation for autonomous vehicles, medical imaging analysis, and augmented reality. Fully Convolutional Networks (FCNs) have been instrumental in advancing per-pixel tasks like semantic segmentation. Unlike traditional Convolutional Neural Networks (CNNs) that produce a single output for the entire input image, FCNs enable dense predictions by preserving spatial information throughout the network

14.2.4 OBJECT TRACKING

Deep learning has also impacted object tracking, which involves continuously following objects across frames in a video sequence.

Utilizing recurrent neural networks and long short-term memory networks to capture temporal dependencies has resulted in increased tracking accuracy. Deep learning-based trackers have been successful in various applications, including surveillance, video analysis, and human-computer interaction.

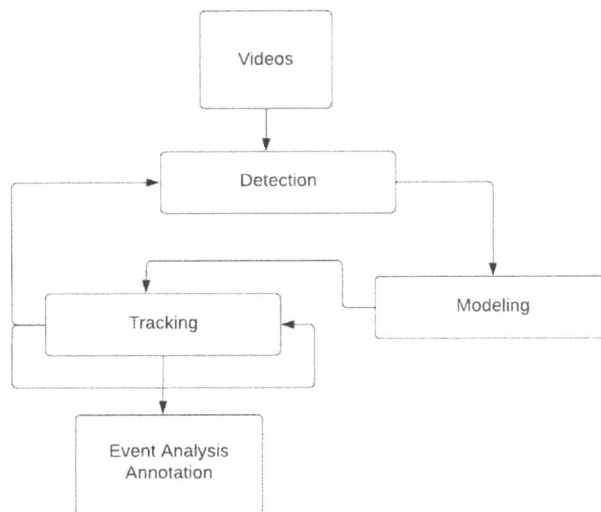

FIGURE 14.6   Object Detection and Tracking [3].

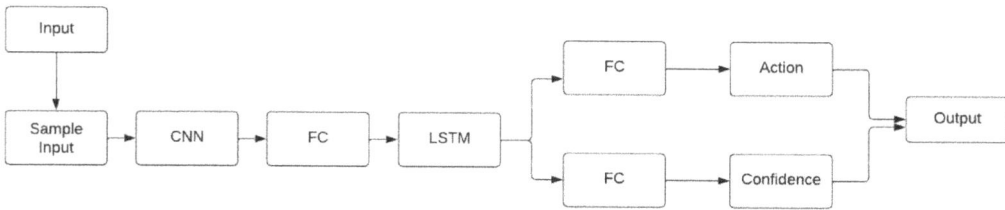

FIGURE 14.7 Pipeline of single object Tracker [4].

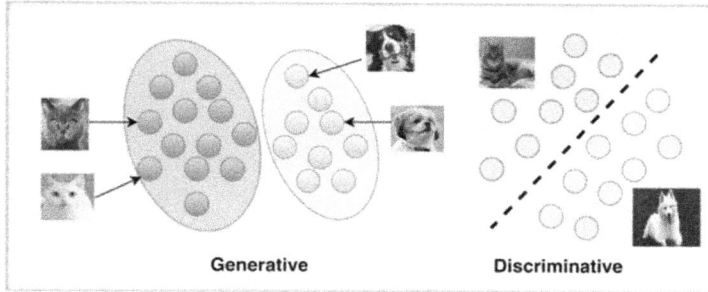

FIGURE 14.8 Generative versus Discriminative [5].

14.2.5 GENERATIVE MODELS

Generative models intend to model the joint distribution of the input features (X) and the corresponding labels (Y) in a dataset. These models learn the underlying data distribution and possess the ability to generate new samples that closely mirror the training data. Generative models can be used for tasks like data generation, unsupervised learning, and modeling complex dependencies in the data. Examples of generative models include:

Gaussian Mixture Models (GMMs): GMMs model the distribution of data as a mixture of several Gaussian distributions.

Hidden Markov Models (HMMs): HMMs are used for sequential data modeling, assuming that the observed data depends on an unobservable (hidden) state.

Variational AutoEncoders (VAEs): VAEs are deep generative models capable of learning a latent representation of input data and generating fresh samples through sampling from the acquired latent space.

Generative Adversarial Networks (GANs): GANs comprise two primary components: a generator network and a discriminator network that compete against each other. The generator network's role is to generate synthetic samples, such as images or text that closely resemble the real samples found in a training dataset. Conversely, the discriminator network acts as a binary classifier, differentiating real samples from generated ones.

Generative models can also be used for discriminative tasks by estimating the conditional probability P(Y|X) using Bayes' rule. However, generative models often have more parameters to estimate due to modeling the joint distribution and their performance in discriminative tasks may be inferior to specialized discriminative models. Deep learning has enabled the development of generative models capable of synthesizing realistic images. Generative adversarial networks and variational autoencoders have the capacity to acquire the inherent distribution of a dataset and produce

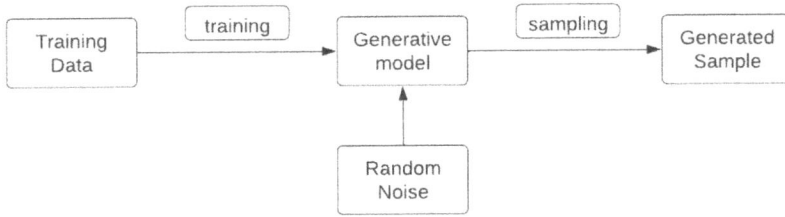

FIGURE 14.9   Generative Modeling Process [6].

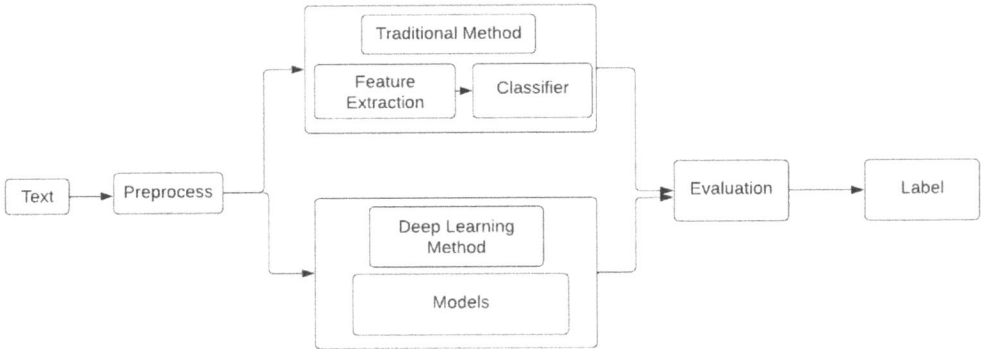

FIGURE 14.10   Sentiment analysis using traditional and Deep Learning Method.

novel samples. This has applications in generating realistic images, data augmentation, and creating synthetic training data.

14.2.6   VISUAL UNDERSTANDING

Deep learning models have facilitated higher-level visual understanding tasks, such as scene understanding, image captioning, and visual question answering. These models can comprehend complex visual scenes, generate descriptions, and answer questions based on the content of an image. This has implications in areas like content-based image retrieval, human-computer interaction, and assistive technologies.

14.3   DEEP LEARNING IN NATURAL LANGUAGE PROCESSING

Deep learning has made remarkable progress in the domain of Natural Language Processing (NLP), revolutionizing various tasks such as language understanding, text generation, sentiment analysis, machine translation, and more. Below are some key applications and techniques in which deep learning has been successful in NLP.

14.3.1   SENTIMENT ANALYSIS AND TEXT CLASSIFICATION

Deep learning models have greatly improved sentiment analysis and text classification tasks. Convolutional neural networks and recurrent neural networks have been employed to autonomously extract features from textual data, capturing both semantic and contextual information. These models have achieved high accuracy in sentiment analysis, document classification, spam detection, and other similar tasks.

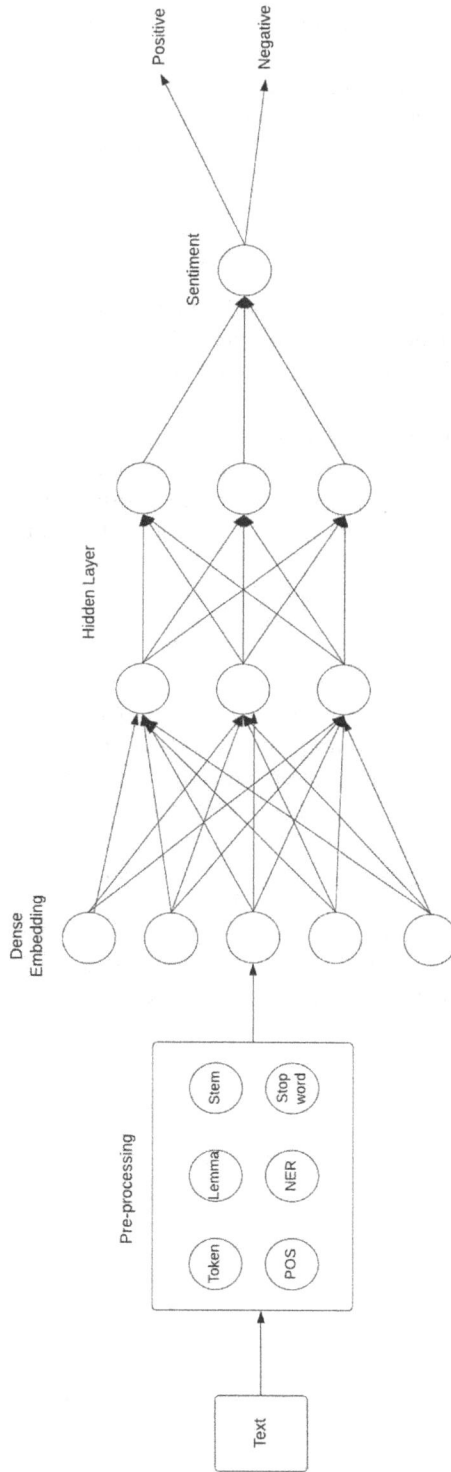

FIGURE 14.11 Sentiment Analysis using Deep Learning.

14.3.2 NAMED ENTITY RECOGNITION AND INFORMATION EXTRACTION

Deep learning has advanced Named Entity Recognition (NER) and information extraction techniques. The scope of NER involves the objective of detecting and classifying named entities, including but not limited to persons' names, organizations, locations, and other specialized terms present within text. Deep learning models, such as bidirectional LSTM-CRFs (Conditional Random Fields), have been successful in accurately identifying and classifying named entities, improving information extraction tasks and applications like question answering systems and knowledge graph construction.

14.3.2.1 Bidirectional LSTM

The long short-term memory network belongs to the family of recurrent neural networks and is designed to process the input sequence in both the forward and backward directions. It captures contextual information by considering the past and future words around a given word.

14.3.2.2 Conditional Random Fields (CRF)

While the bidirectional LSTM network can capture the contextual information of individual words, it doesn't explicitly model the dependencies between neighboring words. The CRF layer is added on top of the LSTM to model these dependencies. It considers the transition probabilities between different named entity labels in a sequence.

14.3.3 TEXT SUMMARIZATION

Deep learning models have made significant contributions to text summarization, where the goal is to generate concise and coherent summaries of long texts. Sequence-to-sequence models, typically based on recurrent neural networks or transformers, have been employed to generate abstractive summaries by learning the mapping from a source text to a summary text. These models have been effective in summarizing news articles, scientific papers, and social media posts.

14.3.4 LANGUAGE TRANSLATION

Deep learning has had a significant impact on language translation in the field of Artificial Intelligence (AI). Traditional approaches to machine translation relied on rule-based methods or statistical models, but deep learning has revolutionized the translation process.

Below, we show how deep learning has transformed language translation in AI.

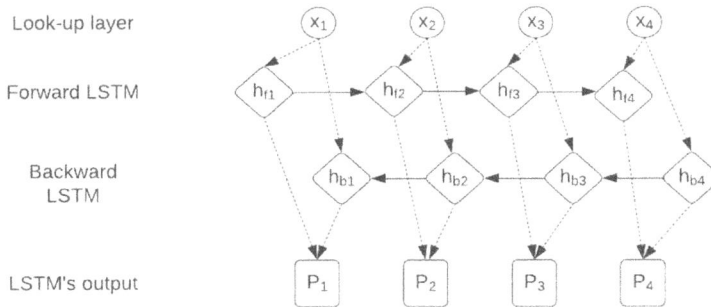

FIGURE 14.12 BiLSTM model structure.

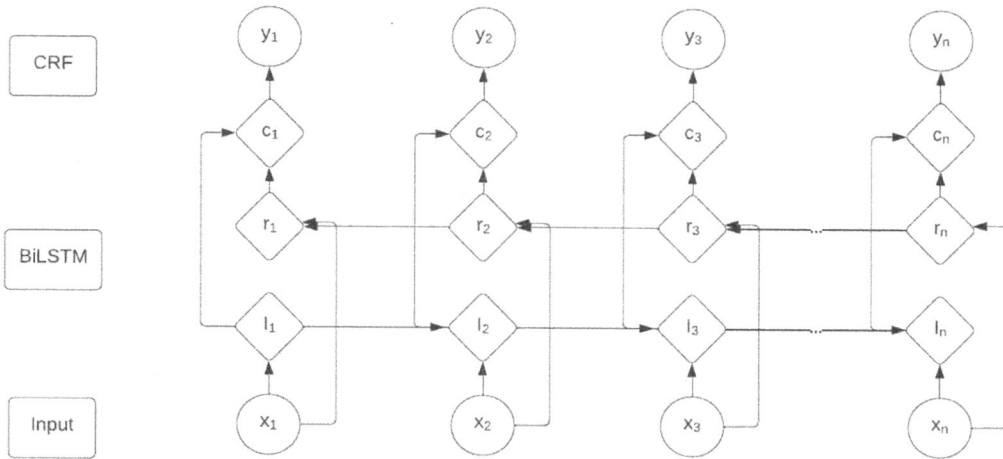

FIGURE 14.13 Architecture of a BiLSTM-CRF model.

14.3.4.1 Neural Machine Translation (NMT)

Deep learning models, particularly sequence-to-sequence models, have been employed for neural machine translation. These models utilize Recurrent Neural Networks (RNNs) or transformer architectures to acquire the mapping between a source language and a target language. NMT models have shown superior performance compared to traditional Statistical Machine Translation (SMT) approaches.

14.3.4.2 End-to-end translation

Deep learning models enable end-to-end translation systems where the entire translation process, from input sentence to output translation, is handled by a single neural network. This eliminates the need for intermediate representations and simplifies the translation pipeline. NMT models have demonstrated improved translation quality and fluency in end-to-end translation.

14.3.4.3 Improved contextual understanding

Deep learning architectures, like Recurrent Neural Networks (RNNs) and transformers, excel at capturing contextual information from the input text. This allows the translation model to consider the entire sentence or document, improving the understanding of word order, syntax, and semantics. Consequently, deep learning-based translation systems can generate more accurate and contextually appropriate translations.

14.3.4.4 Handling long sentences

Deep learning models, especially transformer-based models, have proven effective in handling long sentences, which was a challenge for traditional statistical machine translation approaches. Transformers utilize self-attention mechanisms that allow them to capture long-range dependencies, making them more suitable for translating complex and lengthy sentences.

14.3.4.5 Multilingual translation

Deep learning models have demonstrated the capability to handle multiple languages in a single translation system. Multilingual neural machine translation models can be trained to translate between various language pairs, allowing for more efficient and versatile translation systems. This eliminates the need to build separate models for each language pair.

14.3.5 QUESTION ANSWERING AND DIALOGUE SYSTEMS

Deep learning has advanced question answering and dialogue systems by enabling better understanding of natural language queries and generating appropriate responses. Models like Recurrent Neural Networks (RNNs), transformers, and BERT (Bidirectional Encoder Representations from Transformers) have been used to build question answering systems that can comprehend complex questions and provide accurate answers. In dialogue systems, deep learning models have been employed to generate context-aware responses, leading to more natural and engaging conversations with users.

14.4 AUTOMATED FEATURE EXTRACTION

Deep learning models can automatically learn and extract pertinent features from raw data, thus removing the necessity for manual feature engineering. Traditional machine learning often requires domain experts to hand-craft features, which can be time-consuming and limiting. Deep learning's ability to automatically learn hierarchical representations from data has reduced the dependency on manual feature engineering, allowing AI systems to handle more complex tasks.

14.4.1 END-TO-END LEARNING

Deep learning enables end-to-end learning, where the entire system, from input to output, is learned automatically. Traditional machine learning approaches required human experts to design and select relevant features from the data, a process known as feature engineering. Deep learning models, on the other hand, can learn hierarchical representations directly from raw data, eliminating the need for explicit feature engineering. This automated feature extraction capability has significantly simplified the development process and reduced human effort.

14.4.2 LEARNING DISCRIMINATIVE FEATURES

Deep learning models are designed to learn discriminative features directly from the data, enabling them to capture complex patterns and representations. In traditional feature engineering, human experts often rely on domain-specific knowledge and heuristics to define relevant features. Deep learning models, through their inherent ability to learn from large-scale datasets, can automatically extract features that are relevant and informative for the task at hand. This enables more accurate and robust performance across various domains.

14.4.3 HANDLING HIGH-DIMENSIONAL DATA

Deep learning excels in handling high-dimensional data, such as images, audio, and text, which pose challenges for traditional feature engineering approaches. Deep neural networks can process raw data directly, capturing both local and global patterns. For example, in computer vision, Convolutional Neural Networks (CNNs) automatically learn hierarchical representations of visual features at different spatial scales. This capability allows deep learning models to effectively leverage the rich information contained in high-dimensional data, improving performance in tasks like image classification, object detection, and speech recognition.

14.4.4 TRANSFER LEARNING AND GENERALIZATION

Deep learning enables transfer learning, where models trained on one task or dataset can be leveraged to improve performance on related tasks or datasets. The possibility of this lies in the deep learning models' capacity to learn generic and reusable features from vast-scale datasets. Pre-trained models, such as those trained on ImageNet for computer vision, can serve as a starting point for various downstream tasks, even with limited labeled data. This transfer of learned features and knowledge facilitates faster model development, better generalization, and improved performance on tasks with limited training data.

14.4.5 ADAPTABILITY AND FLEXIBILITY

Deep learning models are highly adaptable and flexible, making them suitable for various domains and applications. They can handle different data modalities, such as images, text, audio, and video, by appropriately designing the network architecture and input representations. Deep learning models can also incorporate additional domain-specific information, such as metadata or side information, into the learning process. This adaptability allows deep learning to be useful in diverse areas, ranging from computer vision and natural language processing to speech recognition and recommendation systems.

14.5 ENHANCED DATA ANALYSIS

Deep learning has enabled more sophisticated and accurate analyses of large and complex datasets. By leveraging deep neural networks, AI systems can discover intricate patterns and relationships within data, leading to improved predictions, recommendations, and decision-making. Deep learning has found applications in various fields, including finance, healthcare, marketing, and cybersecurity.

Overall, deep learning has had a transformative impact on artificial intelligence, pushing the boundaries of what AI systems can achieve in terms of perception, understanding, decision-making, and creative output. Its ability to learn complex representations directly from data has opened up new possibilities and continues to drive advancements in the field of artificial intelligence.

14.6 CHALLENGES AND FUTURE DIRECTION OF DEEP LEARNING ON ARTIFICIAL INTELLIGENCE

Deep learning has its own set of challenges and future directions that impact the broader field of AI. Let's explore some of these challenges and potential future directions:

14.6.1 DATA QUALITY AND QUANTITY

Deep learning models typically rely on extensive amounts of high-quality labeled data for successful training. The process of acquiring and labeling such data can be both time-consuming and costly. Future research will focus on techniques for data augmentation, active learning, and transfer learning to improve data efficiency and quality.

14.6.2 COMPUTATIONAL POWER AND EFFICIENCY

Deep learning models often require significant computational resources, including specialized hardware such as GPUs or TPUs, to train and infer efficiently. Future directions involve developing more efficient algorithms and hardware architectures to reduce the computational requirements of deep learning models and make them more accessible.

14.6.3 DOMAIN ADAPTATION AND TRANSFER LEARNING

Deep learning models trained on one domain may not perform well when applied to a different domain due to distributional shifts. Future research will focus on domain adaptation and transfer learning techniques, enabling models to leverage knowledge from one domain and apply it to another.

14.6.4 CAUSAL REASONING AND EXPLAINABLE AI

Deep learning models excel at correlation-based pattern recognition but struggle with causal reasoning. Future research will focus on developing techniques that enable deep learning models to understand causal relationships, leading to more explainable and reliable AI systems.

14.6.5 GENERALIZATION AND ROBUSTNESS

Deep learning models tend to overfit, leading to good performance on the training data but poor generalization to unseen data. Future research will explore regularization techniques, robust optimization methods, and data augmentation strategies to improve generalization and robustness of deep learning models.

14.6.6 HARDWARE AND SCALABILITY

The computational expense and need for specialized hardware in deep learning models can hinder their accessibility and scalability. Future research will focus on optimizing deep learning algorithms and developing hardware architectures that can efficiently support deep learning computations.

Through the careful consideration of these challenges and the exploration of new frontiers, deep learning will continue to advance the broader field of AI, leading to more powerful, interpretable, robust, and ethical AI systems that can effectively tackle real-world problems.

REFERENCES

[1] Amey, T., & Archit. K. Fundamentals of neural networks. *International Journal for Research in Applied Science & Engineering Technology (IJRASET)* 9(VIII), 2321-9653 (2021).
[2] Brahimi, S., Ben, Aoun, N., Benoit, A., et al. Semantic segmentation using reinforced fully convolutional densenet with multiscale kernel. *Multimedia Tools and Applications* 78, 22077–22098 (2019).
[3] Porikli, F., & Yilmaz, A. Object detection and tracking. (2012).
[4] Ming-xin, J., Chao, D., Zhi-geng, P., Lan-fang, W., & Xing, S. Multiobject tracking in videos based on LSTM and deep reinforcement learning. *Complexity* 2018, 12 (2018), Article ID 4695890.
[5] Taye, M. M. Understanding of machine learning with deep learning: architectures, workflow, applications and future directions. *Computers* 12(5), 91 (2023).
[6] https://www.oreilly.com/library/view/generative-deep-learning/9781492041931/ch01.html
[7] Dang, N. C., Moreno-García, M. N., & de la Prieta, F. Sentiment analysis based on deep learning: a comparative study. *Electronics* 9, (3), 483 (2020).
[8] Kandel, I. & Castelli, M. Transfer learning with convolutional neural networks for diabetic retinopathy image classification. A review. *Applied Science* 2020, 10 (2021).
[9] Dadong, W., Jian-Gang, W., & Xu, K. Deep learning for object detection, classification and tracking in industry applications. *Sensors* 21, 7349 (2021).
[10] Xiaoli, Z., Zheng, C., & Lu, Z. Review of deep learning-based semantic segmentation for point cloud. 7 (2019).
[11] Pathak, A. R., Pandey, M., & Rautaray, S. Application of deep learning for object detection. *Procedia Computer Science* 132, 1706-1717 (2018), ISSN 1877-0509.
[12] Adate, A., Arya, D., Shaha, A., & Tripathy, B. K. 4 impacts of deep neural learning on artificial intelligence research. *Deep Learning: Research and Applications*, edited by Siddhartha Bhattacharyya, Vaclav Snasel, Aboul Ella Hassanien, Satadal Saha & B. K. Tripathy, Berlin, Boston: De Gruyter (2020), pp. 69-84.

[13] Chatterjee, S., Chaudhuri, R., Vrontis, D., et al. Examining the impact of deep learning technology capability on manufacturing firms: moderating roles of technology turbulence and top management support. *Annals of Operations Research* (2022).

[14] Chai, J., Zeng, H., Li, A., & Ngai, E. W. T. Deep learning in computer vision: a critical review of emerging techniques and application scenarios, *Machine Learning with Applications* 6, 100134 (2021), ISSN 2666-8270.

[15] Alzubaidi, L., Zhang, J., Humaidi, A. J., et al. Review of deep learning: concepts, CNN architectures, challenges, applications, future directions. *Journal of Big Data* 8, 53 (2021).

[16] Lauriola, I., Lavelli, A., & Aiolli, F. An introduction to Deep learning in natural language processing: models, techniques, and tools, *Neurocomputing* 470, 443-456 (2022), ISSN 0925-2312.

[17] Hemanth, D. J. Automated feature extraction in deep learning models: A boon or a bane? *2021 8th International Conference on Electrical Engineering, Computer Science and Informatics (EECSI)*, Semarang, Indonesia (2021), pp. 3-3.

[18] Amjoud, A. B., & Amrouch, M. Object detection using deep learning, CNNs and vision transformers: a review. *IEEE Access* 11, 35479–35516 (2023).

15 Stock Prices Prediction of the FMCG Sector in NSE India
An Artificial Intelligence Approach

Subrata Jana, Anirban Sarkar, Bhaskar Nandi,
Arpan Ghoshal, Binay Maji, and Biswadip Basu Mallik

15.1 INTRODUCTION

The Indian economy places Fast-Moving Consumer Goods (FMCG) as the fourth largest industry. The food and beverage industry accounts for 19% of the sector, healthcare for 31%, and household and personal care for the remaining 50%. About 55% of the income share comes from the urban sector, while 45% comes from the rural sector. The growth of the fast-moving consumer goods industry will be led by rising demand in rural areas. It is anticipated that the Indian processed food market would grow from its current value of $263 billion in 2019–20 to reach $470 billion by 2025.

Despite state-wide lockdowns, the Indian Fast-Moving Consumer Goods (FMCG) market increased by 16% in CY21, a nine-year high, on the back of consumption-led growth and value expansion from higher product pricing, notably for basics. Spending on consumable goods and services was up 5.2% annually on average from 2015 to 2020. Fitch Solutions predicts that, after falling by more than 9.3% in 2020 owing to the pandemic's effect on the economy, real household expenditure would rise by 9.1% year over year in 2021. CRISIL Ratings predicts that the FMCG industry's sales growth would increase from 5–6% in FY21 to 10–12% in FY22. As the cost of inputs continues to rise, growth is being driven by higher pricing across the board, increased production volumes, and a renewed interest in luxury goods. There was a year-over-year increase of 12.6% in the domestic FMCG market in Q3 2021.

With a compound annual growth rate (CAGR) of 28.99%, the Indian online grocery business is predicted to generate revenue of about Rs. 1,310.93 billion (US$ 17.12 billion) by 2026. The online grocery market in India is projected to have an 18-fold rise in Gross Merchandise Value (GMV) during the following five years, from $5.5 billion in FY20 to US$37 billion in FY25.

In recent years, the government has made substantial investments to support the FMCG industry. Between April of 2000 and March of 2022, the industry saw robust FDI inflows of US$20.11 billion. In addition, in the Union Budget 2022–23, the Department of Consumer Affairs receives Rs. 1,725 crore (US$222.19 million), while the Department of Food and Public Distribution receives Rs. 215,960 crore (US$27.82 billion). To aid Indian food product names in overseas markets, the government authorised the Production Linked Incentive Scheme for the Food Processing Industry (PLISFPI) in FY 2021–22 with an investment of Rs. 10,900 crore (US$ 1.4 billion).

The Indian e-commerce sector is predicted to grow from USD 38 billion in 2021 to USD 120 billion in 2026, according to a research published jointly by industry association FICCI and property consultancy firm Anarock. In October 2022, Dabur paid Rs. 587.52 crore (US$ 71.81 million) for a 51% share in Badshah Masala Private Limited, based on an enterprise valuation of Rs. 1,152 crore (US$ 140.81 million). HUL's state-of-the-art plant in Sumerpur was officially opened by Uttar Pradesh Chief Minister Mr. Yogi Adityanath in July 2022, with a total investment of Rs. 700 crore

DOI: 10.1201/9781003433309-15

Indian FMCG Market

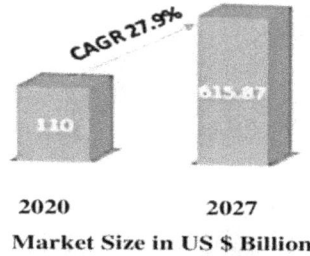

FIGURE 15.1   FMCG Market size in India.

(US$ 88.07 million) scheduled by 2025. Emami entered the Indian pet care market in July 2022 after purchasing a 30% share in Cannis Lupus. Godrej Magic Bodywash, India's first ready-to-mix bodywash, was introduced in July 2022 by Godrej Consumer Products Limited (GCPL) for only Rs. 45 (US$ 0.57).

Companies stand to gain significantly from the government's US$ 1.42 billion Production-Linked Incentive (PLI) programme, which offers them a chance to significantly increase exports. Of the 39 Mega Food Park projects, 22 are fully functional as of February 2021, while another 15 are in the implementation phase.

With the epidemic showing signs of abating, the future prospects of the fast-moving consumer goods industry appear promising. Increases in disposable income and aspiration have driven this growth in rural consumers' spending. There has been a rise in the desire for name-brand goods in rural India. However, as the percentage of the unorganized market in the FMCG sector decreases, the growth of the organised sector is anticipated to expand due to rising brand consciousness and the development of contemporary retail. By 2025, the value of India's fast-moving consumer goods sector is projected to soar to US$ 220 billion, from an anticipated US$ 110 billion in 2020.

Figure 15.1 shows the current and projected FMCG Market size in India

Indian capital and labour markets have relied heavily on the Fast-Moving Consumer Goods (FMCG) sector, which has spawned a number of successful companies and multi-bagger stocks. Some examples include Bajaj Consumer, Patanjali Ayurved Ltd, Parle Agro, Asian Paints Ltd, Himalaya, Healthcare Ltd, ITC FMCG, Amul, Godrej Consumer Products Limited, Jyothy Laboratories, Guruji Products Limited, Dabur India Ltd, Hindu

The social and cultural landscape of India is shifting rapidly. Nearly 115 crore people call India, Pakistan, Bangladesh, or Sri Lanka home. They represent a wide range of socioeconomic backgrounds and are evenly distributed across the country's rural, urban, and suburban sectors and they raise consciousness in the rural market, expand access to education, and so forth. The consumer market has expanded thanks to rising consumer consciousness, lowered barriers to entry, and shifting customer preferences.

Figure 15.2 depicts the components of the FMCG sector in India

Managers, investors, users, and creditors must engage in short-term planning in a variety of domains to ensure the survival and continuation of economic organizations' activities and the success of their investment and financing decisions. Preparedness allows us to respond well when we face downturns in the economy. The efficacy of programs can be enhanced by increasing their capacity for precise and consistent prediction for the simple reason that the forecast plays a significant role in the choices made by users both inside and outside the company. As a result of making sound judgements based on accurate forecasts, efficiency and effectiveness would

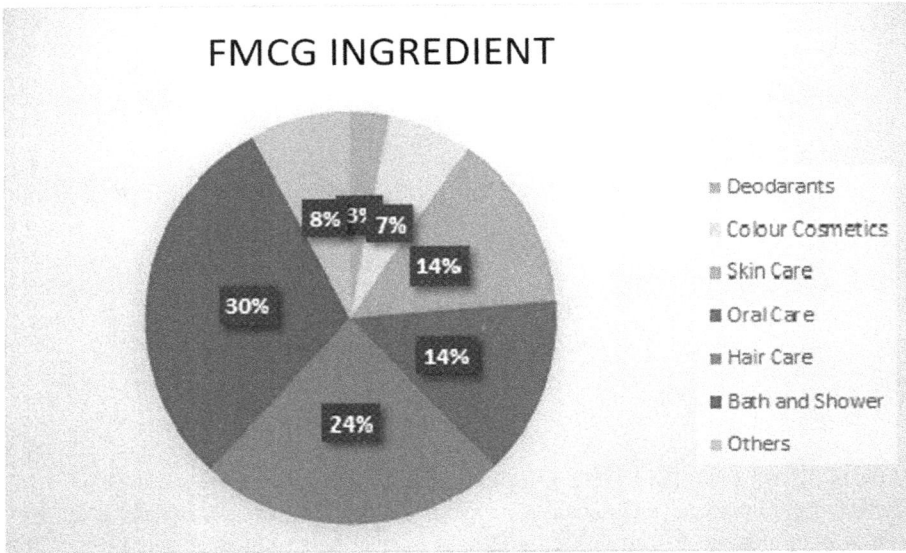

FIGURE 15.2 Components of FMCG Sector in India.

increase (Haghighat, Bakhtiari, Beheshtipoor, 2011). The purpose of prediction is to lessen uncertainty while making choices, hence a good working definition of prediction is estimating future projections. It is true that predictions rarely prove to be 100% accurate, but the degree of error may be minimized by gathering additional data. Predictive value refers to the usefulness of the data in making predictions. Predicting stock prices, earnings, returns on equity, insolvency, and risk are all part of the financial industry's standard fare. Because of the influence that these variables have on the decisions made by activists about NYSE stock prices, accurate stock price forecasting is crucial. Basic transactions in the capital market rely on the presence of connected information; as a result, information is the most valuable asset in the capital market (Economic World Press, 2007). The key to success in these markets is anticipating price movements so that you can adjust your buying and selling accordingly. However, macro characteristics aren't the only thing that has an impact on the stock market (Torabi and Homan, 2010). Future stock values can be anticipated using a number of different methods. Fundamental analysis is a process that considers a wide range of factors (Yaldiz and Yazgol, 2010). At its core it is generally accepted that stock market fluctuations are random, and that stock price behaviour is shaped by the random walk hypothesis (exact date not known). The stock market's value is determined by a number of variables. These include financial ratios, industrial circumstances, and other factors that might have an impact on the economy. Fundamental analysis, which considers the economy, industry, and the firm itself, may be used to forecast stock prices. Predictions in basic analysis are grounded on reality. Traders evaluate current affairs and news stories in light of their expertise and tactics for dealing with market projections (Murphy, 2008). Fundamental analysts think the true intrinsic worth of stocks may be determined and base their work on the premise that all price movements can be traced back to underlying financial and economic factors. Potential investors in the Fast-Moving Consumer Goods (FMCG) sector can boost productivity through stock price forecasting, leading enterprises in this sector to allocate capital where it will be most fruitful. Capital markets that are both dynamic and trustworthy are essential for emerging nations to achieve economic growth and enhance investment incentives. Industry is able to achieve this result because of the sector's ability to absorb and allocate capital and financial resources (Namazi and Kyamehr, 2007). There has been a lot of study devoted to accounting and finance projections during the previous decade. Indicators include expected earnings per share, stock price, earnings management, bankruptcy, and so forth (Peter, Vladimir and Renata, 2013). Statistical and AI-based methods have been extensively explored in this space. Traditional statistical approaches rely heavily on assumptions like linearity, normalcy, independence of predictor variables, and the like, and may not be the best option (Russell and Noroyg, 2012). Previous studies have demonstrated that neural networks outperform the statistical model in terms of prediction (Olson and Masman, 2003, Arabmazar Yazdi, Ahmadi and Abduli, 2006, Kordloyi and Haideri Zare, 2010, Fahimi fard, Salarpoor and Sabouhi, 2011, Makiyan and Mousavi, 2012, 7 Chung et al, 2012). Additionally, the chance of obtaining the best forecast rises if two or more models are merged [7]. We also apply econometric models for estimation and forecasting, alongside neural networks. These strategies include using time series data, cross-section data, and panel data. The use of econometric techniques like panel data is becoming increasingly popular in scientific research (Ashrafzadeh and Mehregan, 2008). Given the importance of stock prices to investors, managers, and creditors, it is reasonable to assume that they employ cutting-edge methods like artificial neural networks and panel data to make stock price forecasts. Panel data and an artificial neural network (an econometric model) are employed to improve efficiency, cut down on expenses, and conduct a thorough analysis in this research.

15.2 THEORETICAL APPROACH

Participants in the financial markets are seeking instruments that can accurately anticipate the future condition of the market in light of the proliferation of information and the expansion of financial markets. Predictive models may be broken down into four categories based on an analysis of prior studies on financial market forecasting:

These are technical analysis, fundamental analysis, econometrics, and intelligent methods.

The integration of fundamental analysis, smart forecasting, and econometrics (panel data) is the focus of this chapter. The premise of fundamental analysis is that stock prices may not perfectly reflect all relevant information at any one moment, and therefore it seeks to identify non-price factors that could lead to a change in the stock's value in the near future, for the simple reason that market prices eventually converge on fundamental values [1,2].

15.2.1 PANEL DATA

Time series data and cross-sectional data are what make up panel data. What this means is that data from cross sections may be tracked over long periods of time. Such information has two dimensions, one connected to the units of measurement in use at any particular moment, and the other related to the passage of time itself. The values of one or more variables are tracked through time in time series data (for example, GDP over the past few years). One or more variables' values are gathered simultaneously for a single data set or a sample in cross sectional data (for example the stock price for the 50 companies listed in a given year). In contrast, measurements in panel data are taken continuously over the same cross-sectional units (such as in a certain industry) (Gajarati, 2010).

15.2.2 NEURAL NETWORKS

When it became clear that the brain's parallel processing design and dynamic state made it fundamentally incompatible with the known conventional processors, research and interest in artificial neural networks began. William James advocated in the 19th century for a mental shift in which stacking altered preconceived notions about how the body and brain worked. Perhaps parallel processing processes and neurons in the brain are two examples. From the family of forward neural networks, function networks are the radial basis. In 1998, Bromhid and Low were the first to present the world with these networks. This kind of network, which has many different uses, is a major rival to the multi-layered Persporon network. As a result of variations in training, neuronal activity, and the use of hidden layers, this technique is distinct from others [5].

15.3 REVIEW OF LITERATURE

To evaluate the efficacy of the neural networks we have the Fuzzy (ANFIS) model for forecasting Bahman business stock price in contrast to linear models, Botshekan (2000) employed a neural networks Fuzzy (AI) approach (ARIMA). In addition, the box-Jenkins technique was utilised to identify the ARIMA model. The index, the most recent stock prices, trade volume, and the price of oil are all input factors. Based on the data, it's clear that Network–Neuro-Fuzzy (ANFIS) is a better tool for predicting stock prices than the linear model (ARIMA). Research by Nazarian et al. [7] indicates that the use of advanced mathematical models with complex structures to predict the market is quite acceptable, especially in light of the growth of financial markets and the importance of these markets and their close relationship to macroeconomic variables. With the daily data of stock price index from 03.25.2009 to 22.10.2011, a reduction model with potential long-term memory, a forward neural network model, and a combination of the two models were used to forecast the volatility of the stock index on the National Stock Exchange in India. Using prediction error as an assessment metric, we find that the feed forward neural network model performs better than the reduction model with the potential for long-term memory. While both of these models are useful, the combined model is more precise. By using ARIMA and artificial neural network models, Schumann and Lohrbach [8] forecast the following day's stock price on the Frankfurt market. For nine years, they've relied on the 13 sets of daily information provided. They have utilised a two-layer deep neural network and the data is of a technical nature. Therefore, we cannot favour both of these approaches simultaneously. Using the asymmetric APARCH model with the two different distributions T-Student and GED, Feransesko and Rakesh (2012) predicted volatility in the stock markets of the five founding members of the Association of Southeast Asian Nations (ASEAN-5). Their goal was to determine whether there was symmetry or asymmetry in the relationship between stock returns and market volatility in the ASEAN-5 market. The model relied on financial turnover and the value of traded shares from domestic firms listed on the stock exchange from January 2, 2002, to January 30, 2012. The t-distribution is used to demonstrate the model's superiority in terms of the measurement error in the prediction, and the results are shown as a graphical representation of this error. Stock price predictions by Peter, Vladimir, and Renata (2013) utilizing a neural network and regression technique that considers both the quantitative data and the hidden emotions (lack of certainty in reports and associated phrases) in the annual report. The ratios of profitability and the variables of technical analysis employed are from 2010 US-listed corporations. The findings demonstrated that neural networks yield superior outcomes, particularly when considering covert emotions conveyed in yearly reports.

15.4 METHODOLOGY

15.4.1 GOAL OF THE STUDY

An econometric model for panel data and a neural network are utilised to accomplish these goals, along with cost savings and deeper analysis. The primary purpose of this research is to compare the predictive power of an econometric model and a neural network in the context of the Fast-Moving Consumer Goods (FMCG) industry.

15.4.2 HYPOTHESIS OF THE STUDY

When compared to an econometric model, how much more accurate are stock price predictions made utilizing fundamental analysis and a neural network for the FMCG industry?

15.4.3 POPULATION

Companies trading on the National Stock Exchange of India in the Fast Moving Consumer Goods (FMCG) industry between 2012 and 2022 make up the population.

15.4.4 METHODS FOR TAKING SAMPLES

Companies that do not meet the following criteria are eliminated from the sample pool using this technique of selection:

- The companies have been approved for listing on India's National Stock Exchange.
- No adjustments to the fiscal year were made within the time frame of the research.
- Basic financial statements from 2012 through 2022 are available for the example firms.

15.4.5 COLLECTION OF DATA AND TOOLS

All of the necessary data and information for this study was obtained from the National Stock Exchange of India's database. Data from the Central Bank's database that is essential for assessing the economy. The TOPSIS approach is used to assess the state of the market.

15.4.6 STEPS OF TOPSIS ALGORITHM

The steps included in this method are provided in the following:

Step 1: Formation of Decision Matrix:
Aggregation of data by Arithmetic" Mean.

Step 2: Formation of Normalized Decision" Matrix:

$$N_m = \frac{x_{ij}}{\sqrt{\sum_{i=1}^{n} x_{ij}^2}} \tag{15.1}$$

Step 3: Formation of Weighted Normalized Decision Matrix:

$$WN_m = w_i * N_m \tag{15.2}$$

Step 4: Determination of the Ideal Best and Ideal Worst Value:

$$P^+ = (a_1^+, a_2^+, \dots, a_o^+) = \left\{ \left(\max_i a_{ij} \mid j \in B \right), \left(\min_i a_{ij} \mid j \in NB \right) \right\}, \tag{15.3}$$

$$P^- = (a_1^-, a_2^-, \dots, a_o^-) = \left\{ \left(\min_i a_{ij} \mid j \in B \right), \left(\max_i a_{ij} \mid j \in NB \right) \right\}, \tag{15.4}$$

Where B represents Benefit Criteria and NB denotes Non- Benefit Criteria.

Step 5: Calculation of the Euclidian Distance from the Ideal best and Ideal Worst value:

$$S_i^+ = \sqrt{\sum_{j=1}^{o} \left(WN_m - a_j^+ \right)^2}, \quad i = 1, 2, \dots, n \tag{15.5}$$

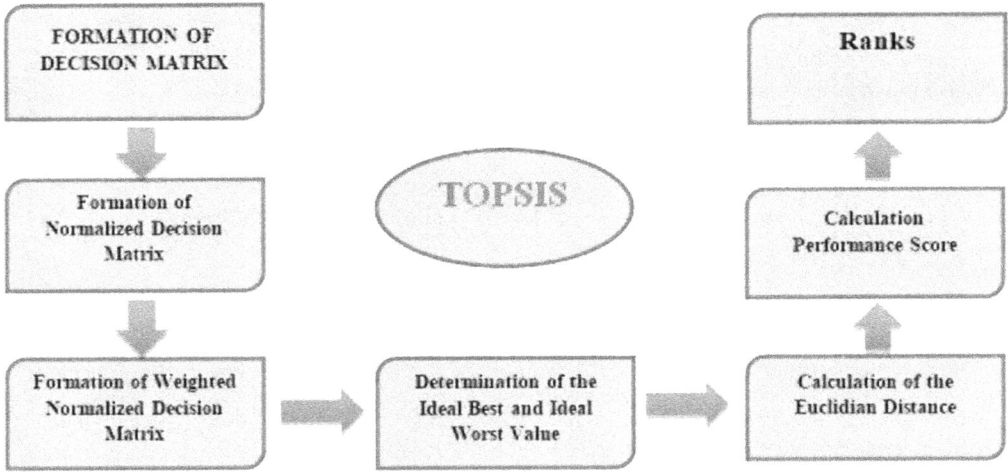

FIGURE 15.3   Diagrammatic Steps of TOPSIS.

$$S_i^- = \sqrt{\sum_{j=1}^{o}\left(WN_m - a_j^-\right)^2}, \quad i = 1, 2, \dots, n \tag{15.6}$$

Step 6: Calculation Performance Score:

$$P_i = \frac{S_i^-}{S_i^+ + S_i^-}; 0 \le P_i \le 1 \tag{15.7}$$

Step 7: Ranks:
Higher value of P_i represents the better alternatives.

The following Figure 15.3 shows diagrammatic steps of TOPSIS

15.4.7 ALGORITHM FOR PANEL DATA

Panel data model can be established as follows:

$$y_{it} = \alpha_{i0} + \alpha_1 x_{1it} + \alpha_2 x_{2it} + \dots + \alpha_k x_{kit} + \epsilon_{it} \tag{15.8}$$

$$\epsilon_{it} = \mu_i + \gamma_t + v_{it}\left(\ \right) \tag{15.9}$$

$$P_i = \frac{S_i^-}{S_i^+ + S_i^-}; 0 \le P_i \le 1 \tag{15.10}$$

Here μ_i and γ_t are undetectable effects of individuals and time respectively and v_{it} is residual for error component and in the matrix will be as follows:

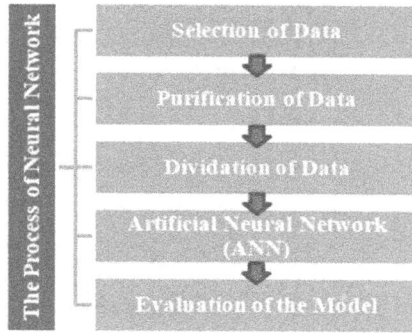

FIGURE 15.4 The Process of Neural Network.

$$y = \alpha_{NT} + X\beta + \epsilon; \quad i = 1, 2, \ldots, N; t = 1, 2, \ldots T \tag{15.11}$$

$$\epsilon = Z_\mu \mu + Z_\gamma \gamma + \nu \tag{15.12}$$

EVIEWS7 software is used to implement for the econometric model of panel data.

15.4.8 NEURAL NETWORK ALGORITHM

Figure 15.4 shows the process of the neural network.

The neural network used consists of an input layer (3 neurons), the middle layer (1 to 20 neurons) and output layer (1 neuron).

15.4.9 METHODS OF DATA ANALYSIS

In this research, we employ a combination of a panel data econometric model and a neural network to foretell the direction of the stock market. So, here we look at the variables and data sets that were employed, as well as the implementation algorithms, procedures, and assessment criteria of the models that were used. Figure 15.5 presents the systematic steps of methodology.

15.5 RESULTS

As aforementioned, the optimal moment to invest is determined by looking at the fundamentals of the market and the economy as a whole, as part of the fundamental research. After that, they do research into various sectors of the economy with promising futures. Once analysts have determined that an investment is viable and that high-efficiency sectors of the economy are operating normally, they turn their attention to an examination of the business itself. Inflation based on consumer price index, open market exchange rate, oil revenue, money supply, and coin rate have all been analysed to determine the effect economic indicators have on the business. Each year, TOPSIS is used to rank industries by a number of metrics, including the following: current ratio, quick ratio, current debt to equity ratio, debt to equity ratio, debt to assets ratio, interest coverage ratio, quick asset turnover ratio, current asset turnover ratio, current asset turnover ratio, tangible fixed asset turnover ratio, gross profit margin, operating profit margin, net profit margin, return on equity, return on assets, price to earnings, price to book value, and the ratio of current assets to total assets. The primary drivers were identified with the help of factor analysis. Correlated variables in a factor

- Goal of the Study
- Hypothesis of the Study
- Population
- Methods for Taking Samples
- Collection of Data & Tools
- Steps of TOPSIS algorithm
- Algorithm for Panel Data
- Neural Network Algorithm
- Methods of Data Analysis

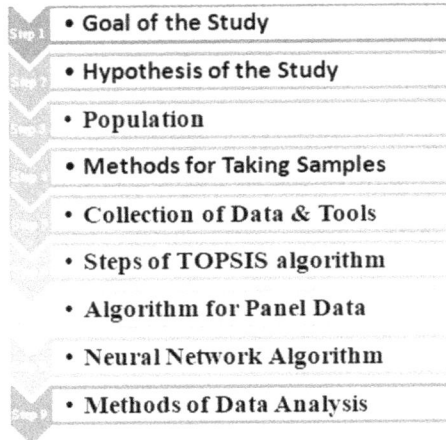

FIGURE 15.5 Systematic Steps of Methodology.

TABLE 15.1
Ranking by TOPSIS method

Year	2012	2013	2014	2015	2016	2017	2018	2019	2020	2021	2022
Ranking	6	4	5	3	7	9	10	3	2	2	10

analysis were grouped together, and eigenvalues were used to determine which variables should be combined (Jelif, 2002). To get a combining variable, eigenvalues must be calculated after finding a set of variables with a correlation of more than 0.5. The matrix of eigenvalues is calculated with EVIEWS7. You may get combining variables (PCs) by multiplying the matrix of eigenvalues by a set of highly correlated variables. TOPSIS use metrics like current ratio, debt-equity ratio, operating profit margin, and net profit margin to categorise sectors based on their financial performance.

Table 15.1 shows the results of applying this methodology to the classification of Stock Exchange industries throughout the ten-year period 2012–2022.

After factor analysis and making appropriate and meaningful patterns the combining variables FR3 (Gross profit margin, Operating profit margin, EPS) were significant as accounting variables in fundamental analysis using panel data. Econometric models are not significant economic variables of the hypothesis model.

Table 15.2 shows the descriptive statistics of the variables in the model hypothesis, including: PS (stock price), and FR3 (gross profit margin, operating profit margin, EPS). Stock price is the dependent variable and the other variables are independent variables. In this table, the number represents the observations for each variable, the mean represents the mean of observations for each variable, the median reflects the middle of observations, mode represents the highest frequency of observations for each variable, the standard deviation represents the deviation of the observation from the mean and variance is the squared deviations from the mean of observations. To test for normality, a standard error of skewness coefficient and a standard error of Kurtosis are used. If the values of the indices are smaller than 2 – or greater than 2 +, the normality is rejected (Momeni & Faal Ghaiyoomi, 2012, 33). The results in Table 15.3 indicate that the variables are normally distributed.

A situation that indicates independent variables is a linear function of other independent variables. To detect linearity two factors of variance inflation factor and tolerance were used. The minimum amount that can be get variance inflation is one and this is in the situation that there is no linearity

TABLE 15.2
Description of Statistics

Model	FR3	PS
Number	10	10
Mean	1.7606	3876.2955
Median	1.4051	4112
Mode	0.45	1214
S.D	1.01627	3335.8764
Variance	1.120	1123467.89

TABLE 15.3
"Standard Error Test"

Model	FR3	PS
Number	10	10
S.E of Skewness	0.382	0.382
S.E of Kurtosis	0.928	0.928

TABLE 15.4
"The Linearity between Independent Variables"

	Model	Tolerance	Variance Inflation Factor
Independent Variable	FR3	0.824	1.121

between the independent variables, whenever the value of this factor is greater than one indicates better linearity between the independent variables. Also, linearity is serious when the amount of variance inflation factor is greater than 10. Regarding the variance inflation factor values in Table 15.4, the linearity between the independent variables is very small and the tolerance level is greater than 0.4, which is good" (Momeni & Faal Ghaiyoomi, 2012).

Table 15.5 represents when the mean residual is zero, which is underpinned by an assumptions of regression.

After considering the assumptions on the regression and preparing variables to build a model for selecting between integrated model and fixed effects model, the Chow test and for selecting fixed effects or random in panel data model Hausman test is used. This test has an asymptotic chi-square distribution and degrees of freedom are equal to the number of explanatory variables (regression). According to the probability column in Table 15.6, the result of the Chow test indicates panel data model is used. The Hausman test is provided for selecting between random and fixed effects.

According to the column of Table 15.7, the probability is more than 0.5, so random effects model is accepted. According to results, the stock price modelling by considering panel data model with fixed effects is summarized in Table 15.8.

As you see in above Table 15.8, the column probability is less than 5% for all variables, so all variables are significant and the model is suitable. The coefficient of determination represents the number of 0.7304, which represents the power produced to justify the model. It means the dependent variable is explained by the independent variables by 73%; 73% change in the stock price depends

TABLE 15.5
Statistical Indexes in relation to the anticipated residual values

	Number of Observation	S.D	Mean	Max	Min
Predicted Value	10	4925.4501	4706.5382	3152.7612	901.6965
Residual	10	2517.0067	0.0000	4647.65924	-7930.7542
Predicted Standardized Value	10	1.000	0.0000	4.021	- 1.002
Standardized Residual	10	0.8912	0.0000	1.824	-4.2040

TABLE 15.6
Test F Chow

Hypothesis	Distribution	Statistic	Degree of Freedom	Probability
	F	16.1505	4.30	0.0024
	Chi-Square	13.4579	4	0.0002

TABLE 15.7
Hausman Test

Hypothesis	Statistic	DOF (Degree of Freedom)	Probability	Result
	0.0000	3	1.000	Random Effects

TABLE 15.8
Stock Price Modelling

Variable	Estimated Coefficient	Statistic t	Probability
C	2510.683	3.9721	0.0003
FR3	233.2248	12.4821	0.0000
$R^2 = 0.730455$;	$\bar{R}^2 = 0.722287$	$F = 0.000$	

on selected variables in the model. Also the calculated probability of F-statistic is less than 5%, so the model is valid. so the selected model for predicting is valid.

$$Price\ Stock = 2510.683 + 233.2248*FR3 + \varepsilon \qquad (15.13)$$

After building a suitable model to forecast stock prices, the research hypothesis was tested using the criteria of prediction error. Table 15.9 shows that as compared to the econometric model, the neural network has a lower criterion. This lends credence to the study's premise that neural networks are more accurate than econometric models at predicting FMCG stocks.

TABLE 15.9
Comparison of the Model Accuracy

		Type of Error		
		RMSE	MAE	MAPE
Model	Neural Network	1096.954	555.899	12.648
	Econometric Model of Panel Data	1659.289	1438.084	49.646

15.6 CONCLUSION

Predicting stock prices on exchanges is a major challenge for market participants. Using fundamental analysis, an econometric model using panel data, and an artificial neural network, this research forecasts FMCG stocks. The econometric model for panel data and artificial neural network were used to improve the efficiency and minimise the expenses associated with fundamental analysis, which often needs a huge amount of data and a significant amount of time for split investors to master all the essential parts. The use of econometric models for panel data to the forecasting of stock prices has been largely overlooked in the literature pertaining to fundamental analysis. Consistent with prior research in the field of neural networks, the results of prediction in both models indicate a higher accuracy of neural network into the econometric model. This is because the variables in the selected models, such as FR3 (gross profit margin, operating profit margin, EPS), were significant. Finally, the selected model is provided to aid investment decisions in light of the outcomes.

REFERENCES

1. Abarbanell, J. S., and Bushee, B. J. "Fundamental analysis, future earnings, and stock prices." *Journal of Accounting Research* 35, no. 1 (1997): 1–24.
2. Abarbanell, J. S., and Bushee, B. J. "Abnormal returns to a fundamental analysis strategy." *Accounting Review* (1998): 19–45.
3. Budhani, N., Jha, C. K., and Sandeep, K. "Application of neural networks in analysis of stock market prediction." *International Journal of Computer Science & Engineering Technology (IJCSET)* 3, no. 4 (2012): 61–68.
4. Ghosh, A. "Analyzing the efficiency of Indian life insurance companies using DEA and SEM." *Turkish Journal of Computer and Mathematics Education (TURCOMAT)* 12, no. 12 (2021): 3897–3919.
5. Haykin, S. *Neural networks: a comprehensive foundation.* Prentice Hall PTR (1998).
6. Jolliffe, I. "A 50-year personal journey through time with principal component analysis." *Journal of Multivariate Analysis* 188 (2022): 104820.
7. Nazarian, R., Alikhani, N. G., Naderi, E. and Amiri, A. "Forecasting stock market volatility: a forecast combination approach." (2013).
8. Schumann, M., and Lohrbach, T. "Comparing artificial neural networks with statistical methods within the field of stock market prediction." In *[1993] Proceedings of the Twenty-Sixth Hawaii International Conference on System Sciences*, vol. 4, pp. 597–606. IEEE (1993).
9. Sudheer, K. P., and Jain, S. K. "Radial basis function neural network for modeling rating curves." *Journal of Hydrologic Engineering* 8, no. 3 (2003): 161–164.
10. Upadhyay, A., Bandyopadhyay, G., and Dutta, A. "Forecasting stock performance in Indian markets using multinomial logistic regression." *Journal of Business Studies Quarterly* 3, no. 3 (2012): 16.

16 Multi-Attribute Decision Modelling

Gurjapna Anand, Priyanka Vashisht, and Simar Preet Singh

16.1 INTRODUCTION

This chapter provides an overview of the concept and application of multi-attribute decision modelling in various domains. Decision-making is a fundamental aspect of human life, and in complex scenarios where multiple factors need to be considered, multi-attribute decision modelling offers a structured approach to facilitate informed and effective decision-making.

Traditional decision-making approaches often focus on single criteria, which may not fully capture the complexities and trade-offs inherent in real-world decision contexts. Multi-attribute decision modelling, on the other hand, enables decision-makers to assess and evaluate alternatives based on multiple criteria simultaneously, considering various factors such as costs, benefits, risks, preferences, and constraints.

The chapter begins by introducing the foundational concepts of multi-attribute decision modelling, including decision criteria, attributes, alternatives, and decision matrices. It explores different types of decision modelling techniques, such as weighted scoring models, the Analytic Hierarchy Process (AHP), and multi-objective optimization. These techniques provide structured frameworks to systematically evaluate and rank alternatives based on their performance across multiple attributes.

Moreover, this chapter discusses the importance of defining decision criteria and attributes, as well as the challenges associated with quantifying and measuring them. It explores methods for gathering and analyzing data, including subjective judgments, expert opinions, surveys, and statistical analysis. The chapter also delves into the use of mathematical and statistical models to derive decision weights, calculate scores, and assess the overall desirability of alternatives.

The application of multi-attribute decision modelling spans various domains, such as business, engineering, healthcare, finance, and environmental management. The chapter presents case studies and examples from these domains to illustrate how multi-attribute decision modelling can be effectively employed to address complex decision problems. It highlights the benefits of using such models, including improved transparency, consistency, and defensibility of decision outcomes.

Additionally, the chapter explores the integration of decision support tools and technologies in multi-attribute decision modelling. It discusses the use of computer-based software applications, visualization techniques, and interactive interfaces to enhance decision-making processes and facilitate collaboration among stakeholders.

By introducing multi-attribute decision modelling, this chapter aims to familiarize readers with the concepts, methods, and applications in this field. It serves as a foundation for further exploration of advanced techniques, such as multi-criteria decision analysis, uncertainty modelling, and decision under risk. The chapter concludes by emphasizing the significance of multi-attribute decision

DOI: 10.1201/9781003433309-16

modelling in addressing complex decision challenges and the need for ongoing research and development in this area.

16.2 BACKGROUND OF THE STUDY

Multi-attribute decision modelling is a field of study that emerged from the recognition that decision-making processes often involve multiple criteria or factors that need to be considered simultaneously. Traditional decision-making approaches, which focus on single criteria or objectives, may not adequately capture the complexities and trade-offs involved in real-world decision contexts. As a result, the need for a structured and systematic approach to decision-making led to the development of multi-attribute decision modelling techniques.

The origins of multi-attribute decision modelling can be traced back to the fields of operations research, management science, and decision analysis, where researchers and practitioners sought ways to support complex decision-making problems. Early works in these fields laid the foundation for multi-attribute decision modelling by introducing concepts such as decision matrices, utility theory, and value-focused thinking.

One of the pioneering approaches in multi-attribute decision modelling is the Analytic Hierarchy Process (AHP), proposed by Thomas L. Saaty in the 1970s. AHP provides a structured method for prioritizing and comparing alternatives based on a hierarchy of criteria and sub-criteria. It allows decision-makers to assign weights to criteria, derive pairwise comparisons, and calculate overall priority scores for alternatives.

Since then, numerous other methods and techniques have been developed to address different aspects of multi-attribute decision modelling. Weighted scoring models, outranking methods, fuzzy [12] logic, and multi-objective optimization are among the approaches that have been widely adopted in various disciplines.

The application of multi-attribute decision modelling has gained prominence in a wide range of domains, including business, engineering, healthcare, finance, public policy, and environmental management. In business, for example, organizations use multi-attribute decision models to evaluate investment opportunities, select suppliers, or assess project proposals. In healthcare, decision models aid in determining treatment options, resource allocation, and healthcare policy development.

The increasing complexity of decision problems, coupled with advances in computing power and data analytics, has contributed to the growing relevance and adoption of multi-attribute decision modelling techniques. These techniques offer decision-makers a systematic and transparent framework to evaluate alternatives, consider multiple criteria, and incorporate stakeholder preferences.

However, challenges persist in the application of multi-attribute decision modelling. These challenges include defining appropriate criteria and attributes, obtaining reliable and accurate data, addressing uncertainty and risk, handling conflicting stakeholder perspectives, and ensuring the validity and reliability of the decision models.

Given the ongoing advancements in technology and the increasing need for effective decision-making in complex environments, further research and development in multi-attribute decision modelling are necessary. This chapter aims to contribute to the existing body of knowledge [2, 13] by providing an overview of the concepts, methods, and applications in multi-attribute decision modelling. By understanding the background and current state of the field, readers can explore and apply these techniques to address real-world decision challenges in their respective domains.

16.3 ABOUT THE CHAPTER TOPIC

Multi-Attribute Decision Modelling is a chapter that focuses on the theory, methods, and applications [15] of decision-making processes that involve multiple criteria or attributes. In today's complex and interconnected world, decision-makers face numerous challenges when evaluating alternatives, as they need to consider a range of factors simultaneously. Multi-attribute decision modelling provides a structured and systematic approach to address these challenges by incorporating multiple criteria and facilitating informed decision-making.

The chapter begins by introducing the fundamental concepts of multi-attribute decision modelling. It explores the basic elements of decision analysis, including criteria, attributes, alternatives, and decision matrices. The chapter discusses the importance of defining decision criteria and attributes accurately, as well as their relevance to the decision problem at hand.

The chapter then presents various decision modelling techniques that can be employed in multi-attribute decision-making. Weighted scoring models, where criteria are assigned weights based on their relative importance, allow decision-makers to rank alternatives by calculating weighted scores. The Analytic Hierarchy Process (AHP), developed by Thomas L. Saaty, enables decision-makers to decompose complex decision problems into hierarchical structures and derive priority weights through pairwise comparisons.

Additionally, this chapter explores other approaches such as outranking methods, fuzzy logic, and multi-objective optimization. These techniques accommodate diverse decision scenarios, handling uncertainty, imprecision, and multiple conflicting objectives.

The practical application [15] of multi-attribute decision modelling is illustrated through case studies and examples across various domains. The chapter highlights its relevance in business and management contexts, including project selection, supplier evaluation, and investment decisions. It also explores applications in fields like healthcare, environmental management, public policy, and engineering, where decision models aid in resource allocation, policy development, and system design.

Moreover, the chapter emphasizes the importance of data collection and analysis in multi-attribute decision modelling. It discusses techniques for gathering and quantifying data, including subjective judgments, expert opinions, surveys, and statistical analysis. The chapter also addresses the challenges associated with data quality, reliability, and uncertainty, and presents approaches to handle these issues effectively.

Lastly, the chapter discusses the benefits and limitations of multi-attribute decision modelling, as well as potential areas for future research and development. It emphasizes the need for ongoing advancements in methodology, integration with emerging technologies, and addressing complex decision problems involving large-scale datasets [7] and dynamic environments.

By exploring the theory, methods, and applications of multi-attribute decision modelling, this chapter aims to provide readers with a comprehensive understanding of this decision-making approach. It equips decision-makers, researchers, and practitioners with the knowledge [2] and tools necessary to navigate complex decision scenarios, make informed choices, and enhance their decision-making processes across various domains.

16.4   METHODOLOGY

The methodology chapter in a book on multi-attribute decision modelling describes the specific methodology and techniques employed to conduct the decision analysis process. It provides a detailed explanation of the steps followed to construct decision models, evaluate alternatives, and derive meaningful insights. The chapter typically includes the following components:

1. Problem identification and formulation: This section outlines the process of identifying and defining the decision problem. It involves understanding the context, objectives, and constraints of the decision, and formulating it in a clear and concise manner.
2. Selection of decision criteria: Here, the chapter discusses the process of selecting the appropriate decision criteria that will be used to evaluate the alternatives. It explores techniques such as brainstorming, literature review, and expert opinions to identify relevant criteria that capture the key dimensions of the decision problem.

3. Data collection: This section focuses on gathering the necessary data to assess the performance of the alternatives against the identified criteria. It discusses various methods of data collection, including surveys, interviews, experiments, and existing databases. The chapter also addresses data quality assurance and validation techniques.

4. Data analysis and preprocessing: This step involves analyzing and preprocessing the collected data to ensure its suitability for the decision modelling process. It includes techniques such as data cleaning, normalization, transformation, and outlier detection to enhance the accuracy and consistency of the data.

5. Criteria weighting: This section explores the methods used to assign weights to the decision criteria based on their relative importance. It discusses both subjective and objective approaches, such as the Analytic Hierarchy Process (AHP), pairwise comparisons, and mathematical optimization techniques.

6. Alternative evaluation and ranking: Here, the chapter explains the process of evaluating and ranking the alternatives based on their performance across the selected criteria. It covers techniques such as weighted scoring, utility theory, outranking methods, and multi-objective optimization to assess and compare the alternatives.

7. Sensitivity analysis: This section emphasizes the importance of conducting sensitivity analysis to examine the robustness of the decision model and its outcomes. It involves varying the criteria weights, altering data inputs, or considering different scenarios to understand the impact on the final rankings and decision recommendations.

8. Validation and verification: The chapter discusses the validation and verification process for the decision model. It explores techniques such as expert reviews, sensitivity testing, and cross-validation to ensure the model's accuracy, reliability, and adherence to sound decision-making principles.

9. Decision support system implementation: This section addresses the implementation of the decision model into a decision support system or software tool [20]. It discusses the integration of the model with relevant technologies, user interface design, and user training to facilitate its practical use.

10. Case studies and application examples: The chapter includes case studies and application examples that demonstrate the methodology in action. These real-world scenarios illustrate how multi-attribute decision modelling has been applied to solve complex decision problems and provide insights into the practical implementation challenges and considerations.

11. By providing a detailed methodology for multi-attribute decision modelling, this chapter guides readers through the systematic process of constructing decision models and evaluating alternatives. It ensures that readers have a clear understanding of the steps involved, the techniques used, and the best practices to follow when applying multi-attribute decision modelling in various decision-making contexts.

16.4.1 WEIGHTED SCORING MODELS

This section focuses on the use of weighted scoring models in multi-attribute decision modelling. Weighted scoring models provide a structured approach to assess and rank alternatives based on multiple criteria [3] or attributes. By assigning weights to the criteria, decision-makers can quantify the relative importance of each criterion and calculate a weighted score for each alternative. This chapter explores the methodology, advantages, and limitations of weighted scoring models in multi-attribute decision modelling.

1. Understanding weighted scoring models:
   • Definition and purpose of weighted scoring models
   • Basic components: criteria, weights, alternatives, and scores

- Importance of criteria weighting in decision-making.
2. Criteria selection and definition:
   - Techniques for identifying and selecting relevant decision criteria
   - Ensuring criteria are well-defined and measurable
   - Incorporating stakeholder preferences in criteria selection.
3. Weight assignment methods:
   - Subjective methods: expert judgment, Delphi method, and nominal group technique
   - Objective methods: analytical approaches, such as the Analytic Hierarchy Process (AHP) and entropy-based methods
   - Sensitivity analysis to assess the impact of varying weights on decision outcomes.
4. Scoring techniques:
   - Different scoring methods: point allocation, performance scales, and fuzzy logic-based scoring [16]
   - Considerations for scaling criteria and establishing score ranges
   - Addressing challenges related to missing or incomplete data in scoring.
5. Aggregation and ranking of alternatives:
   - Calculation of the weighted score for each alternative
   - Aggregating scores across criteria to determine overall performance
   - Ranking alternatives based on their weighted scores.
6. Interpretation and decision-Making:
   - Interpreting the results of weighted scoring models
   - Assessing trade-offs and sensitivity to criteria weights
   - Facilitating decision-making through the ranking of alternatives.
7. Limitations and challenges:
   - Recognizing limitations of weighted scoring models
   - Handling subjective biases in criteria weighting
   - Dealing with interdependencies and interactions among criteria.
8. Real-world applications and case studies:
   - Illustrating the use of weighted scoring models in various domains (for example, project selection, supplier evaluation, resource allocation)
   - Demonstrating the practical implementation of weighted scoring models through case studies and examples.
9. Best practices and guidelines:
   - Providing best practices for effectively implementing weighted scoring models
   - Guidelines for communicating and presenting results to stakeholders
   - Considerations for updating and refining the model over time.

16.4.2 ANALYTIC HIERARCHY PROCESS (AHP)

This section focuses on the Analytic Hierarchy Process (AHP) as a powerful technique in multi-attribute decision modelling. AHP provides a structured approach to handle complex decision problems involving multiple criteria [3] and alternatives. It enables decision-makers to systematically evaluate and prioritize alternatives based on their relative importance and performance across different criteria. This chapter explores the methodology, applications, and benefits of using AHP in multi-attribute decision modelling.

1. Understanding the Analytic Hierarchy Process (AHP):
   - Overview of AHP and its theoretical foundations
   - Hierarchical structure: goals, criteria, sub-criteria, and alternatives
   - The role of pairwise comparisons in AHP

2. Defining the decision hierarchy:
   - Steps involved in constructing the decision hierarchy
   - Identifying the decision goal and criteria
   - Decomposing criteria into sub-criteria and defining their relationships.
3. Pairwise comparisons and priority estimation:
   - Techniques for conducting pairwise comparisons between criteria and alternatives
   - Using scales and linguistic [5] terms for pairwise comparisons
   - Estimating priority weights using the Eigenvector method and consistency checks.
4. Consistency analysis and sensitivity testing:
   - Assessing the consistency of pairwise comparison matrices
   - Calculation of the Consistency Ratio (CR) and its interpretation
   - Sensitivity analysis to evaluate the impact of changes in judgments on the final priorities.
5. Aggregation and synthesis of priorities:
   - Combining priority weights at different levels of the hierarchy
   - Calculation of the overall priority scores for alternatives
   - Synthesizing priorities and deriving the ranking of alternatives.
6. Handling uncertainty and inconsistency:
   - Dealing with uncertain or incomplete judgments in pairwise comparisons
   - Incorporating fuzzy sets [4] and fuzzy logic in AHP for handling imprecise information
   - Addressing conflicts and inconsistencies in pairwise comparison matrices.
7. Applications and case studies:
   - Real-world applications of AHP in various domains (for example, project selection, investment decisions, supplier evaluation)
   - Case studies demonstrating the practical implementation of AHP in multi-attribute decision modelling.
8. Software tools and decision support systems:
   - Overview of software tools [20] and platforms that facilitate AHP implementation
   - Utilizing decision support systems to streamline the AHP process
   - Considerations for selecting appropriate software tools for AHP applications.
9. Advantages and limitations of AHP:
   - Discussing the strengths and benefits of using AHP in multi-attribute decision modelling
   - Recognizing the limitations and potential pitfalls of AHP
   - Addressing common challenges and providing strategies to overcome them.

16.4.3 OUTRANKING METHODS

This section focuses on outranking methods as a valuable approach in multi-attribute decision modelling. Outranking methods provide a systematic way to evaluate and rank alternatives based on their performance across multiple criteria. Unlike traditional scoring models, outranking methods consider the relative importance and preferences of decision criteria, allowing decision-makers to capture the complexity of decision problems. This chapter explores the methodology, applications, and benefits of outranking methods in multi-attribute decision modelling.

1. Understanding outranking methods:
   - Overview of outranking methods and their theoretical foundations
   - Key concepts: outranking relations, preference modelling, and dominance analysis
   - Comparison with other decision modelling approaches (for example, weighted scoring models, AHP).
2. Preference modelling and elicitation:
   - Techniques for eliciting and modelling decision-maker preferences

- Use of preference statements, pairwise comparisons, or value functions
- Incorporating uncertainty and imprecision in preference modelling.

3. Dominance analysis:
   - Definition and identification of dominance relations among alternatives
   - Different types of dominance: strict dominance, weak dominance, and indifference
   - Assessing the dominance of alternatives based on their performance on criteria.

4. Outranking relations and preference indices:
   - Calculation of outranking relations between alternatives
   - Introduction of preference indices and outranking degrees
   - Understanding the significance and interpretation of outranking relations.

5. Multi-Criteria Decision Methods (MCDM):
   - Integration of outranking methods within broader MCDM frameworks
   - Combining outranking with other techniques (for example, AHP, ELECTRE, PROMETHEE)
   - Hybrid approaches and their benefits in complex decision problems.

6. Sensitivity analysis and robustness:
   - Assessing the robustness of outranking results
   - Sensitivity testing for variations in preference models or criteria weights
   - Analyzing the impact of uncertainties and changes in decision inputs.

7. Visualization and communication of results:
   - Techniques for visualizing and presenting outranking results
   - Use of ranking diagrams, spider charts, or decision maps
   - Communicating the decision outcomes to stakeholders effectively.

8. Applications and case studies:
   - Real-world applications of outranking methods in diverse domains
   - Case studies demonstrating the practical implementation of outranking methods
   - Highlighting the advantages and insights gained through outranking approaches.

9. Limitations and challenges:
   - Recognizing the limitations and potential biases of outranking methods
   - Handling incomplete or imprecise preference information
   - Addressing computational challenges in large-scale decision problems.

16.4.4 FUZZY LOGIC IN MULTI-ATTRIBUTE DECISION MODELLING

This section focuses on the application of fuzzy logic in multi-attribute decision modelling. Fuzzy logic provides a flexible and intuitive framework for handling uncertainty, vagueness, and imprecision in decision-making. By incorporating fuzzy sets [4] and fuzzy inference, decision-makers can effectively capture and represent complex and subjective information. This chapter explores the methodology, benefits, and applications of fuzzy logic in multi-attribute decision modelling.

1. Understanding fuzzy logic:
   - Introduction to fuzzy sets [17] and fuzzy logic principles
   - Key concepts: membership functions, fuzzy operations, and linguistic [5] variables
   - Advantages of fuzzy logic in dealing with uncertainty and imprecision.

2. Fuzzy sets and membership functions:
   - Definition and properties of fuzzy sets [7]
   - Designing membership functions to represent linguistic [6] terms
   - Techniques for determining membership functions from data or expert opinions.

3. Fuzzy inference systems:
   - Structure and components of a fuzzy inference system
   - Fuzzy rule-based reasoning and fuzzy IF-THEN rules

- Defuzzification methods for obtaining crisp outputs.
4. Fuzzy logic in criteria evaluation:
    - Modelling criteria with fuzzy sets [19] to capture imprecise or subjective information
    - Fuzzy comparison and aggregation methods for evaluating alternatives
    - Handling linguistic [6] assessments and preferences using fuzzy logic.
5. Fuzzy decision support systems:
    - Integration of fuzzy logic into decision support systems
    - Designing user-friendly interfaces for fuzzy logic-based decision models
    - Utilizing fuzzy logic tools and software for multi-attribute [6] decision modelling.
6. Fuzzy logic and uncertainty management:
    - Handling uncertainty and variability in decision modelling
    - Incorporating fuzzy reasoning in sensitivity analysis and risk assessment
    - Fuzzy approaches for addressing incomplete or vague data.
7. Fuzzy logic in group decision making:
    - Extending fuzzy logic to accommodate group decision-making scenarios
    - Aggregating individual preferences and fuzzy evaluations
    - Consensus-reaching [10] techniques in fuzzy group decision models.
8. Applications and case studies:
    - Real-world applications of fuzzy logic in various domains
    - Case studies demonstrating the practical implementation of fuzzy logic in multi-attribute [6] decision modelling
    - Highlighting the advantages and insights gained through fuzzy logic approaches.
9. Limitations and challenges:
    - Recognizing the limitations and potential pitfalls of fuzzy logic in decision modelling
    - Addressing computational complexity and interpretability issues
    - Dealing with subjectivity and uncertainty in fuzzy logic-based models.

16.4.5 MULTI-OBJECTIVE OPTIMIZATION

This section explores the application of multi-objective optimization in multi-attribute decision modelling. Multi-attribute decision modelling involves evaluating alternatives based on multiple criteria, and multi-objective optimization provides a framework to simultaneously optimize multiple conflicting objectives. By considering trade-offs and Pareto optimality, decision-makers can identify a set of optimal solutions that represent different trade-off possibilities. This chapter discusses the methodology, benefits, and applications of multi-objective optimization in multi-attribute decision modelling.

1. Understanding multi-objective optimization:
    - Introduction to multi-objective optimization and its key concepts
    - Pareto optimality and the concept of trade-offs
    - Different approaches to multi-objective optimization (for example, weighted sum, Pareto dominance).
2. Problem formulation in multi-objective decision modelling:
    - Defining the decision problem and specifying objectives
    - Identifying decision variables and constraints
    - Incorporating multiple criteria into the optimization formulation.
3. Pareto-based multi-objective optimization algorithms:
    - Overview of popular multi-objective optimization algorithms
    - Genetic algorithms, particle swarm optimization, and evolutionary algorithms
    - Handling constraints and diversity preservation in multi-objective optimization.

4. Multi-objective decision making:
   - Interpreting and analyzing Pareto optimal solutions
   - Decision rules and preference-based methods for selecting solutions
   - Visualizing and exploring the Pareto front for decision support.
5. Interactive decision-making methods:
   - Incorporating decision-maker preferences in the optimization process
   - Interactive evolutionary algorithms and preference articulation methods
   - Balancing exploration and exploitation in interactive decision-making.
6. Handling uncertainty in multi-objective optimization:
   - Dealing with uncertain or imprecise input data
   - Robust optimization and stochastic optimization approaches
   - Incorporating uncertainty in the Pareto-based decision-making process.
7. Applications and case studies:
   - Real-world applications of multi-objective optimization in various domains
   - Case studies demonstrating the practical implementation of multi-objective optimization in multi-attribute decision modelling
   - Highlighting the advantages and insights gained through multi-objective optimization approaches.
8. Multi-objective optimization and decision support systems:
   - Integration of multi-objective optimization into decision support systems
   - Utilizing optimization software and tools for multi-objective decision modelling
   - Challenges and considerations in implementing multi-objective optimization in decision support systems.
9. Limitations and future directions:
   - Recognizing the limitations and challenges of multi-objective optimization in decision modelling
   - Addressing computational complexity and scalability issues
   - Emerging trends and future research directions in multi-objective decision modelling.

16.4.6 DATA COLLECTION AND ANALYSIS IN MULTI-ATTRIBUTE DECISION MODELLING

This section focuses on the importance of data collection and analysis in multi-attribute decision modelling. Multi-attribute decision modelling involves evaluating alternatives based on multiple criteria, and the quality and relevance of data play a crucial role in the accuracy and validity of the decision-making process. This chapter explores the methodology, techniques, and best practices for data collection and analysis in multi-attribute decision modelling.

1. Importance of data in multi-attribute decision modelling:
   - Understanding the role of data in decision-making
   - Impact of data quality on the accuracy and reliability of decision models
   - Challenges and considerations in data collection and analysis.
2. Defining decision criteria and variables:
   - Identifying and defining decision criteria and variables
   - Specifying the measurement scales and units for each criterion
   - Ensuring clarity and consistency in criterion definitions.
3. Data collection methods:
   - Overview of data collection methods for multi-attribute decision modelling
   - Surveys, interviews, and questionnaires
   - Data mining, web scraping, and automated data collection.
4. Handling missing data and outliers:

- Strategies for handling missing data in decision models
- Imputation techniques and sensitivity analysis
- Identifying and dealing with outliers in the data.
5. Data preprocessing and transformation:
   - Data cleaning, filtering, and normalization
   - Handling categorical and ordinal data
   - Feature selection and dimensionality reduction techniques.
6. Exploratory data analysis:
   - Descriptive statistics and data visualization techniques
   - Identifying patterns, trends, and relationships in the data
   - Assessing the distribution and variability of the data.
7. Statistical analysis for decision modelling:
   - Statistical techniques for analyzing data in multi-attribute decision modelling
   - Correlation and regression analysis
   - Hypothesis testing and significance analysis.
8. Data integration and aggregation:
   - Combining data from multiple sources or experts
   - Aggregating data and transforming it into decision inputs
   - Handling conflicting or subjective data.
9. Validation and sensitivity analysis:
   - Validating the decision model using data analysis techniques
   - Sensitivity analysis to assess the robustness of the decision model
   - Interpreting and communicating the results of data analysis.
10. Ethical considerations in data collection and analysis:
    - Ethical guidelines and considerations for data collection and analysis
    - Ensuring privacy, confidentiality, and data security
    - Transparency and accountability in data-driven decision-making.

16.4.7 CASE STUDIES IN MULTI-ATTRIBUTE DECISION MODELLING

This section presents real-life case studies that demonstrate the practical application and effectiveness of multi-attribute decision modelling in various domains. Multi-attribute decision modelling involves evaluating alternatives based on multiple criteria, and these case studies provide valuable insights into how decision-makers have utilized this approach to make informed decisions. The case studies highlight the methodology, challenges, and outcomes of multi-attribute decision modelling in real-world scenarios.

1. Healthcare decision-making:
   - Case study on hospital selection for a specific medical procedure
   - Evaluation of hospitals based on criteria such as quality of care, cost, and patient satisfaction
   - Application of multi-attribute decision modelling to assist patients in making informed decisions.
2. Environmental planning and resource allocation:
   - Case study on selecting the optimal location for a renewable energy project
   - Evaluation of potential sites based on criteria like environmental impact, resource availability, and cost-effectiveness
   - Incorporation of stakeholder preferences and sustainability considerations in the decision model.
3. Supplier selection in supply chain management:
   - Case study on choosing suppliers for a manufacturing company

- Evaluation of potential suppliers based on criteria such as quality, price, delivery time, and reliability
- Multi-attribute decision modelling to optimize supplier selection and enhance supply chain efficiency.

4. Project portfolio management:
   - Case study on prioritizing and selecting projects in a portfolio
   - Evaluation of projects based on criteria like strategic alignment, resource requirements, and expected returns
   - Multi-attribute decision modelling to optimize project selection and allocation of resources.

5. Risk assessment and management:
   - Case study on risk assessment in the insurance industry
   - Evaluation of insurance policies based on criteria such as coverage, premium, and claim settlement process
   - Multi-attribute decision modelling to assess and manage risks effectively.

6. Investment decision-making:
   - Case study on selecting investment opportunities in the financial sector
   - Evaluation of investment options based on criteria like risk, return, liquidity, and market conditions
   - Application of multi-attribute decision modelling to optimize investment decisions.

7. Supplier evaluation and performance measurement:
   - Case study on evaluating and monitoring the performance of suppliers in a global supply chain
   - Evaluation of suppliers based on criteria such as quality, delivery performance, and customer service
   - Multi-attribute decision modelling to improve supplier selection and performance management.

8. Transportation and logistics planning:
   - Case study on route optimization and transportation planning for a logistics company
   - Evaluation of different transportation routes based on criteria like cost, delivery time, and fuel efficiency
   - Application of multi-attribute decision modelling to enhance transportation and logistics decision-making.

16.4.8 UNCERTAINTY AND SENSITIVITY ANALYSIS

This section focuses on the incorporation of uncertainty and sensitivity analysis techniques in multi-attribute decision modelling. Multi-attribute decision modelling involves evaluating alternatives based on multiple criteria, and uncertainties in data or model parameters can significantly impact the decision-making process. This chapter explores the methodology, tools, and best practices for handling uncertainty and performing sensitivity analysis in multi-attribute decision modelling.

1. Understanding uncertainty in multi-attribute decision modelling:
   - Sources of uncertainty in decision modelling
   - Types of uncertainty, including aleatory and epistemic uncertainty
   - Implications of uncertainty on decision outcomes.
2. Uncertainty quantification and modelling:
   - Techniques for quantifying uncertainty in decision models

- Probabilistic methods, such as Monte Carlo simulation and probability distributions
- Non-probabilistic methods, such as interval [8] analysis and possibility theory.
3. Probabilistic sensitivity analysis:
   - Evaluating the impact of uncertainty on decision outcomes
   - Methods for propagating uncertainties through the decision model
   - Sensitivity measures, including variance-based sensitivity analysis and tornado diagrams.
4. Scenario analysis and what-if analysis:
   - Exploring different plausible scenarios and their implications
   - Identifying critical scenarios that significantly influence decision outcomes
   - Conducting what-if analysis to assess the effects of specific changes or events.
5. Robust decision making:
   - Incorporating robustness considerations in decision modelling
   - Techniques for handling uncertainty through robust optimization or robust decision rules
   - Balancing performance and robustness in decision-making under uncertainty.
6. Decision trees and Bayesian networks:
   - Utilizing decision trees and Bayesian networks for decision modelling under uncertainty
   - Probabilistic decision trees and Bayesian networks for capturing uncertainties and dependencies
   - Analyzing decision trees and Bayesian networks to assess the effects of uncertainties.
7. Sensitivity analysis techniques:
   - Assessing the sensitivity of decision outcomes to model parameters
   - One-factor-at-a-time sensitivity analysis
   - Global sensitivity analysis using variance-based methods or design of experiments.
8. Decision-making under deep uncertainty:
   - Handling deep uncertainties characterized by limited knowledge or ambiguous information
   - Techniques such as scenario planning, robust decision-making, and adaptive strategies
   - Decision pathways and adaptive management in the face of deep uncertainties.
9. Real-world applications and case studies:
   - Demonstrating the application of uncertainty and sensitivity analysis in multi-attribute decision modelling through case studies
   - Assessing the impact of uncertainties on decision outcomes in various domains
   - Highlighting the insights gained and decision improvements achieved through uncertainty and sensitivity analysis.

16.4.9 INTEGRATION OF MULTI-ATTRIBUTE DECISION MODELLING WITH EMERGING TECHNOLOGIES

This section explores the integration of multi-attribute decision modelling with emerging technologies to enhance the effectiveness and efficiency of decision-making processes. Multi-attribute decision modelling involves evaluating alternatives based on multiple criteria, and the advancements in emerging technologies provide new opportunities to improve decision models and decision support systems. This chapter discusses the integration of various emerging technologies with multi-attribute decision modelling and highlights their benefits, challenges, and potential applications.

1. Artificial intelligence and machine learning:
   - Leveraging AI and machine learning techniques in multi-attribute decision modelling
   - Automated data analysis and pattern recognition for improved decision-making
   - Utilizing machine learning algorithms for predictive modelling and decision support.

2. Big data analytics:
   - Integration of big data analytics in multi-attribute decision modelling
   - Utilizing large and diverse datasets for enhanced decision insights
   - Real-time data processing and decision-making using big data technologies.
3. Internet of Things (IoT) and sensor networks:
   - Incorporating IoT and sensor networks data in multi-attribute decision modelling
   - Utilizing real-time and contextual information for decision support
   - Integration of IoT devices and data streams with decision models.
4. Blockchain technology:
   - Harnessing the capabilities of blockchain in multi-attribute decision modelling
   - Ensuring transparency, immutability, and trust in decision processes
   - Smart contracts and decentralized decision-making systems.
5. Augmented Reality (AR) and Virtual Reality (VR):
   - Enriching decision-making experiences through AR and VR technologies
   - Visualizing and simulating decision scenarios in immersive environments
   - Virtual collaboration and decision-making support using AR and VR tools.
6. Cloud [11] computing and distributed computing:
   - Leveraging cloud computing [1] and distributed computing in multi-attribute decision modelling
   - Scalability, flexibility, and on-demand computing resources for decision support
   - Distributed decision-making systems and collaborative decision modelling.
7. Natural Language Processing (NLP) and text mining:
   - Integration of NLP and text mining techniques in decision modelling
   - Extracting and analyzing unstructured textual data for decision insights
   - Text-based decision support systems and sentiment analysis in decision-making.
8. Explainable AI and decision transparency:
   - Ensuring interpretability and transparency in AI-driven decision models
   - Explainable AI techniques for understanding and validating decisions
   - Ethical considerations and fairness in decision-making with emerging technologies.
9. Case studies and practical applications:
   - Real-world case studies showcasing the integration of emerging technologies with multi-attribute decision modelling
   - Demonstrating the benefits and challenges of using emerging technologies in decision-making processes
   - Highlighting the improvements achieved in decision outcomes through technology integration.

16.4.10 APPLICATIONS OF MULTI-ATTRIBUTE DECISION MODELLING IN BUSINESS AND MANAGEMENT

Real-life applications of multi-attribute decision modelling in business and management are:

1. Supplier selection and evaluation:
   - Businesses often use multi-attribute decision modelling to select and evaluate suppliers based on various criteria such as quality, price, delivery time, and reliability. This helps in making informed decisions about which suppliers to engage with and ensures the procurement process is efficient and effective.
2. Product portfolio management:
   - Multi-attribute decision modelling is employed to manage product portfolios by evaluating and prioritizing potential products based on criteria like market demand,

profitability, resource requirements, and strategic fit. This enables businesses to optimize their product offerings and allocate resources effectively.

3. Project selection and prioritization:
    *   Multi-attribute decision modelling is utilized to select and prioritize projects within organizations. Decision-makers consider criteria such as project feasibility, expected returns, alignment with strategic objectives, and resource availability. This ensures that the most valuable and feasible projects are chosen for execution.
4. Risk assessment and management:
    *   Multi-attribute decision modelling helps businesses assess and manage risks by considering multiple risk factors and their potential impacts. By quantifying risks and evaluating their probabilities and consequences, businesses can make informed decisions about risk mitigation strategies and resource allocation.
5. Marketing campaign optimization:
    *   Businesses employ multi-attribute decision modelling to optimize marketing campaigns by evaluating different marketing channels, messaging strategies, target segments, and budget allocations. This enables organizations to allocate resources effectively, maximize campaign effectiveness, and achieve marketing objectives.
6. Facility location selection:
    *   Multi-attribute decision modelling is used to select the optimal location for new facilities such as warehouses, manufacturing plants, or retail outlets. Factors such as transportation costs, market access, labor availability, and infrastructure are evaluated to make informed decisions regarding facility locations.
7. Vendor and technology selection:
    *   In the context of Information Technology (IT) and software procurement, multi-attribute decision modelling aids in vendor and technology selection. Businesses consider criteria such as functionality, cost, compatibility, scalability, and vendor reputation to choose the most suitable technology solutions and vendors.
8. Financial investment decisions:
    *   Multi-attribute decision modelling is employed in financial investment decisions by considering criteria like risk, return, liquidity, and market conditions. It helps businesses evaluate investment alternatives and optimize their investment portfolios based on their specific financial objectives and risk tolerance.
9. Supply chain optimization:
    *   Multi-attribute decision modelling assists businesses in optimizing their supply chains by evaluating different logistics strategies, transportation modes, inventory levels, and supplier relationships. This enables organizations to improve supply chain efficiency, reduce costs, and enhance customer satisfaction.
10. Performance evaluation and employee appraisal:
    *   Multi-attribute decision modelling is used in performance evaluation and employee appraisal processes. Organizations consider various performance metrics, such as individual goals, skills, teamwork, and organizational values, to objectively assess employee performance and make informed decisions regarding promotions, rewards, and training.

16.4.11 MULTI-ATTRIBUTE DECISION MODELLING IN HEALTHCARE AND PUBLIC POLICY

Multi-attribute decision modelling plays a vital role in healthcare and public policy domains, where complex decisions need to be made considering multiple criteria. This approach enables decision-makers to evaluate alternatives, allocate resources, and prioritize actions based on a comprehensive assessment of various factors. This section explores the applications of multi-attribute decision modelling in healthcare and public policy and highlights its significance in improving decision-making processes.

1. Healthcare resource allocation:
   • Multi-attribute decision modelling is extensively used in healthcare to allocate limited resources effectively. Decision-makers consider criteria such as patient needs, cost-effectiveness, quality of care, and access to healthcare services when making decisions about the allocation of healthcare resources, such as funding, medical equipment, and personnel. By employing multi-attribute decision modelling, healthcare organizations and policymakers can ensure that resources are distributed equitably and efficiently to maximize patient outcomes.
2. Treatment selection and comparative effectiveness:
   • In healthcare, the choice of treatments or interventions can be challenging due to the wide range of options available. Multi-attribute decision modelling enables a systematic evaluation of different treatment options based on criteria such as efficacy, safety, cost, and patient preferences. By considering multiple attributes simultaneously, healthcare professionals and policymakers can make informed decisions about the most appropriate treatment options for specific patient populations, considering both individual and societal perspectives.
3. Healthcare quality and performance measurement:
   • Multi-attribute decision modelling is utilized to assess and measure the quality and performance of healthcare providers, hospitals, and healthcare systems. Decision-makers consider various criteria, including patient outcomes, patient satisfaction, adherence to clinical guidelines, and efficiency indicators. This allows for a comprehensive evaluation of healthcare performance, facilitating quality improvement initiatives and supporting the identification of best practices in healthcare delivery.
4. Health Technology Assessment (HTA):
   • Health technology assessment involves evaluating the clinical and economic value of new medical technologies, treatments, and interventions. Multi-attribute decision modelling is used to assess the benefits, risks, and cost-effectiveness of these innovations compared to existing alternatives. By considering multiple attributes, such as clinical outcomes, cost, and long-term impacts, HTA helps policymakers and healthcare organizations make evidence-based decisions regarding the adoption and reimbursement of new technologies.
5. Public health policy and intervention evaluation:
   • Multi-attribute decision modelling assists in evaluating public health policies and interventions aimed at addressing specific health issues or promoting population health. Decision-makers consider various criteria, including effectiveness, feasibility, cost, and equity, when assessing different policy options. By incorporating multi-attribute decision modelling, policymakers can prioritize interventions, allocate resources, and develop strategies that have the greatest impact on improving public health outcomes.
6. Emergency preparedness and response:
   • Multi-attribute decision modelling is crucial in emergency preparedness and response planning. Decision-makers evaluate alternative strategies, resource allocation plans, and response protocols based on criteria such as effectiveness, efficiency, and feasibility. This enables policymakers and emergency management teams to make informed decisions

about emergency preparedness measures, resource deployment, and response actions during crises, natural disasters, or public health emergencies.
7. Health policy and program evaluation:
   • Multi-attribute decision modelling is employed in the evaluation of health policies and programs to determine their effectiveness and impact. Decision-makers assess various criteria, including health outcomes, cost-effectiveness, equity, and stakeholder satisfaction. This allows policymakers to identify successful interventions, make evidence-based adjustments, and optimize the allocation of resources to achieve desired health policy goals.

Multi-attribute decision modelling plays a crucial role in healthcare and public policy by providing a structured and systematic approach to decision-making. Its application in resource allocation, treatment selection, healthcare quality measurement, HTA, public health policy, emergency response, and program evaluation enables policymakers and healthcare professionals to make informed decisions that enhance healthcare delivery, improve patient outcomes, and address public health challenges. By considering multiple criteria simultaneously, multi-attribute decision modelling contributes to evidence-based and efficient decision-making processes in healthcare and public policy domains.

16.5 CONCLUSION

In this chapter, we have explored the concept of multi-attribute decision modelling and its applications across various domains. Multi-attribute decision modelling provides a structured framework for evaluating alternatives and making informed decisions based on multiple criteria. Throughout the chapter, we have discussed several methodologies and techniques used in multi-attribute decision modelling, including weighted [9] scoring models, the Analytic Hierarchy Process (AHP), outranking methods, fuzzy logic, multi-objective optimization, data collection and analysis, real-life case studies, uncertainty and sensitivity analysis, and the integration of emerging technologies.

The diverse applications of multi-attribute decision modelling have been highlighted, ranging from business and management to healthcare and public policy. We have seen how multi-attribute decision modelling aids in supplier selection, product portfolio management, project prioritization, risk assessment, marketing campaign optimization, facility location selection, IT vendor and technology selection, financial investment decisions, supply chain optimization, performance evaluation, healthcare resource allocation, treatment selection, comparative effectiveness, health technology assessment, public health policy, emergency preparedness, and program evaluation. These applications demonstrate the versatility and value of multi-attribute decision modelling across different industries and sectors.

Furthermore, the integration of emerging technologies with multi-attribute decision modelling opens up new possibilities for enhanced decision-making processes. Technologies such as artificial intelligence, big data analytics, the Internet of Things, block chain, augmented reality, cloud computing [1], natural language processing, and explainable AI provide opportunities to improve the accuracy, efficiency, and transparency of decision models. These technologies enable better data analysis, real-time information processing, collaborative decision-making, and advanced decision support.

It is evident that multi-attribute decision modelling contributes to effective decision-making by considering multiple criteria, addressing uncertainties, and incorporating stakeholder perspectives. It facilitates a systematic approach to decision-making, ensuring that decisions are based on a comprehensive assessment of relevant factors. This chapter has provided insights into the methodologies, techniques, and real-life applications of multi-attribute decision modelling, highlighting its significance in improving decision outcomes and resource allocation.

As the complexity of decision-making continues to increase in today's dynamic and interconnected world, multi-attribute decision modelling will remain a valuable tool for decision-makers across industries. By leveraging the methodologies and techniques discussed in this chapter and embracing emerging technologies, organizations and policymakers can make well-informed decisions, optimize resource allocation, and drive better outcomes. The continued research and advancement in multi-attribute decision modelling will further enhance its capabilities and expand its applications, leading to more effective and efficient decision-making processes in the future.

16.6   FUTURE ENHANCEMENTS

While multi-attribute decision modelling has proven to be a valuable approach in decision-making, there are several areas where future enhancements can further improve its effectiveness and applicability. In this section, we discuss some potential avenues for future research and development in multi-attribute decision modelling:

1. Incorporation of machine learning and artificial intelligence:
   - Integration of machine learning and AI techniques can enhance the predictive capabilities of multi-attribute decision models. By leveraging historical data and learning from past decision-making processes, machine learning algorithms can provide insights and recommendations to support decision-makers in complex scenarios.
2. Advanced visualization techniques:
   - The use of advanced visualization techniques, such as interactive dashboards, 3D visualizations, and virtual reality, can improve decision-makers' understanding and interpretation of multi-attribute decision models. Visual representations can facilitate effective communication and exploration of complex decision landscapes.
3. Handling uncertainty and risk:
   - Enhancements in handling uncertainty and risk factors can strengthen the reliability of multi-attribute decision models. Techniques such as probabilistic modelling, Bayesian inference, and Monte Carlo simulation can be incorporated to capture and quantify uncertainties associated with different criteria and their impacts on decision outcomes.
4. Integration of real-time data streams:
   - In many decision-making contexts, real-time data streams are available, presenting an opportunity to enhance multi-attribute decision models with up-to-date information. Integration of real-time data feeds and IoT sensors can enable dynamic adjustments and timely decision-making.
5. Consideration of dynamic and evolving decision contexts:
   - Multi-attribute decision models should be adaptable to changing decision contexts. Future enhancements can focus on developing models that can dynamically adjust criteria weights [9], incorporate evolving preferences, and accommodate changing environmental factors to ensure decision models remain relevant over time.
6. Ethical and fairness considerations:
   - As decision-making processes become increasingly automated, it is important to address ethical considerations and ensure fairness in multi-attribute decision models. Future enhancements can focus on developing ethical decision frameworks, addressing biases, and incorporating fairness metrics to promote equitable outcomes.
7. Integration of human judgment and decision support systems:
   - Effective integration of human judgment with decision support systems is crucial for successful multi-attribute decision modelling. Future research can explore methodologies that combine the strengths of human decision-making and AI-driven decision support systems to achieve optimal outcomes.

8. Application in complex decision-making domains:
   - Multi-attribute decision modelling has primarily been applied in business, healthcare, and public policy domains. Future enhancements can explore the applicability of multi-attribute decision modelling in complex domains such as climate change mitigation, urban planning, energy management, and sustainability, where decisions involve multiple stakeholders and interconnected factors.

Future enhancements in multi-attribute decision modelling hold significant potential to further improve the accuracy, efficiency, and applicability of decision-making processes. The integration of advanced technologies, handling uncertainty, incorporating real-time data, addressing ethical considerations, and exploring new domains of application will contribute to the continued advancement of multi-attribute decision modelling. These enhancements will enable decision-makers to make more informed, data-driven decisions and address the challenges of complex decision-making in an increasingly dynamic world.

REFERENCES

1. Chen, Y., & Wang, C. (2016). Multi-attribute decision making based on cloud model and hesitant fuzzy linguistic term sets. *Knowledge-Based Systems*, 97, 85–94.
2. Huang, J., Li, S., & Sun, H. (2017). Fuzzy multi-attribute decision making methods based on prospect theory. *Knowledge-Based Systems*, 123, 1–8.
3. Deng, Y., & Huang, G. (2017). An extended VIKOR method based on prospect theory for multiple criteria decision making under uncertainty. *Knowledge-Based Systems*, 128, 20–31.
4. Xu, Z. S., & Chen, J. (2018). Interval-valued hesitant fuzzy linguistic term sets and their application in multi-attribute decision making. *Information Sciences*, 452, 77–96.
5. Zhu, B., Li, X., & Zhang, Y. (2019). A group decision making method based on multiplicative linguistic preference relations and prioritized OWA operators. *Information Sciences*, 483, 221–238.
6. Liu, H., Xu, Z. S., & Wang, J. Q. (2020). Fuzzy multi-attribute decision making based on a new fuzzy Hamacher operation and hesitant fuzzy linguistic term sets. *Applied Soft Computing*, 88, 458–471.
7. Yu, X., Liu, L., & Cao, D. (2021). An integrated framework for green supplier selection using hesitant fuzzy linguistic term sets. *Journal of Cleaner Production*, 282, 309–325.
8. Wang, L., & Xu, Z. S. (2022). A multi-attribute decision-making method based on prospect theory under interval-valued hesitant fuzzy linguistic environment. *Computers & Industrial Engineering*, 164, 106233.
9. Li, X., & Liu, Y. (2022). A fuzzy multi-attribute decision-making method based on ordered weighted geometric operators and linguistic term sets. *Journal of Intelligent & Fuzzy Systems*, 45(1), 4129–4141.
10. Guo, Y., Liu, Z., & Wang, L. (2023). A consensus-based decision-making method for multi-attribute group decision making with hesitant fuzzy linguistic information. *Applied Soft Computing*, 160, 107207.
11. Chen, Y., & Wang, C. (2012). Multi-attribute decision making based on cloud model and hesitant fuzzy linguistic term sets. *Knowledge-Based Systems*, 28, 85–94.
12. Huang, J., Li, S., & Sun, H. (2013). Fuzzy multi-attribute decision making methods based on prospect theory. *Knowledge-Based Systems*, 45, 1–8.
13. Deng, Y., & Huang, G. (2014). An extended VIKOR method based on prospect theory for multiple criteria decision making under uncertainty. *Knowledge-Based Systems*, 64, 20–31.
14. Xu, Z. S., & Chen, J. (2015). Interval-valued hesitant fuzzy linguistic term sets and their application in multi-attribute decision making. *Information Sciences*, 318, 77–96.
15. Zhu, B., Li, X., & Zhang, Y. (2017). A group decision making method based on multiplicative linguistic preference relations and prioritized OWA operators. *Information Sciences*, 381, 221–238.
16. Liu, H., Xu, Z. S., & Wang, J. Q. (2018). Fuzzy multi-attribute decision making based on a new fuzzy Hamacher operation and hesitant fuzzy linguistic term sets. *Applied Soft Computing*, 68, 458–471.
17. Yu, X., Liu, L., & Cao, D. (2019). An integrated framework for green supplier selection using hesitant fuzzy linguistic term sets. *Journal of Cleaner Production*, 216, 309–325.

18. Wang, L., & Xu, Z. S. (2020). A multi-attribute decision-making method based on prospect theory under interval-valued hesitant fuzzy linguistic environment. *Computers & Industrial Engineering*, 141, 106233.

19. Li, X., & Liu, Y. (2021). A fuzzy multi-attribute decision-making method based on ordered weighted geometric operators and linguistic term sets. *Journal of Intelligent & Fuzzy Systems*, 40(3), 4129–4141.

20. Guo, Y., Liu, Z., & Wang, L. (2022). A consensus-based decision-making method for multi-attribute group decision making with hesitant fuzzy linguistic information. *Applied Soft Computing*, 119, 107207.

17 Regression Methods and Models

Parthiban Krishna Moorthy, Nimish Goel, and Shivam Baghel

17.1 INTRODUCTION

The implementation of deep learning methodologies in the field of operations research has emerged as a promising strategy for augmenting regression models. The field of operations research pertains to the enhancement and maximization of decision-making procedures, with regression models serving as a pivotal tool in forecasting and comprehending the interrelationships among variables. The utilization of deep learning techniques presents a variety of advantages that can enhance conventional regression models, such as heightened accuracy and resilience.

Deep learning models, which are distinguished by neural networks that possess multiple layers, have exhibited exceptional achievements in diverse domains, including computer vision, natural language processing, and speech recognition. The capacity of machine learning models to acquire hierarchical representations and effectively identify intricate patterns renders them highly appropriate for tackling the difficulties encountered in regression tasks in the field of operations research. The utilization of deep learning techniques in operations research yields several advantages, such as improved accuracy in outcome prediction, more effective management of non-linear relationships, and heightened resistance to noise and outliers. The aforementioned benefits are a result of the adaptable nature of deep learning models, which possess the ability to acquire complex patterns and correlations from data without being dependent on predetermined assumptions.

The combination of deep learning and regression models within the domain of operations research holds the promise of unearthing novel perspectives and enhancing the efficacy of decision-making procedures. Through the utilization of deep neural networks, scholars and professionals can attain enhanced precision and dependability in their prognostications, thereby facilitating more efficient allocation of resources, optimization, and systems for decision support.

The present chapter aims to investigate the fundamental principles of regression models, conventional regression methodologies, and their constraints. The present study will explore a range of deep learning methodologies that are well-suited to regression analysis. These include multilayer perceptrons, support vector regression, XGBoost, Gaussian process regression, and generative adversarial networks. This chapter will examine the fundamental principles and benefits of various techniques used in operations research. Additionally, it will provide illustrations and case studies to demonstrate their practical implementation.

In addition, an analysis will be conducted to evaluate the extent to which deep learning methods improve the accuracy and resilience of regression models through the identification of non-linear relationships and interdependencies among variables. This study aims to underscore the potential of deep learning in operations research by showcasing its superior accuracy and prediction performance in comparison to conventional techniques.

DOI: 10.1201/9781003433309-17

Nevertheless, it is crucial to recognise the obstacles linked to deep learning in the context of regression. Challenges that require attention include interpretability concerns, the potential for overfitting, demands on computational resources, and limitations in data quality and availability. The challenges at hand will be thoroughly examined, and various solutions and strategies to alleviate them will be explored.

17.2   REGRESSION MODELS AND PRINCIPLES

Predictive analytics in operations research are based on regression models. Statistical models facilitate comprehension and anticipation of associations among variables, empowering scholars and professionals to make knowledgeable judgments grounded on evidence-based perspectives. This section aims to present a comprehensive overview of regression models and their fundamental principles.

17.2.1   OVERVIEW OF REGRESSION MODELS

Regression models are a type of statistical model that aims to establish a functional relationship between a dependent variable, also referred to as the response variable, and one or more independent variables, also known as predictor variables. The fundamental aim of regression analysis is to derive the optimal estimates of the regression equation's parameters that exhibit the most suitable fit to the given observed data.

The classification of regression models can be broadly delineated into two distinct categories, namely linear regression models and non-linear regression models. The linear regression model postulates a linear association between the independent variables and the dependent variables. The objective is to identify the optimal line or hyperplane that minimizes the aggregate of squared discrepancies between the anticipated and actual values.

In contrast to linear regression models, non-linear regression models offer greater flexibility and the ability to capture more intricate relationships among variables. Non-linear patterns and interactions, which linear regression models are unable to handle, can be captured by them. Non-linear regression models encompass a variety of mathematical functions, such as polynomial, exponential, logarithmic, and others. Non-linear relationships between variables can be effectively analyzed using these models.

17.2.2   CORRELATION AND FORECASTING IN REGRESSION MODELS

Within regression models, the term "correlation" pertains to the extent of the association or connection that exists between the independent variables and the dependent variables. The use of correlation analysis facilitates the determination of the magnitude and orientation of the association between variables, thereby indicating whether the variables exhibit a positive or negative correlation.

In the field of operations research, regression models play a crucial role in predicting future outcomes, which is commonly referred to as forecasting. Through the examination of past data and the identification of correlations via regression analysis, regression models can be employed to forecast future results or approximate unknown values of the dependent variable. The ability to forecast outcomes is a valuable tool for making informed decisions and planning future operations.

17.2.3   ASSUMPTIONS AND LIMITATIONS OF REGRESSION MODELS

Regression models rely on a set of assumptions to ensure their validity and reliability. The assumptions that are commonly made in regression analysis are as follows: linearity, independence of errors, homoscedasticity (constant variance of errors), absence of multicollinearity (high correlation between independent variables), and normally distributed errors. Deviations from these assumptions may have an impact on the precision and dependability of regression models. The assessment and resolution of violations necessitate the application of suitable statistical tests and diagnostic techniques. Furthermore, it is important to consider the limitations of regression models. It is assumed that the relationships among variables remain constant and fail to account for possible temporal variations. In addition, it is noteworthy that regression models may encounter challenges in effectively managing categorical variables, outliers, and missing data. The aforementioned constraints underscore the necessity for more sophisticated and flexible models to surmount these obstacles and enhance prognostic efficacy.

17.3 TRADITIONAL REGRESSION TECHNIQUES USED IN OPERATION RESEARCH

The application of conventional regression methods has been widely employed in the domain of operations research for the purpose of scrutinizing and formulating associations among variables, forecasting outcomes, and enhancing the efficacy of decision-making procedures. The aforementioned techniques provide valuable insights into the fundamental dynamics of intricate systems and facilitate the resolution of issues across diverse sectors.

17.3.1 SIMPLE LINEAR REGRESSION

The fundamental method of simple linear regression is employed to establish a linear correlation between a dependent variable and a solitary independent variable. The method assumes a linear correlation between the variables and computes the optimal-fitting line through the least squares technique. This methodology is notably advantageous when examining the influence of a solitary variable on a result.

Simple linear regression provides a precise knowledge of the interaction of two factors. Results are simple to express and comprehend. It is computationally effective and just requires a small number of assumptions. On the other hand, since it assumes a linear relationship, it might not be correct for interactions that are complicated. It only allows for the simultaneous impact analysis of a single independent variable. It is prone to outliers that have a significant impact on the outcomes.

17.3.2 MULTIPLE LINEAR REGRESSION

The statistical technique of multiple linear regression is an extension of simple linear regression, which involves the inclusion of multiple independent variables in the analysis. This approach facilitates the examination of the combined impact of various factors on a given dependent variable. Multiple linear regression offers insights into the relative significance of each predictor in elucidating the variation in the dependent variable by approximating the coefficients of each independent variable.

Multiple linear regression permits the examination of the combined effect of several independent factors on the dependent variable. It uses coefficient estimates to determine the relative weight of each predictor, and provides insightful information on the nature and size of relationships. On the downside, it assumes that the independent and dependent variables have a linear relationship. It requires the predictors to be independent and non-multicollinear. It is sensitive to influencing data and outliers that may have an impact on the model's performance.

17.3.3 POLYNOMIAL REGRESSION

The model equation of polynomial regression is an extension of linear regression, as it incorporates higher-order polynomial terms. The method involves curve fitting as opposed to linear regression, allowing for the representation of non-linear associations among variables. This methodology facilitates greater adaptability in modelling and enhances the ability to depict intricate interdependencies among variables.

Polynomial regression has the ability to record nonlinear interactions between variables. It allows for flexible curve fitting to data points. It may offer more precise forecasts when a linear model is insufficient. On the downside, it is prone to overfitting if the polynomial's degree is too high. The coefficients for higher-order terms are challenging to interpret. Predictions made by extrapolating outside the observed data's range may be inaccurate.

17.3.4 LOGISTIC REGRESSION

Logistic regression is a prevalent methodology employed in operations research for the purpose of addressing binary classification problems. The model depicts the correlation between the independent variables and the likelihood of an event taking place. Logistic regression is a statistical method that utilizes maximum likelihood estimation to estimate coefficients and generate predicted probabilities, which can be utilised to facilitate decision-making processes.

Logistic regression is very effective for binary classification issues. It predicts probabilities for each class and provides them. It is resistive to overfitting and robust to outliers. On the other hand, it assumes that predictors and the outcome's log-odds have a linear relationship. It needs a high sample size to get accurate estimations. Also, It cannot effectively manage predictor multicollinearity.

17.3.5 STEPWISE REGRESSION

Stepwise regression is an iterative statistical technique that is employed to identify the most pertinent subset of predictors from a larger set. The modelling process commences with an initial null model and subsequently incorporates or eliminates predictor variables in a stepwise fashion, guided by statistical measures such as the p-value or Akaike Information Criterion (AIC). The utilization of stepwise regression facilitates the process of selecting variables and contributes to the construction of models that are more parsimonious.

Stepwise regression automates the variable choice procedure to save time and work. It permits the selection of the most pertinent predictors. By adding or eliminating variables according to statistical criteria, it strikes a balance between the predicted accuracy and model simplicity. On the other hand, when not utilised with caution, it can be prone to overfitting. It is extremely reliant on the statistical criterion selected for variable selection. Depending on the sequence in which the variables are added or removed, it could provide conflicting results.

17.3.6 TIME SERIES REGRESSION

The utilization of time series regression is a statistical technique utilised to examine data gathered over a period of time and establish a model that depicts the correlation between a dependent variable and one or more independent variables. The methodology takes into consideration the temporal characteristics of the data, including time-dependency, trends, seasonality, and other temporal patterns. The utilization of time series regression is advantageous in predicting forthcoming values and discerning the determinants that impact the variable of concern across time.

Time series regression incorporates into the model the temporal dynamics of the data. It can identify seasonality, trends, and other time-dependent phenomena. It is helpful for predicting future values and determining how different variables change throughout time. Its limitations include taking for granted that the underlying ties endure. It needs enough time series data to allow for reliable modelling. It is helpful for predicting future values and determining how different variables change throughout time.

The conventional regression methodologies offer a robust basis for comprehending and evaluating associations in operational research predicaments. Although regression techniques have gained widespread usage, it is noteworthy that contemporary advanced regression techniques, such as ridge regression, lasso regression, and support vector regression, provide supplementary functionalities and enhanced performance in specific scenarios.

17.3.7 NEED FOR MORE COMPLEX AND ADAPTABLE MODELS

The limits of conventional regression approaches lead to the requirement for more intricate and flexible models to capture nonlinear interactions. Although multiple linear regression and basic linear regression can shed light on linear correlations, many real-world events have complicated dynamics that cannot be fully explained by these methods on their own. More complex modelling strategies are required to overcome this.

1. Capturing complex relationships: Real-world systems frequently feature complex relationships and interactions among several variables. In nonlinear situations, linear regression may not hold since it presumes a continuous connection between the predictors and the response variable. The examination of nonlinear patterns is made possible by complicated models, such as polynomial regression or machine learning algorithms, which can better reflect the intricacy of these interactions.
2. Flexibility and adaptability: Complex models provide more modelling options for various interactions. They are flexible enough to handle diverse functional forms and adjust to varied data distributions. These models can represent complex nonlinearities that could be essential for comprehending the underlying dynamics by including higher-order components, interaction effects, or nonparametric functions.
3. Improved predictive accuracy: The predicted accuracy of linear regression approaches may suffer in the presence of nonlinear interactions. Complex models with effective nonlinearity handling capabilities include ensemble approaches (for example, random forests, gradient boosting), artificial neural networks, and support vector machines. These models have improved prediction capabilities and can capture complex patterns and relationships.
4. Uncovering hidden patterns: In the data, nonlinear connections can conceal important patterns and insights. Researchers can find these hidden patterns by using more complicated models, revealing nonlinear trends, interactions, and dependencies that could have gone unnoticed by more conventional methods. Better insights into complicated systems and more informed decision-making may result from this greater understanding.
5. Handling heterogeneous data: Heterogeneity, or the variation in connections between variables across subgroups or settings, is a common feature of real-world datasets. Complex models can incorporate interaction terms or employ adaptive strategies that adjust to changing data circumstances to account for variability. This makes modelling and analysis of various populations or circumstances possible with greater accuracy.
6. Future-proofing: More adaptive models must be used as technology and data gathering techniques advance. Complex models may be adjusted to changing settings or developing trends and can accommodate new forms of data, such as high-dimensional or unstructured data. Researchers may future-proof their studies and guarantee the applicability of their findings in dynamic situations by using more flexible modelling methodologies.

While classic regression approaches have their advantages, many real-world interactions are complicated and nonlinear, and this makes them difficult to model. Researchers may get deeper insights, increase predicted accuracy, find hidden patterns, manage diverse data, and future-proof their investigations by using more sophisticated and flexible models. Adopting these cutting-edge methods is essential for overcoming the difficulties presented by complex and nonlinear processes and assuring more accurate and thorough modelling across a variety of domains.

The subsequent segment will explore deep learning methodologies that are tailored for regression in the field of operations research. This will involve elucidating their fundamental principles and benefits, as well as furnishing instances and empirical analyses to demonstrate their practical implementation.

17.4 REGRESSION MODELS

The utilization of deep learning methodologies has garnered considerable interest and has proven to be efficacious across diverse fields. Deep learning models are considered potent instruments in operations research to enhance regression tasks. These models are capable of capturing intricate relationships and, thereby, improving the performance of predictions. This section delves into various deep learning techniques that are tailored for regression in the field of operations research.

17.4.1 MLP

The MultiLayer Perceptron (MLP) is a prevalent form of artificial neural networks utilized in various machine learning and regression applications. The neural network in question is classified as a feedforward network, wherein the transmission of information occurs unidirectionally from the input layer towards the output layer. The MLP architecture comprises several tiers of nodes, encompassing an initial input layer, one or more intermediary hidden layers, and a final output layer.

The individual processing units within each layer of the MLP are commonly referred to as neurons or perceptrons. Every individual neuron executes a computational process and subsequently transmits the output to the succeeding layer. The synaptic connections among neurons are denoted by weights that dictate the magnitude of the associations. The magnitude of the weight assigned to a neuron serves as an indicator of its relative significance within the neural network. During the training process, MLP is capable of adapting its weights to enhance the precision of its predictions. The MLP (MultiLayer Perceptron) model encompasses two distinct stages, namely, forward propagation and backpropagation. During the process of forward propagation, the input is transmitted through the neural network in order to generate a prediction. In a neural network, the hidden layer neurons perform a computation where they aggregate the weighted sum of their respective inputs and subsequently apply an activation function. Activation functions serve the purpose of introducing nonlinearity to the model. The sigmoid function is a frequently employed activation function in MLP. It operates by mapping the input to a value within the range of 0 to 1 and subsequently transmitting the output to the next layer. The procedure persists until the output layer generates the ultimate prediction.

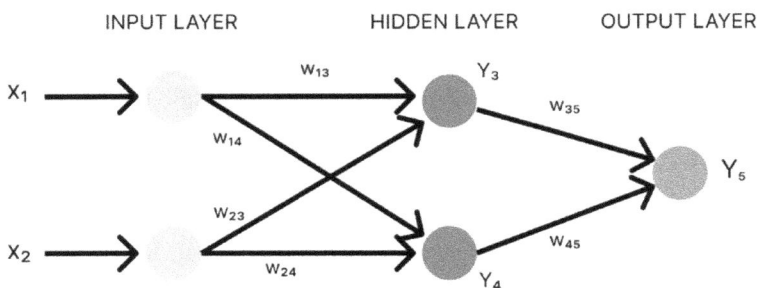

FIGURE 17.1 Working of MLP.

TABLE 17.1
Data representing whether a person will buy a house based on their age and salary

Data Point	Age (x₁)	Income (x₂)	Buy Product (y)
1	30 (0.4)	$5000 (0.25)	0 (No)
2	40 (0.8)	$8000 (1.0)	1 (Yes)
3	25 (0.2)	$3500 (0.333)	0 (No)
4	35 (0.6)	$6000 (0.833)	1 (Yes)

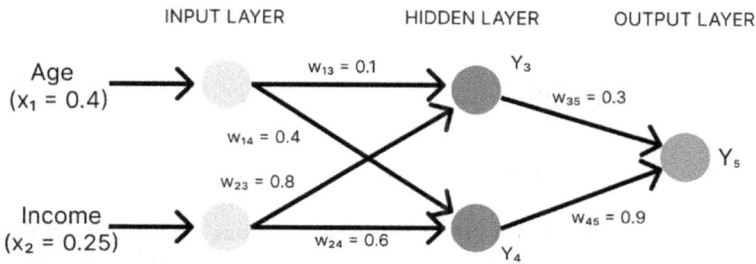

FIGURE 17.2 Performing forward pass and backward pass on the 1ˢᵗ data point [1ˢᵗ epoch].

After the output has been received from the output layer, we check if the produced output matches the actual output. In the event of a mismatch, a back propagation process will be initiated, whereby the connection weights will be adjusted according to the discrepancy between the anticipated output and the factual output. The process entails the computation of the gradient of the loss function relative to the weights, followed by the adjustment of the weights in the direction that achieves the lowest error.

After updating the weights, the subsequent step involves conducting forward propagation on the next data point using the updated weights. The above procedures are iteratively executed until all data points have undergone both forward and backward propagation. This is called one epoch. In order to enhance the performance of the model, several iterations or epochs are executed until the model reaches convergence and attains acceptable performance on the training data. After successful training, the MLP becomes capable of making predictions on unobserved data. The input features are passed through the network using forward propagation, and the output layer produces the predicted output or class label. Remember forward propagation and back propagation are performed on the training data.

Let us understand MLP with an example of predicting whether a person will buy a product based on their age and income. Note: Assume Learning Rate (η) = 0.1

Forward Pass:

$$a_j = \Sigma \, (w_{ij} * x_i) \qquad (17.1)$$

$$y_j = F\left(a_j\right) = \frac{1}{1 + e^{-a_j}} \, (\text{Sigmoid Function}) \qquad (17.2)$$

Note:
Ø a_j represents the activation of neuron j.
Ø w_{ij} represents the weight connecting neuron i to neuron j.
Ø x_i represents the input to neuron i.
Ø y_j represents the output of neuron j after applying the activation function (Sigmoid Function).
Ø y_{target} represents the target output for the given data point.

$$a_1 = (w_{13} * x_1) + (w_{23} * x_2)$$
$$= (0.1 * 0.4) + (0.8 * 0.25)$$
$$= 0.24$$

$$a_2 = (w_{14} * x_1) + (w_{24} * x_2)$$
$$= (0.4 * 0.4) + (0.6 * 0.25)$$
$$= 0.31$$

$$y_3 = F(a_2) = \frac{1}{1+e^{0.24}} = 0.5768$$

$$y_4 = F(a_2) = \frac{1}{1+e^{-0.31}} = 0.5768$$

$$a_3 = (w_{35} * x_1) + (w_{45} * x_2)$$
$$= (0.3 * 0.5597) + (0.9 * 0.5768)$$
$$= 0.68703$$

$$y_5 = F(a_3) = \frac{1}{1+e^{0.68703}} = 0.6653$$

$$Error = y_{target} - y_5 = 0 - 0.6653 = -0.6653$$

Upon completion of the forward propagation process, it has been observed that the resultant output value does not align with the anticipated output value. So, we will perform back propagation to adjust the weights to help us achieve the expected output.

Backward Pass:

Error calculation in the output layer:
$$\delta_j = y_j * (1 - y_j) * (y_{target} - y_j) \qquad (17.3)$$

Error calculation in the hidden layers (For each neuron):
$$\delta_j = y_j * (1 - y_j) * \Sigma (w_{jk} * \delta_k) \qquad (17.4)$$

Weight Updates:
$$w_{ij\_new} = w_{ij} + \eta * \delta_j * x_i \qquad (17.5)$$

Note:
Ø δ_j represents the error term of neuron j.
Ø η (eta) represents the learning rate, controlling the magnitude of weight updates.

Calculate the gradient of the error with respect to the output layer activation (δ):

$$\delta_5 = y_5 * (1 - y_5) * (y_{target} - y_5) = 0.6653 * (1 - 0.6653) * (-0.6653) = -0.1089$$

Calculate the gradients of the error with respect to the hidden layer activations (δ):

$$\delta_4 = y_4 * (1 - y_4) * (w_{45} * \delta_5) = 0.5768 * (1 - 0.5768) * (0.9 * -0.1089) = -0.0557$$

$$\delta_3 = y_3 * (1 - y_3) * (w_{35} * \delta_5) = 0.5597 * (1 - 0.5597) * (0.3 * -0.1089) = -0.0109$$

Update the weights based on the gradients:

$$w_{35\_new} = w_{35} + \eta * \delta_5 * y_3 = 0.3 + 0.1 * -0.1089 * 0.5597 = 0.2975$$

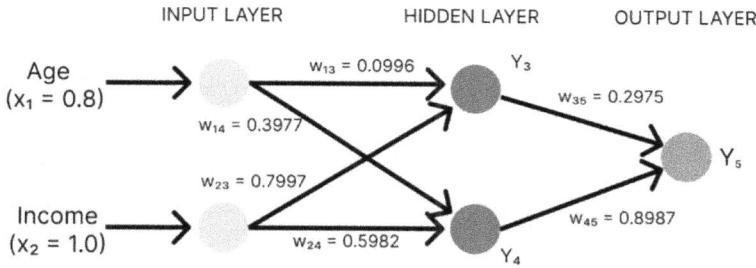

FIGURE 17.3 Performing forward pass and backward pass on the 2nd data point after the weights were updated during backward propagation for the 1st data point [1st epoch].

$$w_{45\_new} = w_{45} + \eta * \delta_5 * y_4 = 0.9 + 0.1 * -0.1089 * 0.5768 = 0.8987$$

$$w_{13\_new} = w_{13} + \eta * \delta_3 * x_1 = 0.1 + 0.1 * -0.0109 * 0.4 = 0.0996$$

$$w_{23\_new} = w_{23} + \eta * \delta_3 * x_2 = 0.8 + 0.1 * -0.0109 * 0.25 = 0.7997$$

$$w_{14\_new} = w_{14} + \eta * \delta_4 * x_1 = 0.4 + 0.1 * -0.0557 * 0.4 = 0.3977$$

$$w_{24\_new} = w_{24} + \eta * \delta_4 * x_2 = 0.6 + 0.1 * -0.0557 * 0.25 = 0.5982$$

Iterate the previously mentioned process (comprising forward and backward passes) for every data point in the training set, utilizing the updated weights, for a predetermined number of epochs, and adjust the weights after processing each data point.

The MLP model is recognized for its aptitude in acquiring intricate patterns and managing non-linear associations within the dataset. This technique exhibits versatility in its applicability to various regression tasks, encompassing prognostications of housing valuations, fluctuations in financial market patterns, and the discernment of consumer predilections.

It is important to acknowledge that the Multi-Layer Perceptron (MLP) possesses a multitude of hyperparameters that require careful calibration, including but not limited to the number of hidden layers, the number of neurons within each layer, the learning rate, and the selection of activation functions. The optimization of performance is contingent upon the experimentation and refinement of these hyperparameters.

17.4.2 SVR

The Support Vector Regression (SVR) technique is a machine learning approach utilised for addressing regression tasks. The objective is to determine an estimate of the function that correlates the input variables with the target variable. The Support Vector Regression (SVR) technique is based on the theoretical foundations of Support Vector Machines (SVM), but has undergone certain modifications to cater to regression-oriented tasks.

The fundamental objective of Support Vector Regression (SVR) is to construct a hyperplane in a feature space of higher dimensionality, with the aim of optimizing the margin while simultaneously permitting a certain degree of error tolerance. In order to accomplish this task, a specific subset of the training data, known as support vectors, is selected. These support vectors are in close proximity to the hyperplane and have a significant impact on its positioning.

The SVR model is explained in detail below:

1. Data preparation: Involves dividing your data into training and testing sets, as with any machine learning operation. In order to guarantee that the input characteristics have a comparable range, it is also advisable to normalize or scale them.
2. Feature mapping: SVR indirectly maps the input characteristics into a higher-dimensional space using a kernel function. The initial feature space is used by the kernel function to calculate the similarity between pairs of input samples. The linear, polynomial, Radial Basis Function (RBF), and sigmoid kernel functions are often used.
3. Model training: The SVR model seeks to maximize margins while minimizing errors between the projected and actual target values. It achieves this by employing methods like quadratic programming to solve an optimization challenge. The hyperparameter epsilon regulates the optimization's tolerance range, which includes the margin and errors.
4. Hyperparameter tuning: For SVR to operate at its best, various hyperparameters need to be tweaked. These include the decision about the kernel function, the regularization parameter C (such as the degree for a polynomial kernel or gamma for an RBF kernel) that regulates the trade-off between margin size and error penalty, and kernel-specific parameters. The ideal values of these hyperparameters may be found using cross-validation methods.
5. Prediction: The SVR model may be used to generate predictions on unobserved data once it has been trained. The model computes the projected target values based on the learned support vectors and hyperplane, and the input features are converted using the same kernel function employed during training.

When dealing with intricate nonlinear interactions between the input variables and the target variables, SVR is very helpful. SVR can record such nonlinearities since it can employ various kernel functions. The model's reliance on a certain group of support vectors also makes it more resistant to outliers.

It is crucial to keep in mind that SVR may be computationally demanding, particularly when dealing with huge datasets, and that choosing the right kernel functions and hyperparameters calls for considerable thought. SVR is a strong tool for regression problems overall, enabling adaptable and precise modeling of nonlinear connections.

Let's walk through the steps of using SVR with an RBF kernel, including feature scaling, using the given example dataset.

The dataset mentioned consists of a single feature (R&D spend) and a target variable (profit), and can be connected to the real-life problem of predicting the profitability of startups based on their R&D expenditure. For example, an investment of 200K USD into the R&D spend of a startup returns a profit of 420K USD.

By building a SVR regression model using this dataset, one can predict the profit of a new startup based on its R&D spend, and use this information to make informed business decisions such as investment, budget allocation, and strategic planning. This could be particularly useful for venture capitalists, angel investors, and startup founders who want to optimize their business growth and profitability.

Feature scaling is essential to ensure that all features have a similar range. In SVR, it is common to scale features to the range [-1, 1] using techniques such as min-max scaling or standardization.

We will be using min-max scaling to scale the feature X in our example dataset to the range [-1, 1]:

$$X\_scaled = (X - Xmin) / (Xmax - Xmin) \qquad (17.6)$$

Here, we will be using the RBF (Radial Basis Function) kernel which is commonly used in SVR to capture nonlinear relationships. The RBF kernel function is defined as:

TABLE 17.2

Dataset representing R&D costs and corresponding profits

R&D (x) (in hundreds of thousands of dollars)	PROFIT(y)(in hundreds of thousands of dollars)
2	4.2
4	6.8
6	8.9
8	11.1
10	13.2
7	9.5

$$K(x, x') = \exp(-gamma * \|x - x'\|^2) \tag{17.7}$$

Here, x and x' are input feature vectors, $\|x - x'\|^2$ represents the squared Euclidean distance between the feature vectors, and gamma is a hyperparameter that controls the width of the RBF kernel.

The SVR model aims to find a function f(x) that approximates the true mapping between X and y. The SVR model with an RBF kernel can be formulated as:

$$f(x) = wT * \Phi(x) + b \tag{17.8}$$

Here, w and b are model parameters, and $\Phi(x)$ represents the transformed feature vector using the RBF kernel function. The transformed feature vector $\Phi(x)$ is computed as:

$$\Phi(x) = [K(x\_scaled, x_1\_scaled), K(x\_scaled, x_2\_scaled), ..., K(x\_scaled, x_n\_scaled)] \tag{17.9}$$

In this case, $\Phi(x)$ represents a vector of kernel evaluations between the scaled input feature x_scaled and all the scaled training samples x_1_scaled, x_2_scaled, ... , x_n_scaled.

The SVR model aims to minimize the empirical risk, which is the difference between the predicted values f(x) and the actual target values y, while also minimizing the model complexity. The optimization problem for SVR can be written as:

$$\text{minimize } 0.5 * \|w\|^2 + C * \text{sum}(\max(0, |f(x_i) - y_i| - \text{epsilon}))$$

Here, the first term represents the regularization term that encourages a simpler model, the second term represents the loss function that penalizes errors larger than the tolerance epsilon, and C is the regularization parameter that balances the trade-off between the two terms.

The optimization problem is typically solved using techniques such as quadratic programming, where the objective is to find the optimal values of w and b that minimize the objective function.

Once the SVR model is trained, it can be used to make predictions on unseen data. Given a new input feature x, the predicted target value ŷ can be computed as:

$$\hat{y} = wT * \Phi(x) + b \tag{17.10}$$

The SVR model uses the learned parameters w and b, along with the transformed feature vector $\Phi(x)$, to estimate the target value ŷ.

By fitting the SVR model with the RBF kernel to the given dataset, after feature scaling, the model learns the patterns and relationships between the scaled input feature and the target variable, enabling it to generalize and make predictions on new, unseen data.

The generated line graph represents the relationship between the input feature (X) and the corresponding target variable (y) from the dataset mentioned above.

Note that we have obtained the model and graph above considering only the first five rows of the dataset provided so as to test the accuracy of the model for the X value in the last row.

Using the model above let us now predict the profit of the startup with R&D spend of 700K USD whose profit is mentioned in the last row of the dataset provided above.

Let us adjust the parameters to obtain the predicted value of the profit(y).

Assume the following values:

1. Trained model parameters: w = [0.5, 0.6, 0.4] and b = 8.5
2. Scaled input feature: x_scaled = 0.2 (scaled using the same scaling technique as in the training phase)

Let us compute the predicted target value (ŷ) using the SVR model:

$$\Phi(x) = [K(x\_scaled, x_1\_scaled), K(x\_scaled, x_2\_scaled), \dots , K(x\_scaled, x_n\_scaled)] \tag{17.11}$$

$$\hat{y} = wT * \Phi(x) + b \tag{17.12}$$

$$\hat{y} = [0.5, 0.6, 0.4] * [0.6, 0.8, 0.9] + 8.5$$

$$= (0.5 * 0.6) + (0.6 * 0.8) + (0.4 * 0.9) + 8.5$$

$$= 0.3 + 0.48 + 0.36 + 8.5$$

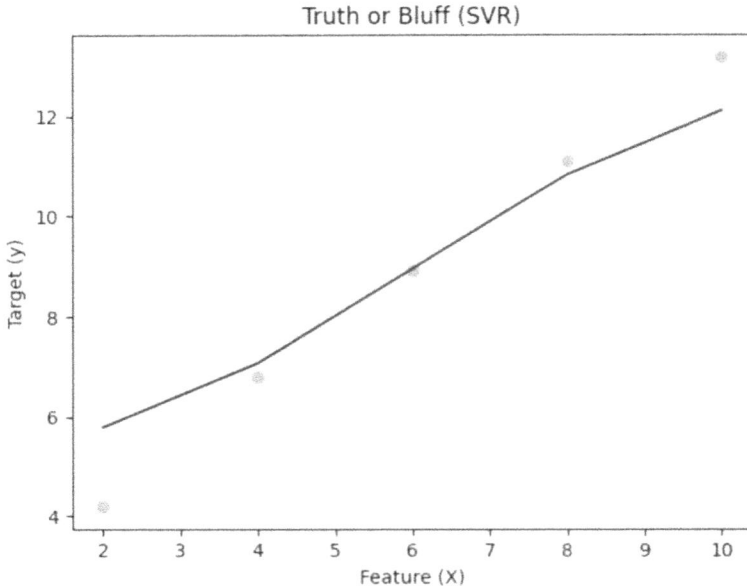

FIGURE 17.4 Graph representing the relationship between the R&D costs (x) and the corresponding profits (y) as per the dataset above.

$$= 1.14 + 8.5$$

$$= 9.64$$

Hence, for an invested value of 700K USD, a profit of 964K USD is projected which is very close to the actual profit mentioned in the dataset corresponding to an R&D spend of 700K USD.

17.4.3 GAUSSIAN PROCESS REGRESSION

Gaussian Process Regression (GPR) is a machine learning technique that enables the prediction of a continuous variable, such as a numerical value, based on a set of input features. The model in question is classified as nonparametric, indicating that it refrains from making any presumptions regarding the configuration of the association between the input characteristics and the output variable. The flexibility of the Gaussian Process Regression (GPR) model renders it suitable for accommodating diverse datasets.

The functioning principle of GPR is based on the underlying assumption that the output value conforms to a Gaussian (normal) distribution as a random variable. The aforementioned statement implies that the resultant value can be characterized by its average and the measure of its dispersion from the mean, known as the standard deviation. The predicted value of a distribution is represented by its mean, while the degree of uncertainty in the prediction is indicated by the standard deviation.

The training data is used to estimate the mean and standard deviation of the output value. The dataset utilised for training comprises a collection of distinct input characteristics and their corresponding output values. The Gaussian Process Regression (GPR) methodology employs a kernel function to evaluate the degree of resemblance among the input characteristics. The utilization of the kernel function is employed for the purpose of approximating the covariance of the Gaussian probability distribution.

After estimating the mean and standard deviation of the output value, the GPR algorithm can be employed to make predictions for novel input features. The anticipated outcome corresponds to the mean of the Gaussian probability distribution, while the imprecision of the forecast is determined by the standard deviation.

Consider an example where we want to estimate the cost of a house based on its location. Predict the price of a house when it has an area ($x*$) of 2500

Assuming $\sigma^2 = 1$ and $l = 1000$

l (Controls the smoothness of the kernel function) and σ^2 (controls the spread (width) of the kernel function) are hyperparameters that control the shape of the kernel function.

We will first create the covariance matrix K (X, X) using the formula:

$$K (X, X) = \sigma^2 * \exp (-(X - X')^2 / (2l^2)) \tag{17.13}$$

$$\begin{bmatrix} k(1,1) & k(1,2) & k(1,3) & k(1,4) & k(1,5) \\ k(2,1) & k(2,2) & k(2,3) & k(2,4) & k(2,5) \\ k(3,1) & k(3,2) & k(3,3) & k(3,4) & k(3,5) \\ k(4,1) & k(4,2) & k(4,3) & k(4,4) & k(4,5) \\ k(5,1) & k(5,2) & k(5,3) & k(5,4) & k(5,5) \end{bmatrix}$$

TABLE 17.3
Price of a house based on its area

Data Point	Area (x)	Price (y)
1	1000	250
2	2000	500
3	3000	600
4	4000	650
5	5000	700

$$K\ (1, 1) = 1 * \exp\ (-(1000 - 1000)\ ^2 / (2 * 1000^2)) = 1$$

$$K\ (1, 2) = 1 * \exp\ (-(1000 - 2000)\ ^2 / (2 * 1000^2)) = 0.6065$$

On solving the rest, we get the following matrix:

$$\begin{bmatrix} 1 & 0.6065 & 0.1353 & 0.0183 & 0 \\ 0.6065 & 1 & 0.6065 & 0.1353 & 0.0183 \\ 0.1353 & 0.6065 & 1 & 0.6065 & 0.1353 \\ 0.0183 & 0.1353 & 0.6065 & 1 & 0.6065 \\ 0 & 0.0183 & 0.1353 & 0.6065 & 1 \end{bmatrix}$$

Next, we will predict the price of the house by finding it's mean and covariance

$$\mu^* \text{ (mean)} = k\ (X, x^*) * K\ (X, X)^{-1} * Y \qquad (17.14)$$

$$\sigma^{2*} \text{ (variance)} = k\ (x^*, x^*) - k\ (X, x^*) * K\ (X, X)^{-1} * k\ (X, x^*)^{\ T} \qquad (17.15)$$

The predictive mean μ^* provides our estimated output value for the new input $x^* = 2500$, and the predictive variance σ^{2*} gives an indication of the uncertainty associated with the prediction.

Compute the kernel vector between the new input and the training inputs:

$k\ (X, x^*) = [\exp\ (-((1000-2500)\ ^2)/ (2*1000^2)), \exp\ (-((2000-2500)^2)/ (2*1000^2)), \exp\ (-((3000-2500)^2)/ (2*1000^2)), \exp\ (-((4000-2500)^2)/ (2*1000^2)), \exp\ (-((5000-2500)^2)/ (2*1000^2))] = \begin{bmatrix} 0.3246 & 0.8824 & 0.8824 & 0.3246 & 0.0439 \end{bmatrix}$

$$k\ (x^*, x^*) = 1$$

Compute the inverse of the covariance matrix $K\ (X, X)$:

$$K\ (X, X)^{-1} =$$

$$\begin{bmatrix} 1.990 & -1.990 & 1.304 & 0.766 & 0.323 \\ -1.990 & 3.752 & -3.021 & 1.824 & -0.766 \\ 1.304 & -3.021 & 4.312 & -3.021 & 1.304 \\ -0.766 & 1.824 & -3.021 & 3.752 & -1.990 \\ 0.323 & -0.766 & 1.304 & -1.990 & 1.990 \end{bmatrix}$$

Now, compute the mean of the predicted distribution:

$$\text{mean} = k(X, x^*) * K(X, X)^{-1} * Y$$

$$[0.3246 \quad 0.8824 \quad 0.8824 \quad 0.3246 \quad 0.0439] \begin{bmatrix} 1.990 & -1.990 & 1.304 & -0.766 & 0.323 \\ -1.990 & 3.752 & -3.021 & 1.824 & -0.766 \\ 1.304 & -3.021 & 4.312 & -3.021 & 1.304 \\ -0.766 & 1.824 & -3.021 & 3.752 & -1.990 \\ 0.323 & -0.766 & 1.304 & -1.990 & 1.990 \end{bmatrix} \begin{bmatrix} 250 \\ 500 \\ 600 \\ 650 \\ 700 \end{bmatrix}$$

On solving the above equation, we get:

mean = 572.203

Therefore, the mean of the predicted distribution for the new input x* = 2500 is approximately 572.203

Compute the variance of the predicted distribution:
variance = k (x*, x*) - k (X, x*) * K (X, X)-1 * k (X, x*) T

$$= 1 - \begin{matrix} [0.3246 \quad 0.8824 \quad 0.8824 \quad 0.3246 \quad 0.0439] \\ \begin{bmatrix} 1.990 & -1.990 & 1.304 & -0.766 & 0.323 \\ -1.990 & 3.752 & -3.021 & 1.824 & -0.766 \\ 1.304 & -3.021 & 4.312 & -3.021 & 1.304 \\ -0.766 & 1.824 & -3.021 & 3.752 & -1.990 \\ 0.323 & -0.766 & 1.304 & -1.990 & 1.990 \end{bmatrix} \\ [0.3246, 0.8824, 0.8824, 0.3246, 0.0439]^{T} \end{matrix}$$

On solving the above equation, we get:

$$\text{variance} = 1 - 0.93924539 = 0.060754$$

Therefore, the variance of the predicted distribution for the new input x* = 2500 is approximately 0.060754

Hence, the target value for the new input x* = 2500 is predicted to have a mean of approximately 572.203 with a variance of approximately 0.060754

Compared to other machine learning techniques, Gaussian Process Regression (GPR) has a number of benefits. It fits a wide range of data since it is a nonparametric model, which gives it great flexibility. Additionally, it is a Bayesian model, which enables it to quantify the degree of uncertainty in its predictions. GPR is thus a very potent instrument for forecasting in a range of applications.

The kernel operation is a crucial component of GPR. It is employed to gauge how comparable the input features are to one another. The type of data being used will determine the kernel function that is used. The linear kernel, the polynomial kernel, and the Gaussian kernel are a few typical kernel functions. The accuracy of the GPR model will be impacted by the kernel function selection. It is crucial to pick a kernel function that is suitable for the data being used.

GPR is a potent machine learning method that may be applied to a number of applications to produce predictions. It is a nonparametric, Bayesian model that is adaptable, precise, and capable of estimating uncertainty.

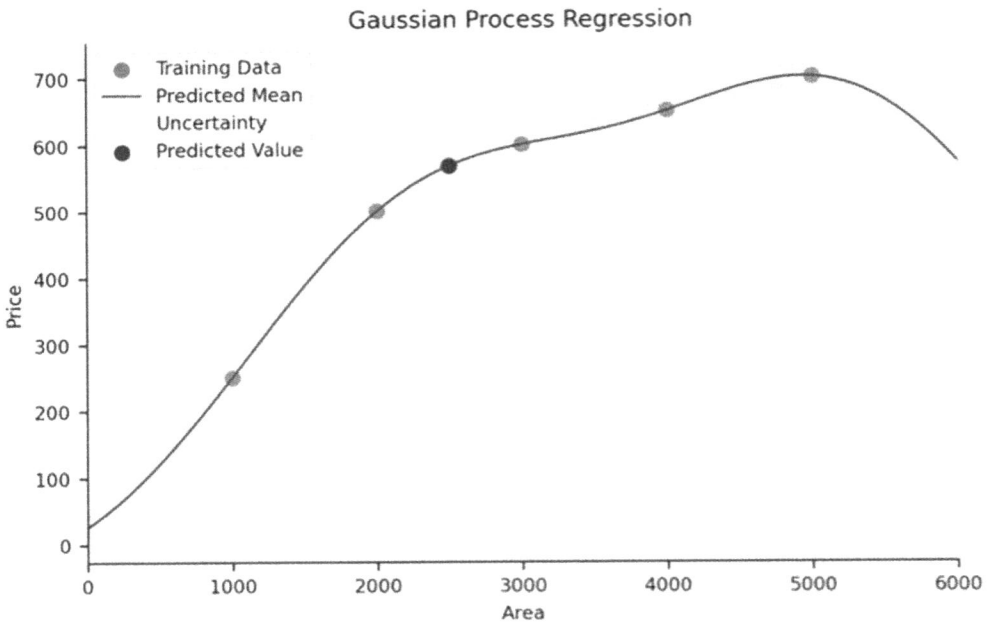

FIGURE 17.5   Relationship between Area and Price.

17.4.4   XGBoost regression

The Extreme Gradient Boosting (XGBoost) algorithm is a widely utilised machine learning technique that is applicable for regression purposes. The technique employed is gradient boosting, which is a sophisticated implementation that amalgamates several weak prediction models, usually decision trees, to produce a robust predictive model. XGBoost is recognised for its effectiveness, capacity to handle large datasets, and precision.

The objective of XGBoost regression is to construct a regression model that can precisely forecast a continuous numerical response. The following is an explanation of the functioning of XGBoost regression:

1. Ensemble of weak predictors: The XGBoost algorithm generates a collection of feeble predictors, commonly referred to as "base learners," which are typically decision trees, to form an ensemble. The base learners are constructed in a sequential manner, whereby each subsequent learner endeavors to rectify the errors committed by its predecessors.
2. Objective function: The XGBoost algorithm formulates an objective function to evaluate the efficacy of the model and direct the training procedure. In regression tasks, the mean Squared Error (MSE) or its variant is a frequently employed objective function.
3. Tree boosting: The XGBoost algorithm utilizes a boosting methodology, whereby every fundamental learner is trained on the residuals, which are the dissimilarities between the factual and anticipated values, of the preceding fundamental learners. Subsequent trees are directed towards capturing the residual patterns and enhancing the overall performance of the model.
4. Regularization: In order to mitigate the issue of overfitting, the XGBoost algorithm employs regularization techniques. The model comprises of various parameters, such as the maximum depth, minimum child weight, and learning rate, which are utilised to regulate the intricacy of the individual trees and the overall model.
5. Gradient-based optimization: The XGBoost algorithm employs the technique of gradient descent optimisation to determine the most suitable values for the parameters of the model.

The iterative process of minimizing the loss involves the computation of the gradients of the objective function relative to the model's parameters, which are subsequently updated.

6. Tree pruning: Upon completion of the construction of all base learners, XGBoost conducts a pruning process to eliminate superfluous nodes and leaves within each individual tree. This process facilitates the streamlining of the model and enhances its capacity for generalisation.

7. Prediction: Upon completion of the training process, the XGBoost algorithm is capable of generating predictions through the consolidation of predictions derived from each of its constituent base learners. The ultimate forecast is the aggregate of these distinct prognostications, typically adjusted by a learning rate.

XGBoost regression is advantageous due to its capability to effectively handle intricate and high-dimensional data, manage missing values, and offer feature importance analysis. Furthermore, XGBoost exhibits proficient parallel processing capabilities, rendering it appropriate for extensive datasets.

It is noteworthy to acknowledge that XGBoost may exhibit overfitting tendencies if not adequately fine-tuned and regularized. Achieving optimal performance may necessitate meticulous hyperparameter selection and parameter tuning through methods such as cross-validation.

In general, XGBoost regression is a robust algorithm utilised for regression tasks, renowned for its precision, effectiveness, and adaptability in managing diverse data types.

Let's illustrate the XGBoost regression using an example.

Suppose we have a dataset with N training examples, denoted as $\{(x_1, y_1), (x_2, y_2), \dots, (x_N, y_N)\}$, where x_i represents the feature vector for the ith house and y_i represents its corresponding sale price.

1. Objective function:

The objective function for XGBoost regression can be defined as follows:

$$obj = \Sigma L(y_i, \bar{y}_i) + \Sigma \Omega(fk) \tag{17.16}$$

where L is the loss function that measures the discrepancy between the true label y_i and the predicted label $\bar{y}_i$, and Ω is the regularization term that penalizes complex models with too many trees (fk) to avoid overfitting.

2. Training process:

The training process involves building an ensemble of weak prediction models (decision trees) that minimize the objective function.

At each iteration t, the model aims to find the optimal prediction function F_t by adding a new decision tree, h_t, to the ensemble:

$$F_t = F_{t-1} + \eta * h_t, \tag{17.17}$$

where η is the learning rate that controls the contribution of each tree and F_{t-1} is the ensemble prediction from the previous iteration.

3. Gradient calculation:

To compute the gradients of the loss function with respect to the predicted values, we have:

$$g_i = \partial L(y_i, \bar{y}_i) / \partial \bar{y}_i \tag{17.18}$$

where g_i represents the gradient for the ith training example.

4. Update leaf values:

For each leaf node in the newly added tree h_t, we compute the optimal weight w_j by minimizing the objective function:

$$w_j = - (\Sigma g_i / (\Sigma h_i + \lambda)) \tag{17.19}$$

where h_i represents the number of instances that belong to the jth leaf, and λ is the regularization parameter that controls the complexity of the trees.

5. Tree pruning:

After constructing the tree, pruning techniques can be applied to remove unnecessary nodes and improve model generalization.

6. Prediction:

To make predictions on new data, the final prediction function F can be obtained by summing up the predictions from all the trees in the ensemble.

This is a simplified overview of the XGBoost regression process using mathematical formulations. The actual implementation may involve additional optimizations and techniques. The goal is to iteratively build an ensemble of decision trees that minimize the objective function and provide accurate predictions for the target variable.

Let's apply the XGBoost regression formulations to an example dataset. Suppose we have the following training and test sets:

Training Set:

Test Set:
To apply the XGBoost regression formulations on the given dataset, let's go through the steps using the mathematical formulations:

1. Objective function:

Assuming we use the Mean Squared Error (MSE) as the loss function, the objective function can be defined as:

$$obj = \Sigma(yi - ȳi)2 + \Omega(fk) \tag{17.20}$$

2. Training process:

Initialize the XGBoost regression model and set hyperparameters such as maximum depth, learning rate, and the number of estimators (trees).
The model will iteratively build decision trees that minimize the objective function using the training set.

3. Gradient Calculation:

Compute the gradients of the loss function (MSE) with respect to the predicted values for each training example. For example, for the first training example:

TABLE 17.4
Dataset representing x variables that determine the y variable (Sale Price)

Bedrooms	Square Footage	Location	Age	Sale Price($)
3	1500	A	10	200000
4	2000	B	5	300000
2	1200	A	8	180000

TABLE 17.5
Dataset representing x variables that determine the y variable (Sale Price)

Bedrooms	Square Footage	Location	Age	Sale Price($)
3	1800	B	7	220000

$$g_1 = \partial L(y_1, \bar{y}_1) / \partial \bar{y}_1 \tag{17.21}$$

$$= 2 * (\bar{y}_1 - y_1) = 2 * (\bar{y}_1 - 200\,000) = 2 * (\bar{y}_1 - 200\,000).$$

4. Update leaf values:

For each leaf node in the newly added tree, calculate the optimal weight based on the gradients and regularization. For example, if a leaf has three training examples in it:

$$w_{leaf} = - (\Sigma g_i / (\Sigma h_i + \lambda)) = - (g_1 + g_2 + g_3) / (h_1 + h_2 + h_3 + \lambda) \tag{17.22}$$

5. Tree pruning:

After constructing the tree, pruning techniques can be applied to remove unnecessary nodes and improve model generalization.

6. Prediction:

 - Once the training process is complete, use the trained XGBoost regression model to make predictions on the test set.
 - Sum up the predictions from all the trees in the ensemble to obtain the final prediction for each test example.

Note that the actual computations involve more iterations, regularization terms, and complex optimizations. The steps provided here give a general understanding of the process based on the mathematical formulations.

Let's assume that we have trained an XGBoost regression model with 3 trees, a learning rate of 0.1, and the following weights for the leaves of each tree:

Tree 1:
Leaf 1: weight = 0.2
Leaf 2: weight = -0.1

Tree 2:
Leaf 1: weight = 0.3
Leaf 2: weight = 0.1
Leaf 3: weight = -0.2

Tree 3:
Leaf 1: weight = -0.3
Leaf 2: weight = 0.4
Leaf 3: weight = 0.2

Now let's proceed with the calculations:

1. Prepare the test example:

   Extract the features of the test example:
   Bedrooms = 3, Square Footage = 1800, Location = B, Age = 7.

2. Use the trained XGBoost model to make predictions:

   Feed the test example's features into the trained model:

$$F = F0 + \eta * h1 + \eta * h2 + \eta * h3 \qquad (17.23)$$

Calculate the predictions for each tree:

Tree 1 prediction: h_1 = 0.2 (as the test example falls into Leaf 1 of Tree 1)
Tree 2 prediction: h2 = 0.1 (as the test example falls into Leaf 2 of Tree 2)
Tree 3 prediction: h3 = 0.4 (as the test example falls into Leaf 2 of Tree 3)

Calculate the final prediction:

Predicted Sale Price = $F = F_0 + \eta * h_1 + \eta * h_2 + \eta * h_3$.

Let's assume the initial prediction (bias) for the XGBoost regression model is 250000, and the learning rate (η) is 0.1. Using the given weights and test example, we can calculate the final prediction:

$$\text{Predicted Sale Price} = 250000 + 0.1 * 0.2 + 0.1 * 0.1 + 0.1 * 0.4$$

$$= 250000 + 0.02 + 0.01 + 0.04$$

$$= 250000.07$$

Hence, with the given features of the test data, the model predicts a price of 250000.07$ which is sufficiently close to its actual value.

WHY XGBoost? In recent years, XGBoost has garnered substantial attention owing to its exceptional performance, accuracy, and versatility in addressing a diverse array of machine learning tasks. The software provides proficient and adaptable implementations of gradient boosting algorithms, facilitating the development of robust ensemble models that can proficiently capture intricate patterns in data. XGBoost has gained popularity among practitioners in diverse fields due to its capacity to manage voluminous datasets, perform feature importance analysis, and incorporate regularization techniques. This has made it a preferred option for both novice and seasoned users. The popularity of the platform is bolstered by its active community and ongoing development, which offer users a plethora of resources and support. The amalgamation of rapidity, precision, adaptability, and communal backing has propelled XGBoost to attain a prominent status and extensive usage as a machine learning library in contemporary times.

17.4.5 GENERATIVE ADVERSARIAL NETWORK

The field of operations research has witnessed the emergence of generative adversarial networks as potent regression models. Generative Adversarial Networks (GANs) utilize the dynamic interplay between the generator and discriminator networks to produce synthetic data samples that exhibit a high degree of similarity to authentic data. This capability has significant implications for operations research, as it enables the development of various applications.

The architecture in question is a neural network comprising a generator network and a discriminator network as its primary components. Generative Adversarial Networks (GANs) are utilized in the field of generative modeling, where the objective is to generate novel data instances that bear resemblance to a pre-existing training dataset.

The fundamental concept underlying Generative Adversarial Networks (GANs) is the concurrent training of two networks in a competitive fashion. The generator network is responsible for producing synthetic data samples, whereas the discriminator network is tasked with discerning between authentic and artificially generated data samples. The primary goal of Generative Adversarial Networks (GANs) is to facilitate the training of the generator network to generate synthetic samples that are virtually indistinguishable from authentic data. Meanwhile, the discriminator network endeavors to classify the samples with precision, distinguishing between genuine and generated ones.

To understand the working and architecture of a GAN, let us consider an illustrative scenario wherein the objective is to forecast the price of a residential property based on a range of determinants.

The Generator network (G): Responsible for generating synthetic house data by mapping a random noise vector, denoted as z, to a high-dimensional space. The function $G(z; \theta_G)$ is utilized to represent it, with θ_G denoting the generator's parameters.

The Discriminator network (D): Is a component in the architecture of certain machine learning models. The primary objective of the discriminator network is to differentiate between authentic and synthesized house data samples. The function $D(x; \theta_D)$ is utilized to represent the discriminator, with x serving as the input house data and θ_D representing the parameters of the said discriminator.

The primary aim of Generative Adversarial Networks (GANs) is to facilitate the joint training of the generator and discriminator networks. The formulation of the objective function can be expressed as a minimax game played between the two networks. The mathematical expression V(G, D) is defined as the expected value of the logarithm of the discriminator function D(x) plus the expected value of the logarithm of the complement of the discriminator function evaluated at the generator function G(z).

$$V(G, D) = E[\log(D(x))] + E[\log(1 - D(G(z)))] \tag{17.24}$$

Regarding the scenario of predicting house prices:

The mathematical expression $E[\log(D(x))]$ denotes the anticipated log-likelihood of the discriminator accurately categorizing authentic housing data samples.

The mathematical expression $E[\log(1 - D(G(z)))]$ denotes the anticipated log-likelihood of the discriminator accurately categorizing artificially generated housing data.

The generator's objective is to minimize the given function by producing synthetic house data that can deceive the discriminator. On the other hand, the discriminator's objective is to maximize the same function by accurately distinguishing between authentic and synthetic house data.

To train the GAN model we first initialize: The parameters θ_G and θ_D are initialized randomly. Then in the training loop, a mini-batch of actual house data, denoted as x, is sampled from the

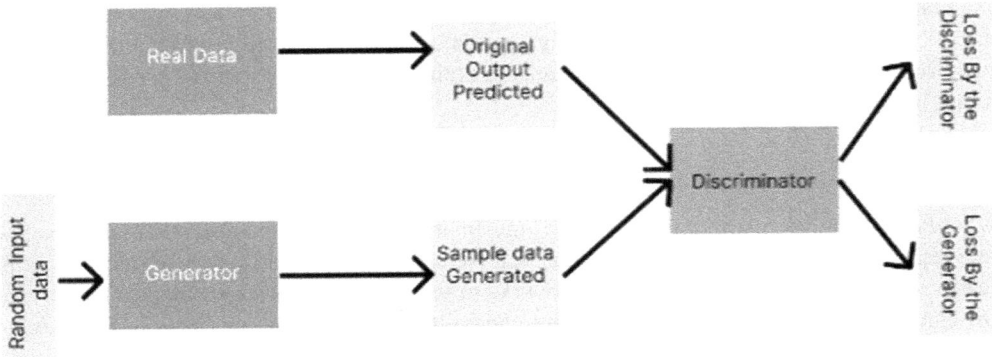

FIGURE 17.6 Working of GAN.

dataset. The next step involves sampling a mini-batch of noise vectors, denoted as z. The synthetic house data can be produced by utilizing the generator function G (z; θ_G). The discriminator is updated through the minimization of its loss function. The equation for L_D is the negative expected value of the logarithm of D(x). The expression being evaluated is the logarithm of one minus the output of the generator network G when given input z, which is then passed through the discriminator network D. The generator can be updated by minimizing its loss function, which is defined as L_G = -E[log(D(G(z)))].

Within the context of predicting house prices, the variables L_D and L_G denote the losses sustained by the discriminator and generator networks, correspondingly. These losses are determined by the networks' capacity to differentiate between authentic and synthesized house data.

Through an iterative process of optimizing the generator and discriminator based on their respective loss functions, the Generative Adversarial Network (GAN) acquires the ability to generate synthetic house data that exhibits a high degree of similarity to authentic house data, thereby enhancing the overall predictive performance.

It is noteworthy that in the context of predicting house prices, the GAN-generated synthetic house data can serve as a valuable supplement to the original dataset. The utilization of this augmented dataset can facilitate the training of alternative regression models, including linear regression and neural networks, thereby enhancing the accuracy of house price prediction. The Generative Adversarial Network (GAN) serves as a means of augmenting data, thereby generating supplementary samples that effectively capture the fundamental data distribution. This process ultimately enhances the predictive models' generalization capabilities.

Generative Adversarial Networks (GANs) are models that require a substantial amount of data for optimal performance. Therefore, it is crucial to utilize data of superior quality during the training process. The optimization settings of Generative Adversarial Networks (GANs), including the learning rate and batch size, can exert a notable influence on the efficacy of the model. The process of exploring various configurations to identify the optimal combination for a given application is deemed crucial. Generative Adversarial Networks (GANs) are trained through a minimax game that involves the generator and the discriminator. If the generator exhibits excessive power, it has the potential to overpower the discriminator and generate output that is implausible or lacks coherence. Monitoring the performance of both the generator and discriminator during the training process is crucial in order to avoid undesired outcomes.

Generative Adversarial Networks (GANs) are a potent instrument for producing authentic and innovative data. Nevertheless, the process of training them can be arduous and demands meticulous attention to particulars. By considering the aforementioned factors, one can enhance the likelihood of achieving favorable outcomes while implementing GANs in personal projects.

17.5   ENHANCING REGRESSION MODELS WITH DEEP LEARNING

The utilization of deep learning techniques presents notable improvements over conventional regression models within the field of operations research. Deep learning models have the ability to capture intricate non-linear relationships and interactions between variables through the use of neural networks. This results in enhanced precision and robustness in regression tasks. The present section delves into the manner in which deep learning techniques augment regression models and the prospective advantages that ensue.

17.5.1   CAPTURING NON-LINEAR CONNECTIONS AND INTERACTIONS

The principal benefit of using deep learning methodologies in regression analysis lies in their capacity to apprehend non-linear associations among variables. Conventional regression models, such as linear regression, posit that there are linear associations between the predictors and the response variable. Nevertheless, in numerous practical situations, the associations may exhibit a significant degree of non-linearity.

Deep learning architectures, such as the MultiLayer Perceptron (MLP) and neural networks featuring multiple hidden layers, possess the ability to acquire intricate associations between input variables and the output variable. Deep learning models are capable of capturing complex non-linear patterns that may be overlooked by conventional regression models through the use of activation functions and interconnected neurons across multiple layers.

In addition, deep learning models demonstrate exceptional proficiency in capturing the complex interrelationships between variables. Through the process of extracting hierarchical and non-linear combinations of input variables, high-level feature representations can be learned in an automated manner. The aforementioned capability facilitates the ability of deep learning models to discern intricate interrelationships and interdependencies that are instrumental in achieving precise prognostications in regression assignments.

17.5.2   IMPROVED ACCURACY AND PREDICTION PERFORMANCE

The utilization of deep learning techniques has exhibited enhanced precision and predictive capabilities in contrast to conventional regression methodologies. Deep learning models are capable of achieving higher accuracy by modelling the underlying complexity of the data due to their ability to capture non-linear relationships and interactions.

Deep learning algorithms have the ability to acquire feature representations from unprocessed data, thereby obviating the necessity for human-driven feature engineering. The aforementioned benefit holds significant value in the field of operations research, where datasets are often characterized by their vastness, heterogeneity, and the presence of features with high dimensionality. Deep learning models have the ability to automatically extract pertinent features and dynamically acquire representations that enhance performance in regression tasks.

Furthermore, deep learning models exhibit proficiency in effectively managing datasets of significant magnitude. Artificial intelligence systems possess the capability to effectively handle extensive quantities of data and acquire knowledge from substantial training sets, thereby facilitating their ability to generalize proficiently and provide precise forecasts on unobserved data.

17.5.3   CHALLENGES AND CONSIDERATIONS

Deep learning techniques provide substantial improvements to regression models in the field of operations research. However, it is important to acknowledge and address the various challenges and considerations associated with their implementation.

The interpretability of deep learning models is frequently perceived as a significant challenge due to their reputation as black boxes, which hinders the ability to comprehend their decision-making mechanisms. The interpretability of deep learning models can pose a challenge in comprehending the rationale behind their predictions, thereby constraining their utilization in contexts where interpretability holds significant importance. Scholars are currently investigating methods to enhance the interpretability of deep learning models.

Overfitting is a phenomenon that can occur in deep learning models due to their high flexibility and ability to learn intricate patterns in data. The pliability of the model may result in overfitting, a phenomenon in which the model becomes excessively tailored to the training data and is unable to effectively extrapolate to novel data. The implementation of regularization techniques, such as dropout and weight decay, can effectively address the issue of overfitting and enhance the overall generalisation performance of deep learning models.

The combination of deep learning and operations research presents significant prospects for enhancing decision-making, streamlining processes, and attaining more precise predictions across diverse fields. Further investigation is warranted to enhance and advance the utilization of deep learning within regression models, thereby broadening their influence on methodologies and practices in the field of operations research.

17.6   CHALLENGES OF DEEP LEARNING FOR REGRESSION

Although deep learning techniques provide several advantages for improving regression models in operations research, there are various challenges associated with their implementation. This section endeavors to tackle the aforementioned challenges and offers valuable perspectives on how to effectively alleviate them.

17.6.1   INTERPRETABILITY ISSUES

The absence of interpretability stands out as a key obstacle to employing deep learning for regression. Deep learning models are frequently regarded as opaque due to their ability to acquire intricate representations and mappings among variables. Comprehending the fundamental elements that propel the forecasts can prove challenging, particularly when handling intricate neural networks that possess numerous layers.

The significance of interpretability is paramount in numerous operations research applications, given that decision-makers require a comprehensive understanding of the rationale behind the model's forecasts. Scholars are currently investigating techniques to improve the interpretability of deep learning models. Various methodologies, including attention mechanisms, layer-wise relevance propagation, and feature visualization, can offer valuable insights into the decision-making process of a model and facilitate the elucidation of the interconnections between input variables and the output.

17.6.2   OVERFITTING AND GENERALIZATION

Overfitting represents a significant obstacle that deep learning models may encounter when utilised for regression tasks. Deep neural networks exhibit a high degree of flexibility and possess a considerable number of parameters, thereby leading to the possibility of models that tend to memorise the training data rather than acquire significant patterns. As a result, models that are overfitted exhibit inadequate generalisation to novel data, resulting in suboptimal performance in practical situations. Regularization techniques are frequently utilised to address the issue of overfitting. Methods such as dropout, which involves the random deactivation of neurons during the training process, and weight decay, which introduces a penalty term to the loss function, are effective in promoting model regularization and mitigating the risk of excessive complexity. The implementation of appropriate validation and testing protocols, including the utilization of distinct validation sets and cross-validation techniques, is imperative for evaluating the overall performance of the model and determining optimal hyperparameters.

17.6.3 COMPUTATIONAL RESOURCES AND TRAINING TIME

The training of deep learning models, particularly those with intricate architectures and multiple layers, necessitates substantial computational resources. The process of training deep neural networks can be characterized by its high computational demands and lengthy time requirements, especially in the context of extensive datasets. Rapid decision-making and real-time forecasting can present difficulties in the context of operations research scenarios.

In order to tackle these obstacles, scholars and professionals may utilize the advancements in hardware infrastructure, such as potent GPUs (Graphics Processing Units) and TPUs (Tensor Processing Units), that are explicitly engineered to expedite deep learning computations. Furthermore, the use of transfer learning methods, which involve the refinement of pre-existing models for particular tasks, can effectively curtail training duration by capitalizing on the knowledge acquired from prior tasks.

17.6.4 DATA AVAILABILITY AND QUALITY

Deep learning algorithms exhibit superior performance when trained on extensive volumes of high-quality data. Acquiring datasets of this nature can present difficulties in the field of operations research, as the data available may be restricted or contain inaccuracies. Inadequate information may result in models that do not effectively capture fundamental patterns, leading to suboptimal outcomes.

Effective resolution of data-related issues necessitates meticulous preprocessing of data, engineering of features, and utilization of data augmentation methodologies. Scholars may investigate various methodologies, including synthetic data generation, data imputation, and feature extraction, to surmount data constraints and amplify the efficacy of deep learning models in regression assignments.

17.6.5 ETHICAL CONSIDERATIONS AND BIAS

The presence of biases in the training data can render deep learning models vulnerable to producing biased predictions and decisions. In the field of operations research, it is of utmost significance to consider aspects of fairness and ethical considerations.

In order to address potential biases, it is imperative to ensure that training datasets are diverse and representative, encompassing a range of demographic groups while avoiding both under-representation and over-representation of any particular group. Conducting routine audits and monitoring predictive models can aid in the detection and mitigation of biases, thereby promoting equitable and just outcomes.

17.7 CONCLUSION

This chapter delves into the utilization of deep learning methodologies to augment regression models in the field of operations research. Deep learning presents various benefits over conventional regression techniques, such as its capacity to capture non-linear relationships and interactions, enhanced accuracy and prediction performance, and its effectiveness in handling extensive datasets.

The utilization of deep learning methodologies, including multilayer perceptron, support vector regression, XGBoost, Gaussian process regression, and generative adversarial networks, can potentially augment the accuracy and resilience of regression models in the field of operations research. The aforementioned techniques facilitate the representation of intricate associations, flexibility in handling non-linear data, and automated extraction of features from unprocessed data. Nevertheless, the application of deep learning to regression tasks presents certain difficulties. The issue of interpretability persists as deep learning models are frequently perceived as opaque entities. Current endeavors are being made to enhance the comprehension of deep learning models. Various techniques such as attention mechanisms, layer-wise relevance propagation, and feature visualization are being investigated as potential remedies. Overfitting represents a significant obstacle that necessitates attention when employing deep learning methodologies. The implementation of regularization techniques, such as dropout and weight decay, in conjunction with appropriate validation and testing protocols can effectively address the issue of overfitting and enhance the overall generalisation performance.

The challenges posed by the computational resources necessary for training deep learning models are noteworthy. The challenges posed by lengthy training times in machine learning are being mitigated by the progress made in hardware infrastructure, including GPUs and TPUs, which facilitate accelerated training. Furthermore, the implementation of transfer learning methods can utilize pre-existing models to decrease the duration of training and enhance effectiveness.

The crucial factors to be considered in deep learning for regression are the availability and quality of data. The efficacy of deep learning models may be impeded by inadequate or disruptive data. The utilization of preprocessing techniques, data augmentation, and meticulous feature engineering can effectively tackle data-related obstacles and improve the overall performance of machine learning models. The incorporation of deep learning models in operations research necessitates careful consideration of ethical considerations and biases. The implementation of diverse and representative training datasets, coupled with regular monitoring of the model's predictions, can serve as a means of mitigating biases and promoting fair and equitable outcomes. Notwithstanding the obstacles, the incorporation of deep learning methodologies into regression models for operations research exhibits significant promise. Continued progress in the field will be propelled by further research and advancements in interpretability, regularization, computational efficiency, data handling, and ethical considerations.

To conclude, the integration of deep learning techniques can provide significant improvements to regression models within the field of operations research. Deep learning models can offer more accurate and resilient predictions by utilizing their capacity to capture non-linear relationships, extract significant features, and manage vast datasets. Despite the existing challenges, continuous research and innovation are expected to facilitate the extensive integration and further development of deep learning within the domain of operations research regression models.

REFERENCES

1. Aggarwal, A., Mittal, M., and Battineni, G. "Generative adversarial network: An overview of theory and applications." *International Journal of Information Management Data Insights* 1, no. 1 (2021): 100004.
2. Karlik, B., and Olgac, A. V. "Performance analysis of various activation functions in generalized MLP architectures of neural networks." *International Journal of Artificial Intelligence and Expert Systems* 1, no. 4 (2011): 111–122.
3. Tolstikhin, I. O., et al. "MLP-mixer: An all-MLP architecture for vision." *Advances in Neural Information Processing Systems* 34 (2021): 24261–24272.
4. Quinonero-Candela, J., and Rasmussen, C. E. "A unifying view of sparse approximate Gaussian process regression." *The Journal of Machine Learning Research* 6 (2005): 1939–1959.
5. Goodfellow, I. J., et al. "Generative adversarial networks." *Communications of the ACM* 63, no. 11 (2020): 139–144.

6. Schulz, E., Speekenbrink, M., and Krause, A. "A tutorial on Gaussian process regression: Modelling, exploring, and exploiting functions." *Journal of Mathematical Psychology* 85 (2018): 1–16.
7. Abdulaal, A., et al. "Comparison of deep learning with regression analysis in creating predictive models for SARS-CoV-2 outcomes." *BMC Medical Informatics and Decision Making* 20 (2020): 1–11.
8. LeCun, Y., Bengio, Y., and Hinton, G. "Deep learning." *Nature* 521, no. 7553 (2015): 436–444.
9. Qiu, X., Zhang, L., Ren, Y., Suganthan, P. N., and Amaratunga, G. "Ensemble deep learning for regression and time series forecasting." In *2014 IEEE Symposium on Computational Intelligence in Ensemble Learning (CIEL)*, pp. 1–6. IEEE (2014).
10. Xu, Z., Gao, Y., and Jin, Y. "Application of an optimized SVR model of machine learning." *International Journal of Multimedia and Ubiquitous Engineering* 9, no. 6 (2014): 67–80.
11. Wang, A., Lam, J. C., Song, S., Li, V. O., and Guo, P. "Can smart energy information interventions help householders save electricity? An AVR machine learning approach." *Environmental Science & Policy* 112 (2020): 381–393.
12. Zhang, X., Yan, C., Gao, C., Malin, B. A., and Chen, Y. "Predicting missing values in medical data via XGBoost regression." *Journal of Healthcare Informatics Research* 4 (2020): 383–394.
13. Pesantez-Narvaez, J., Guillen, M., and Alcañiz, M. "Predicting motor insurance claims using telematics data—XGBoost versus logistic regression." *Risks* 7, no. 2 (2019): 70.
14. Snee, R. D. "Validation of regression models: Methods and examples." *Technometrics* 19, no. 4 (1977): 415–428.
15. Fahrmeir, L., Kneib, T., Lang, S., and Marx, B. D. "Regression models." In *Regression: Models, Methods and Applications*, pp. 23–84. Berlin, Heidelberg: Springer Berlin Heidelberg (2022).
16. LaValley, M. P. "Logistic regression." *Circulation* 117, no. 18 (2008): 2395–2399.
17. Ostrom, C. W. *Time Series Analysis: Regression Techniques. No. 9.* Sage (1990).
18. Johnsson, T. "A procedure for stepwise regression analysis." *Statistical Papers* 33, no. 1 (1992): 21–29.

18 The Machine Learning Pipeline

Algorithms, Applications, and Managerial Implications

Anjali Munde

18.1 WHAT IS MACHINE LEARNING?

The focus of machine learning, a branch of AI, lies in the advancement of techniques and models that enable computers to learn and make predictions or judgments without relying on explicit programming. It involves the development of computational systems that can autonomously learn and enhance their performance through experience, empowering them to tackle intricate tasks and adapt to dynamic situations. At its core, machine learning is centered around the concept of training models on data. These models are designed to identify patterns, relationships, and structures within the data, enabling them to make predictions or take actions based on new, unseen examples. By analyzing large and complex datasets, machine learning algorithms can uncover valuable insights, detect anomalies, classify objects, and make accurate predictions.

18.1.1 HISTORY OF MACHINE LEARNING

Researchers embarked on exploring the concept of machine learning and decision-making during the mid-20th century, marking the inception of the history of ML. The 1940s–1950s: Early development: The foundation of machine learning can be traced back to the work of pioneers such as Alan Turing, who proposed the idea of universal computing machines capable of learning. The invention of the perceptron by Frank Rosenblatt in 1957, an early type of artificial neural network, marked a significant advancement in machine learning.

The 1960s–1970s: Early AI and rule-based systems: The development of early Artificial Intelligence (AI) systems focused on rule-based expert systems. These systems utilized a set of predefined rules and logical reasoning to solve problems. Allen Newell and Herbert Simon's creation of the General Problem Solver (GPS) in 1961 represented a significant milestone in the field of AI, showcasing a system with remarkable problem-solving abilities.

The 1980s–1990s: Knowledge-based systems and statistical methods: ML research in the 1980s emphasized the integration of knowledge-based systems and statistical methods. The introduction of DT algorithms, such as ID3 (Iterative Dichotomiser 3) by Ross Quinlan in 1986, provided a framework for learning from labeled data and making predictions. SVM, developed by Vladimir Vapnik and Corinna Cortes in the 1990s, became popular for classification and regression tasks.

The late 1990s–2000s: Rise of big data and ensemble methods: With the advent of the internet and the exponential growth of data, ML faced new challenges in processing and analyzing large-scale datasets. Boosting and bagging, ensemble learning techniques, gained prominence in the late 1990s. Boosting methods, such as AdaBoost, combined weak learners to create strong classifiers.

 DOI: 10.1201/9781003433309-18

The field of reinforcement learning also saw significant advancements, with notable contributions from researchers like Andrew Ng and Stuart Russell.

The 2010s–Present: Deep Learning (DL) and breakthroughs: Deep learning, a subfield of machine learning that centers on artificial neural networks comprising multiple layers, witnessed notable advancements that propelled the field forward. Convolutional Neural Networks (CNN) revolutionized computer vision tasks, achieving remarkable accuracy in image recognition and object detection. Recurrent Neural Networks (RNN) and Long Short-Term Memory (LSTM) networks made significant progress in natural language processing and speech recognition. The availability of large, annotated datasets, increased computational power, and improved algorithms contributed to the rapid advancement and adoption of deep learning.

Today, ML has become pervasive in numerous domains, including healthcare, finance, autonomous vehicles, recommendation systems, and more. It continues to evolve, incorporating advancements in DL, reinforcement learning, generative models, and interpretability techniques. With ongoing research and advancements, ML is poised to transform industries, improve decision-making processes, and contribute to the development of innovative technologies.

18.1.2 Principles of machine learning

The core principles of ML revolve around its ability to automatically identify patterns, make predictions, and extract meaningful insights from complex and large-scale datasets. These principles form the foundation of machine learning algorithms and methodologies. Here are the key core principles:

1. Pattern recognition: ML algorithms are made to automatically identify and extract links and patterns from datasets. They can spot hidden structures and patterns that people might not be able to see, leading to the discovery of important ideas.
2. Prediction and forecasting: ML models have the ability to make predictions and forecasts based on patterns identified in historical data. By learning from past examples, these models can generalize their knowledge to predict future outcomes or trends with a certain level of accuracy.
3. Automation: ML enables automation of tasks that would typically require human intervention. Once trained, ML models can perform complex computations and analyses on large datasets without explicit programming, reducing the need for manual effort and intervention.
4. Adaptability and generalization: ML algorithms have the capacity to adapt and generalize their learning from one set of data to new, unseen data. This ability allows models to make accurate predictions or classifications on previously unseen examples, demonstrating their capacity to handle novel situations.
5. Scalability: ML algorithms can handle large-scale datasets with millions or even billions of data points. They can efficiently process and analyze massive amounts of data, enabling the extraction of insights and patterns that would be challenging or impossible to identify through manual analysis.
6. Iterative learning: ML often involves an iterative learning process. Initially, models are trained using a subset of the available data, and their performance is assessed. The evaluation results guide the refinement of the models, followed by retraining and subsequent evaluation iterations. This iterative process allows for continuous improvement and optimization of the model's performance.
7. Interpretability and explainability: While some ML models are highly complex and difficult to interpret, efforts are being made to develop interpretable and explainable models. Interpretable models allow humans to understand and explain the underlying factors contributing to a prediction or decision, enhancing transparency and trust in the system.

8. Always learning and adaptation: ML models can be created to continuously learn from and adjust to the availability of fresh data. This enables models to stay up-to-date with evolving patterns and trends in the data, ensuring their relevance and accuracy over time.

These core principles collectively contribute to the power and capabilities of machine learning, enabling the extraction of valuable insights, accurate predictions, and automation of complex tasks. Machine learning algorithms built upon these advancements serve as the fundamental building blocks for various industries, including finance, healthcare, image identification, and natural language processing. They enable the development and application of advanced solutions in these domains.

18.2 DATA PRE-PROCESSING, FEATURE ENGINEERING, AND MODEL SELECTION IN ML

Data pre-processing, feature engineering, and model selection are critical steps in the ML pipeline that significantly impact the performance and effectiveness of ML models. Data pre-processing involves transforming and preparing the raw data before it is fed into the ML model. It is crucial because of the following:

1. Data cleaning: It addresses missing values, outliers, and inconsistencies in the dataset. Cleaning the data ensures that the ML model is not biased or misled by erroneous data points.
2. Data transformation: It involves scaling or normalizing the data to bring all features to a similar range. This assists in avoiding the dominance of some features over the learning process due to their greater magnitude.
3. Feature selection: It entails locating and choosing pertinent factors that significantly affect the target variable. Removing irrelevant or redundant features reduces noise and simplifies the learning process.
4. Handling categorical variables: Creating numerical representations of categorical variables (for example, one-hot encoding) allows ML algorithms to interpret and utilize these variables effectively.
5. Handling imbalanced data: When the dataset has imbalanced class distributions, pre-processing techniques such as oversampling or under sampling can address the issue and improve the model's ability to learn from minority classes.

By performing effective data pre-processing, the quality and reliability of the dataset are improved, ultimately resulting in enhanced performance of machine learning models and more precise predictions. Subsequently, feature engineering entails the creation of new features or the transformation of existing ones to improve the representation of the data. This process aims to extract meaningful information and enhance the predictive power of machine learning models. It holds significant importance for the following reasons.

1. Feature extraction: It involves deriving new features from the existing ones that capture additional information. For example, extracting time-based features from timestamps or deriving ratios from numerical attributes can provide valuable insights to the model
2. Feature encoding: It involves encoding categorical variables into numerical representations that can be processed by ML algorithms. This transformation enables the model to understand and utilize categorical information effectively.
3. Dimensionality Reduction (DR): It seeks to cut down on features while keeping the most important data. Methods such as Principal Component Analysis (PCA) or feature selection algorithms can be employed to achieve computational complexity reduction and mitigate overfitting. Feature engineering enhances data representation, allowing the model to capture

intricate patterns and improve its predictive performance. Further, the model selection involves choosing the most appropriate ML algorithm or model architecture for the given task. It is crucial for the following reasons:

1. Performance and accuracy: Different ML models have varying strengths and weaknesses. Selecting the right model can lead to higher accuracy and better performance on a specific task.
2. Interpretability: Some models, such as decision trees or linear regression, provide interpretability, allowing stakeholders to understand and explain the model's predictions. Other models, like deep neural networks, offer higher accuracy but may lack interpretability.
3. Computational efficiency: Model selection also considers the computational resources required by different models. For large-scale datasets or real-time applications, selecting models that balance accuracy and computational efficiency is crucial.
4. Generalization: The selected model should generalize well to unseen data and avoid overfitting or underfitting. Model selection techniques, such as cross-validation or hyperparameter tuning, aid in identifying models that generalize well.

Choosing the right model for the task ensures optimal performance, scalability, interpretability, and generalizability. In summary, data pre-processing, feature engineering, and model selection are vital steps in the ML pipeline. They enhance data quality, improve feature representation, and enable the selection of the most suitable ML model for the task at hand, ultimately leading to improved model performance, accuracy, and interpretability.

18.3   TYPES OF MACHINE LEARNING

There are two main forms of machine learning based on the learning procedure and the availability of labeled data:

1. SML: In Supervised Machine Learning (SML), the model is trained using labeled data, where the input examples are associated with corresponding desired outputs or labels. The model learns to make predictions by generalizing from the provided labeled examples. The goal is to teach the model a mapping function that can precisely forecast the result for fresh input samples. SML algorithms include:
   a. Classification: Models predict discrete class labels or categories. Examples include spam detection, sentiment analysis, and image classification.
   b. Regression: Models predict continuous numeric values. Examples include price prediction, demand forecasting, and stock market analysis.
2. UML: With no stated output or label, the model in UML is trained on unlabeled data. Finding hidden patterns, structures, or correlations in the data is the goal. UML algorithms include:
   a. Clustering: Based on their intrinsic similarities or differences, algorithms group comparable cases together. Examples include customer segmentation and image segmentation.
   b. DR: Techniques are employed to decrease the quantity of input features while retaining the critical data. Examples include PCA and t-SNE (t-Distributed Stochastic Neighbor Embedding).
   c. Anomaly detection: Algorithms identify unusual or anomalous instances in the dataset. Examples include fraud detection and network intrusion detection.

In addition to these main types, there are other specialized areas within machine learning, including:

RL: By optimizing a reward signal, RL includes training models to make decisions in a dynamic environment. The environment is interacted with, actions are taken, and feedback in the form of rewards or punishments is received. The model gains the ability to act in a way that maximizes cumulative rewards over time. RL algorithms are commonly used in robotics, gaming, and autonomous systems.

Semi-supervised learning: In semi-supervised machine learning, a combination of labeled and unlabeled data is used for training. The goal is to improve the performance of the model by leveraging a large amount of unlabeled data along with a smaller portion of labeled data. By utilizing the additional information from the unlabeled data, the model can learn more effectively and make more accurate predictions.

These types of machine learning provide a diverse set of tools and techniques for solving various problems and extracting valuable insights from data. The selection of the appropriate type of machine learning, whether supervised, unsupervised, or semi-supervised, depends on several factors including the nature of the task, the availability of data, and the desired outcomes. The specific requirements and characteristics of the problem at hand guide the decision-making process to determine the most suitable approach.

18.3.1 SUPERVISED MACHINE LEARNING

Indeed, Supervised Machine Learning (SML) is a subset of machine learning where the model is trained using labeled data. The labeled data consists of input examples paired with their corresponding labels or expected outputs. By learning from these labeled examples, the model can make predictions or classifications for new, unseen data. The objective is for the model to develop a mapping function that can precisely forecast the result for brand-new, untried input data.

SML requires a dataset where each example has a known input-output pair. The input represents the features or attributes, and the output represents the corresponding label or target variable. During the training phase, the model in supervised machine learning gains knowledge from the labeled data by adjusting its internal parameters or weights. It does so in order to minimize the difference or discrepancy between the predicted outputs generated by the model and the actual labels provided in the training data. This iterative process of updating the model's parameters helps it learn and improve its ability to accurately predict or classify new, unseen data instances. Once trained, the model can use the honed mapping function to predict or categorize fresh, unknown input data. In supervised learning, there are two primary approaches: regression and classification. Regression involves predicting a continuous value or real number, while classification entails predicting a discrete class or label. To evaluate and validate Supervised Machine Learning (SML) models, various techniques are employed, including:

- Training and test sets: To assess the model's performance on unseen data, training, and test sets are typically created from the labeled data. The model is trained on the training set, where the input examples are coupled with their corresponding labels. Once the training phase is complete, the model's performance is evaluated on the test set, which consists of data instances that the model has not seen before. By examining how well the model predicts the labels for the test set, we can gauge its ability to generalize and make accurate predictions on new, untried data. This process helps us estimate the model's performance in real-world scenarios and assess its effectiveness in handling unseen data.
- CV: The data is split up into many subsets and the model is iteratively trained and tested on different combinations of the subsets via Cross-Validation (CV) procedures like k-fold cross-validation to obtain more precise performance estimates.
- Evaluation metrics: Various evaluation metrics, such as accuracy, precision, recall, F1 score for classification, Mean Squared Error (MSE), Root Mean Squared Error (RMSE), or R-squared for regression, are used depending on the job.

Sentiment analysis, spam detection, picture categorization, fraud detection, recommendation systems, and many other uses for SML are common. In order to discover patterns and relationships and generate precise predictions or classifications based on unobserved data, it depends on labeled data.

18.3.1.1 Classification

In SML, the goal of classification involves teaching the model to forecast discrete class labels or categories for fresh, unstudied input data. It involves using labeled data to train a model, where each sample is connected to a particular class or category. The trained model may then classify new cases into one of the established classes based on the patterns and correlations found in the training data. A dataset with labeled examples is needed for classification, and each example must include input features and the associated class label. The input features represent the characteristics or attributes of the instance, while the class label represents the category to which it belongs. The input features should be appropriately represented to capture the relevant information for classification. This may involve pre-processing steps such as feature scaling, normalization, or transformation to ensure fair comparison and effective learning.

Several algorithms are commonly used for classification tasks. Some popular ones include:

- Logistic Regression (LR): It mimics the relationship between the properties of the input and the likelihood of belonging to a particular class.
- DT: These base their conclusions on features and their thresholds using a tree-like structure. They are capable of handling both numerical and category characteristics.
- RF: Multiple decision trees are used in this ensemble method to increase accuracy and decrease overfitting.
- SVM: In a high-dimensional feature space, SVMs locate the ideal hyperplane that divides several classes.
- Naive Bayes (NB): This method assumes that characteristics are conditionally independent given the class and calculates the probabilities of each class using the Bayes theorem.

The following assessment metrics are employed to gauge the efficiency of classification models:

- Accuracy: It calculates what portion of the total number of cases were correctly categorised.
- Precision: The effectiveness of the model to avoid false positives is determined by calculating the ratio of true positive predictions to all predicted positives.
- Recall (sensitivity): The model's capacity to identify all positive cases is indicated by the ratio of true positive predictions to the total actual positives.
- F1 Score: In order to provide a fair assessment of the model's performance, it combines recall and precision.

Lastly, model optimization can be completed in which fine-tuning the classification model involves selecting appropriate hyperparameters, such as the learning rate, regularization parameters, or tree depth, to improve performance. Techniques like CV and Grid Search (GS) can aid in finding the optimal hyperparameter values.

Classification is widely used in various domains, including spam detection, sentiment analysis, image recognition, document classification, customer segmentation, and disease diagnosis. It allows for automated decision-making and can provide valuable insights based on the predicted class labels.

18.3.1.2 Regression

Regression in SML is a task where the model learns to predict continuous numeric values as the output variable based on input features. It entails using labeled data to train a model, with each example having input attributes and a matching numeric target variable. The trained model can then make predictions on new, unseen instances to estimate or approximate a continuous value. A dataset containing labeled examples is needed for regression, and each example must have input features and the corresponding numeric target variable. The input characteristics are the independent variables or traits, whereas the target variable is the dependent variable that we desire to anticipate. Similar to classification, the input features need to be properly represented to capture the relevant information for regression. Pre-processing steps such as feature scaling, normalization, or transformation might be necessary to ensure fair comparison and effective learning.

Various algorithms are commonly used for regression tasks. Some popular ones include:

- Linear regression: The link between the target variable and the input information is modeled using a linear function.
- Polynomial regression: It goes beyond linear regression by capturing more intricate correlations between the features and the target variable using higher-order polynomial functions.
- Support Vector Regression (SVR): Similar to SVMs in classification, SVR finds the best hyperplane that approximates the continuous target variable.
- Random forest regression: It uses an ensemble of decision trees to produce predictions and employs the random forest technique for regression tasks.

Regression model performance is assessed using a variety of evaluation indicators, such as:

- MSE: The average squared deviation between the target values' actual values and the expected values is calculated. Lower values signify better performance.
- RMSE: It is the square root of the MSE and provides a more comprehensible statistic in the original units of the target variable.
- R-squared: It determines how much of the target variable's volatility the model can accommodate. Higher values represent a better match.

Similar to classification, optimizing regression models involve selecting appropriate hyperparameters and tuning the model to improve performance. Techniques like CV and GS can help in finding the optimal hyperparameter values.

In many different industries, including finance, economics, medicine, and engineering, regression is frequently employed. Predictions may be made, values can be estimated, and the relationship between variables can be understood. Price prediction, demand forecasting, sales analysis, housing price estimation, and stock market prediction are a few examples of regression activities.

18.3.2 Unsupervised Machine Learning

In UML, the model learns patterns, structures, or relationships from unlabeled data without having an output or goal variable in mind. Unlike SML, there is no labeled data available to guide the learning process. Instead, the model explores the data on its own to discover hidden insights or patterns. UML relies on unlabeled data, where there are no pre-defined output labels or target variables associated with the input examples. The model explores the inherent structure or distribution of the data to find meaningful patterns or relationships.

Compared to SML, evaluating UML models can be more difficult because there are no specified labels. Depending on the precise task and aims, different evaluation metrics are used. When clustering, measures like the Davies-Bouldin index or the silhouette coefficient can be employed to rate the quality of the clusters. On the basis of their capacity to accurately identify anomalies and reduce false positives, anomaly detection models can be assessed.

Customer segmentation, market basket analysis, anomaly detection, picture and document clustering, and recommendation systems are just a few of the disciplines where UML is used.

Finding hidden patterns or insights that can then be applied to decision-making or feature engineering for supervised learning tasks is helpful in exploratory data analysis.

18.3.2.1 Clustering

In UML, clustering is a common activity where the objective is to put similar instances together based on their proximity or intrinsic similarity. It entails dividing up the data into discrete clusters or subgroups without being aware of the class names beforehand. Clustering algorithms look for organic structures or patterns in the data to help academics grasp the underlying links and get new insights. Numerous clustering techniques are available, each with unique advantages and presumptions. Several frequently employed clustering techniques are:

1. K-means clustering: Assigning instances to the closest centroid or cluster center based on a distance measure, often Euclidean distance, it is a centroid-based clustering algorithm.
2. Hierarchical clustering: By repeatedly merging or dividing clusters based on similarity, it creates a hierarchy of clusters. Either agglomerative (bottom-up) or divisive (top-down) processes can occur.
3. Density-based clustering: Algorithms like DBSCAN (Density-Based Spatial Clustering of Applications with Noise) form clusters based on density-connected regions in the data space.
4. Gaussian Mixture Models (GMM): By assuming that the data points are generated from a combination of Gaussian distributions, GMM aims to estimate the parameters of these distributions.

Distance measurements are frequently used by clustering algorithms to determine how similar or dissimilar data items are. Common distance measurements include the Euclidean distance, the Manhattan distance, and cosine similarity. The distance metric to be utilised depends on the clustering method being used and the type of data being used. Finding the ideal number of clusters in the data is one of the difficulties in clustering. Techniques like the elbow method, silhouette analysis, or domain knowledge can be used for this. The right number of clusters can be automatically determined by algorithms like DBSCAN in situations when the number of clusters is unknown or fixed. Clustering evaluation metrics assess the quality and coherence of the generated clusters. While there is no one-size-fits-all evaluation metric for clustering, common metrics include the silhouette coefficient, Davies-Bouldin index, and Rand index. These metrics provide insights into the compactness, separation, and consistency of the clusters. Clustering finds applications in various domains, including customer segmentation, document clustering, image segmentation, social network analysis, and anomaly detection. It helps in identifying homogeneous groups, discovering patterns, and supporting decision-making processes. Clustering in unsupervised learning is a powerful technique for exploratory data analysis, pattern recognition, and gaining insights from unlabeled data. By grouping similar instances together, it provides a foundation for further analysis and decision-making in various domains.

18.3.2.2 Dimensionality reduction

DR is a UML approach that is used to decrease the number of input characteristics or variables while preserving crucial data. It is especially helpful when working with high-dimensional data when there are many features and it might be difficult to accurately visualise and analyse the data. The goal of dimensionality reduction techniques is to reduce the dimensions of the data while maintaining any inherent structure or relationships.

DR can be categorized into two broad approaches: feature selection and feature extraction.

1. Feature selection: Based on certain criteria, such as relevance, importance, or statistical measurements, feature selection algorithms choose a subset of the original features. It seeks to

locate the most illuminating characteristics that significantly influence the underlying patterns or relationships.
2. Feature extraction: By combining or altering the original features, feature extraction techniques produce new, derived features. These techniques seek to identify the most important data or patterns in a smaller-dimensional space.

A common DR approach is PCA. It pinpoints the data's primary directions or components that account for the most volatility. The PCA effectively decreases the dimensionality while maintaining as much information as possible by projecting the data onto these major components.

t-SNE is a nonlinear DR technique that is particularly useful for visualization. It maps high-dimensional data to a lower-dimensional space while preserving the local structure and relationships between neighboring data points. It is commonly used for visualizing clusters or patterns in complex datasets.

Autoencoders are neural network-based models used for unsupervised learning and dimensionality reduction. They are made up of an encoder, which reduces the dimensions of the data, and a decoder, which extracts the original data from the compressed representation. Autoencoders capture the key features and reduce dimensionality by perfecting the reconstruction of the data.

DR offers several advantages in UML:

- Reducing computational complexity: Processing highly dimensional data can be computationally expensive. Dimensionality reduction reduces the computational burden by reducing the number of features.
- Enhancing interpretability: It is simpler to see and understand the patterns or relationships in the data when the data is shrunk to a lower-dimensional space.
- Mitigating the curse of dimensionality: High-dimensional data often suffers from the curse of dimensionality, where the data becomes sparse, and the performance of learning algorithms deteriorates. Dimensionality reduction helps to overcome this problem by capturing the essential information.

It is important to note that dimensionality reduction involves a trade-off between preserving information and reducing dimensionality. While dimensionality reduction can simplify the data representation, it may also result in some loss of information or detail. The choice of dimensionality reduction technique and the appropriate number of dimensions depends on the specific task, dataset, and the trade-offs that are acceptable in the given context.

Dimensionality reduction in unsupervised learning plays a vital role in simplifying complex data, improving computational efficiency, and aiding in data exploration and visualization. It helps in uncovering the underlying structure and patterns present in high-dimensional data, leading to more effective analysis and understanding.

18.3.3 REINFORCEMENT LEARNING

Machine learning has a subset called Reinforcement Learning (RL) that aims to teach an agent how to make intelligent decisions through interactions with its environment. An agent gradually learns to operate in RL so as to maximize a reward signal. It is predicated on the notion of making mistakes and learning from them, much like how both people and animals learn from their experiences. Here are some key aspects of RL:

1. Agent: The agent is the learner or decision-maker in the RL framework. It interacts with the environment, observes its state, and takes actions based on a policy to maximize its long-term rewards.

2. Environment: The external system that the agent communicates with is represented by the environment. It can be anything, from a straightforward simulation to intricate real-world situations. Based on the agent's behaviors, the environment gives it feedback in the form of incentives or punishments.

3. State: The state is an image of the surroundings at a specific moment. It records the crucial data the agent needs to make decisions. The agent observes the current state and selects actions based on it.

4. Action: Actions are the decisions made by the agent to interact with the environment. The agent's objective is to discover the best mapping between states and actions in order to maximise the cumulative reward over time.

5. Reward: A scalar signal indicating the environment's approval of the agent's activities is the reward. Learning a policy that maximizes the cumulative reward or projected return is the agent's goal.

6. Policy: A mapping from states to actions makes up the policy. It establishes the agent's behavior and chooses what should be done in response to a specific situation. Learning an ideal strategy that maximizes the anticipated long-term benefit is the aim of RL.

7. Exploration and exploitation: Exploration and exploitation must be balanced in RL. Exploration refers to the agent's exploration of the environment to gather information about different actions and their consequences. Exploitation refers to the agent's use of the acquired knowledge to exploit the actions that have resulted in higher rewards in the past.

8. Value function: A certain state or state-action pair's expected cumulative reward is estimated by the value function. It provides a measure of the desirability of being in a certain state or taking a certain action. The agent learns to estimate the value function to guide its decision-making.

9. Algorithms: RL algorithms are used to learn the optimal policy or value function. Common RL algorithms include Q-learning, SARSA, and Deep Q-Networks (DQN). These algorithms use various techniques such as value iteration, policy iteration, and function approximation to learn and improve the agent's decision-making ability.

Robotics, gaming, recommendation systems, autonomous vehicles, and control systems are just a few of the fields where RL is used. Its ability to enable agents to learn from and adapt to complicated contexts makes it a potent paradigm for making decisions in fluid and unpredictable situations.

18.3.4 SEMI-SUPERVISED MACHINE LEARNING

SML and UML are used in a learning paradigm known as SSML. It uses a smaller amount of labeled data along with a bigger amount of unlabeled data to improve the performance of predictive models. In SSML, the objective is to use unlabeled data to extract relevant information and enhance the generalisation capability of the model. SSML assumes that labeled data is costly, time-consuming, or challenging to get. It is dependent on a limited quantity of labeled data that contains both the input features and the labels for the matching output features. SSML takes advantage of the bigger pool of unlabeled data that is readily available, where only input features are present without any corresponding output labels.

By identifying the data's underlying structure or patterns, the unlabeled data is used to speed up learning. It aids in the learning of representations, the discovery of undiscovered links, and the enhancement of model generalisation. A fundamental tenet of SSML is that the data's underlying structure or decision boundaries must match the labeling information. The program seeks to make use of this consistency to forecast the unlabeled data. Both SML and UML concepts are incorporated into SSML. Typical strategies include:

- Self-training: It entails building a model from the sparsely labeled data and utilizing it to predict outcomes from the unlabeled data. The labeled set is then supplemented with the confident predictions, which are utilised to iteratively retrain the model.
- Co-training: Co-training uses multiple views or representations of the data to learn from the unlabeled examples. Each view is associated with a separate classifier, and the classifiers are trained on the labeled data. Following their agreement, they successively exchange and modify the labels for the unlabeled occurrences.
- Transductive learning: In the specific dataset it was trained on, transductive learning seeks to forecast the labels of the instances that are unlabeled. It doesn't apply to fresh, unheard-of situations.
- Graph-based methods: The structure of the data is used by graph-based algorithms to propagate labels from labeled to unlabeled instances. Graph-based algorithms leverage the similarity or affinity between data points to estimate labels.

Evaluating SSML algorithms can be challenging since the true labels of the unlabeled data are generally unknown. Evaluation metrics typically consider the accuracy or error rate on the labeled data and the performance on a separate labeled test set.

When there is a lot of unlabeled data accessible but it is difficult or expensive to obtain labeled data, SSML can be used. It has been successfully used in areas such as natural language processing, image recognition, and fraud detection, where labeled data may be scarce but unlabeled data is abundant. By effectively utilizing both labeled and unlabeled data, SSML offers opportunities for improving model performance and reducing the need for extensive labeling efforts.

18.4 POPULAR MACHINE LEARNING ALGORITHMS

There are numerous ML algorithms available, each with its own characteristics, strengths, and limitations. Here are some commonly used machine learning algorithms:

18.4.1 DECISION TREES

DT are adaptable ML techniques that may be applied to both regression and classification tasks. Based on a sequence of binary judgments on individual characteristics, they divide the feature space into regions. DT can handle categorical and numerical data and are easily interpretable and understandable. But DT can overfit, particularly if the tree gets too deep and complicated. Pruning and establishing a maximum depth are two strategies that can be used to address this problem. Customer segmentation, medical diagnosis, and credit scoring are examples of typical DT use cases.

18.4.2 SUPPORT VECTOR MACHINE

SVM is a powerful technique for both classification and regression tasks. It looks for a hyperplane in a high-dimensional feature space that maximizes the separation between data points of different classes. Through the use of kernel functions, SVM can handle both linear and nonlinear decision boundaries. SVM works well with high-dimensional data and is capable of good generalization. SVM can, however, be computationally expensive, particularly when working with huge datasets. SVM is frequently used in bioinformatics, text categorization, and image classification.

18.4.3 Neural network

Deep neural networks in particular have drawn a lot of interest and been incredibly successful in a number of fields. They are modeled after the organisation and function of the human brain and are composed of interconnected artificial neurons stacked in layers.

A process known as training involves changing the weights of the connections between neurons, which allows NN to understand intricate patterns and relationships in data. They can successfully describe nonlinear interactions and manage large-scale, high-dimensional data. NN can be computationally demanding and require a lot of labelled data for training. They have had success with natural language processing, recommendation systems, and picture and speech recognition.

18.4.4 Ensemble methods (random forests and gradient boosting)

Ensemble approaches, which frequently produce better results, integrate several different individual models to create predictions. RF is made up of several decision trees, each of which is trained on a different subset of the data. The forecasts from all the trees are then integrated to get the final prediction. GB is an ensemble strategy that sequentially creates weak learners (usually decision trees), with each learner attempting to fix the errors of the prior learners. In addition to handling complex data well, ensemble methods are robust and have a good handle on noise and outliers. They perform better in generalisation and are less prone to overfitting. Numerous applications, including anomaly detection, object recognition, and click-through rate prediction, have seen success with ensemble approaches.

18.5 PERFORMANCE EVALUATION OF MACHINE LEARNING MODELS

The effectiveness of ML models must be evaluated in order to determine whether or not they are appropriate for use in practical applications. It involves assessing the model's capacity to generate precise forecasts or estimates based on the available data. Here are some typical metrics for measuring the effectiveness of ML models:

1. Accuracy: Accuracy is used to gauge the overall level of accuracy of the model's predictions. It is the percentage of events that were accurately predicted for all examples in the dataset. Despite being a frequently used statistic, accuracy may not be suitable for datasets with imbalances where the classes are represented differently.
2. Precision and recall: Recall and precision are two metrics that are widely used in binary classification problems. Precision is the proportion of correctly anticipated positive instances out of all cases that were accurately predicted as positive, while recall measures the proportion of correctly predicted positive instances out of all occurrences that were truly positive. Particularly when there is concern about the imbalance between classes, these measurements are useful.
3. F1 Score: The harmonic mean of recall and precision is known as the F1 score, which is a balanced assessment of a model's performance. As a result of its ability to combine precision and recall into a single metric, it is typically used when there is an imbalance across classes.
4. MSE: A common evaluation metric for issues involving regression is MSE. What is measured is the average squared difference between the expected and actual values. Lower MSE values, with 0 being the ideal fit, indicate better performance.
5. R-squared: The percentage of variation in the dependent variable that can be anticipated from the independent variables is calculated using the R-squared formula. Higher scores, which range from 0 to 1, imply a better fit of the model to the data.

6. Area Under the ROC Curve (AUC-ROC): AUC-ROC is a common evaluation statistic for binary classification issues. It evaluates the model's ability to distinguish between positive and negative cases by comparing the real positive rate against the false positive rate. Higher AUC-ROC values are indicative of better discrimination abilities.

7. Mean Average Precision (MAP): MAP is a frequent statistic in information retrieval and ranking issues. It evaluates the model's average accuracy across various retrieval queries. It gives an indication of the model's ability to prioritize relevant instances while accounting for precision and recall.

8. CV: The performance of the model is assessed using the CV technique, which folds the dataset into multiple subsets. Once the model has been iteratively trained and evaluated on numerous folds, the performance metrics are averaged across all folds. Thanks to CV, the performance of the model can be estimated more precisely, which also aids in minimizing overfitting-related issues.

When evaluating the performance of machine learning models, it is imperative to consider the particular task, the nature of the data, and the application requirements. Different evaluation methods might be better suitable for different situations. It is also essential to evaluate the model's performance using independent test data that wasn't used during training in order to obtain an unbiased evaluation.

To ensure the robustness and generalizability of machine learning models, model validation techniques, including cross-validation, are commonly employed. These techniques help assess how well the model performs on unseen data and mitigates issues like overfitting. Here are some techniques for model validation:

1. Train-test split: For this technique, the dataset is divided into a training set and a test set. The model is created on the training set, and the evaluation is done on the test set. The model's performance metrics on the test set can be used to predict how well the model will do when presented with fresh data. Making sure that the test set appropriately depicts the distribution of real-world data is essential.

2. K-fold cross-validation: The dataset is divided into nearly equal-sized subgroups or folds using the cross-validation resampling technique. The model is evaluated K times, with each evaluation using a different fold as the test set and the remaining folds as the training set. The performance of the model is then calculated by averaging the metrics over all iterations. K-fold cross-validation is particularly useful when the dataset is limited and needs to be used efficiently.

3. Stratified cross-validation: The class distribution is maintained in each fold thanks to a K-fold cross-validation variant known as stratified cross-validation. This approach is useful when working with datasets that are unbalanced and have unequal representation of the classes. It guarantees that each fold contains a representative distribution of the different classes.

4. Leave-One-Out Cross-Validation (LOOCV): An exception to K-fold cross-validation is LOOCV, where K is the number of instances in the dataset. One instance is left out as the test set in each iteration, and the model is trained using the other instances. LOOCV offers a thorough assessment of the model's performance, although it can be computationally costly for large datasets.

5. Stratified shuffle split: This method keeps the class distribution while randomly shuffling the dataset and dividing it into training and test sets. It is especially helpful when a rapid estimation of the model's performance is needed and the dataset is huge.

6. Nested cross-validation: When model selection or hyperparameter adjustment are involved, nested cross-validation is performed. It entails two cross-validation loops: an outer loop

to evaluate the model's performance on unobserved data, and an inner loop to select a model or adjust hyperparameters on the training set. This method helps avoid overfitting the hyperparameters to the test set and provides a more precise evaluation of the model's performance.

7. Holdout validation: In holdout validation, an additional validation set is set aside in addition to the training and test sets. After being trained on the training set and evaluated on the test set, the model is further validated on the validation set. With the use of this technique, the model can be adjusted before being used, and its performance can be assessed against yet another independent dataset.

By employing these model validation techniques, ML practitioners can assess the performance, robustness, and generalizability of their models. It helps in selecting the best model, identifying potential issues, and making informed decisions about model deployment and improvement.

Additionally, feature selection and DR methods are essential for overcoming the dimensionality curse and enhancing the effectiveness and interpretability of machine learning models. Let's examine their significance in more detail now:

1. Curse of dimensionality: When working with high-dimensional data, problems and difficulties are referred to as "the curse of dimensionality." The amount of data needed to effectively cover the feature space grows exponentially as the number of features or dimensions increases. As a result, there are more sparse data points, more sophisticated computations, and a greater possibility of overfitting.

2. Feature selection: The goal of feature selection is to choose the most useful and pertinent features from the initial list of features. We can decrease the dimensionality of the data, increase computing effectiveness, and lessen the chance of overfitting by choosing a subset of the characteristics. The removal of superfluous, pointless, or distracting components aids in improving the performance of the model.

3. Benefits of feature Selection:
   - Improved model performance: Feature selection enhances the performance of the model by eliminating noise and irrelevant information by concentrating on the most informative features. It improves the model's capacity to discover pertinent trends and generate reliable forecasts.
   - Faster training and inference: Faster training and inference times are obtained by reducing the dimensionality of the data through feature selection. The computational load is lighter with fewer features, enabling more effective model development and implementation.
   - Enhanced interpretability: Feature selection helps simplify the model by removing less important features. This results in a more interpretable model since the focus is on the most relevant and influential features. Interpretability is crucial in domains where understanding the model's decision-making process is essential, such as healthcare or finance.

4. Dimensionality reduction: The goal of dimensionality reduction approaches is to maintain the most important information while transforming high-dimensional data into a lower-dimensional representation. By lowering computing complexity and noise while preserving important patterns and structures in the data, it aids in overcoming the curse of dimensionality.

5. Benefits of dimensionality reduction:
   - Computational efficiency: By reducing the dimensionality of the data, dimensionality reduction techniques alleviate computational complexity, allowing models to train and make predictions more efficiently. This is especially important when working with large-scale datasets.

- Noise reduction: Data with high dimensions frequently includes noise or unimportant elements. Techniques for dimensionality reduction can assist eliminate or lessen the effect of noisy features, resulting in a clearer and more useful representation of the data.
- Visualization and interpretability: Dimensionality reduction can reduce the dimensions of the data, making it easier to visualise. This makes it possible to comprehend and interpret the data and its underlying structures better.
- Generalization: By lowering the danger of overfitting and enhancing the model's ability to generalize to new data, dimensionality reduction approaches aid in enhancing the generalisation capabilities of machine learning models.

Both feature selection and dimensionality reduction techniques contribute to enhancing the efficiency, interpretability, and generalizability of machine learning models. They help in mitigating the curse of dimensionality, improving model performance, and simplifying the learning process by focusing on the most relevant features or transforming the data into a more manageable representation.

18.6 REAL-WORLD APPLICATIONS OF MACHINE LEARNING

ML has found applications across diverse domains, including operations research, digital marketing, and quantitative finance. Here are some real-world examples of how machine learning is used in these fields:

1. Operations research:
   - Supply chain management: By anticipating demand patterns, maximizing inventory levels, and enhancing logistics and distribution procedures, ML algorithms can improve supply chain operations.
   - Resource allocation: ML techniques help optimize resource allocation in various domains, such as workforce scheduling, transportation planning, and energy management, to improve efficiency and reduce costs.
   - Predictive maintenance: ML models can predict equipment failures or maintenance needs, allowing for proactive maintenance scheduling and minimizing downtime.
2. Digital marketing:
   - Customer segmentation: Customers can be divided into groups based on their behavior, preferences, and demographics using ML algorithms, allowing for more precise marketing campaigns and tailored recommendations.
   - Predictive analytics: ML models can analyze customer data to predict customer behavior, such as churn prediction, purchase propensity, or lifetime value estimation, enabling companies to make data-driven marketing decisions.
   - Ad targeting and optimization: ML is used in programmatic advertising to target the right audience, optimize ad placement, and personalize ad content, resulting in more effective ad campaigns and higher conversion rates.
3. Quantitative finance:
   - Algorithmic trading: ML techniques are widely used in algorithmic trading to analyze historical data, identify patterns, and make predictions for automated trading decisions. This includes strategies such as trend detection, pattern recognition, and risk management.
   - Fraud detection: By examining patterns and anomalies in massive amounts of data, ML models are used to identify fraudulent activity in financial transactions, enabling quick identification and prevention of fraudulent behavior.

- Risk assessment: ML algorithms can assess credit risk, market risk, and portfolio risk by analyzing historical data and market trends, helping financial institutions make informed decisions and manage risk effectively.

These are just a few instances of how operations research, digital marketing, and quantitative finance are using machine learning. Machine learning is used in many different ways across a wide range of industries, including healthcare, manufacturing, cybersecurity, recommendation systems, natural language processing, and others. Due to its ability to examine massive amounts of data, identify patterns, and produce predictions, it is a useful tool for decision-making and optimizing across a range of fields.

18.7   CHALLENGES OF MACHINE LEARNING

ML brings numerous benefits and opportunities, but it also poses ethical considerations and challenges that need to be addressed. Here are some key areas of concern:

1. Privacy: ML often relies on large amounts of data, raising concerns about data privacy. There is a risk of personal and sensitive information being mishandled, shared without consent, or used for purposes other than those originally intended. Safeguarding data privacy and ensuring proper consent and data anonymization is crucial to maintain trust and protect individual privacy.
2. Bias and fairness: Biases existing in the training data for ML models may reflect them unintentionally. Biased information or biased algorithms can provide discriminatory results, which reinforce and maintain societal biases. To reduce bias and make sure that ML applications are fair and equitable, it is crucial to carefully curate and pre-process training data.
3. Transparency and explainability: Deep neural networks are only one example of ML algorithms that frequently function as "black boxes," making it challenging to comprehend and analyse their decision-making processes. Concerns concerning responsibility, trust, and potential biases are brought up by this lack of openness. It is essential to provide interpretable models and methods to illuminate the reasoning behind the model's predictions, especially in key fields like healthcare or the legal system.
4. Accountability and responsibility: As ML models are deployed in various applications, determining accountability and responsibility can be challenging. When an ML model makes a decision or prediction with real-world consequences, it is important to establish clear lines of accountability and ensure that decisions can be traced back to responsible parties. Ethical guidelines and regulations can help define the responsibilities of developers, users, and stakeholders.
5. Data Quality and representativeness: For ML models, training data's representativeness and correctness are essential. Because of inadequate, biased, or unfair data, predictions may be incorrect or unfair. It is crucial to make sure that the data used to train the models is diverse, representative, and of high quality in order to lessen biases and improve the overall performance and fairness of the models.
6. Security and adversarial attacks: The manipulation or deception of ML models by hostile actors using carefully constructed inputs is a threat to their vulnerability. Such attacks may result in grave repercussions like system failure or misclassification. To reduce these risks, effective security mechanisms and procedures are required, such as data encryption, anomaly detection, and adversarial training.

Addressing these ethical considerations and challenges requires a multi-faceted approach involving developers, policymakers, researchers, and stakeholders. It involves establishing clear ethical guidelines, ensuring transparency and accountability, fostering diverse and inclusive data representation, implementing privacy-preserving measures, and continuously monitoring and evaluating the ethical impact of machine learning applications. By prioritizing ethical considerations, we can harness the potential of machine learning while safeguarding individual rights and societal well-being.

REFERENCES

1. Carbonneau, R. , Laframboise, K. & Vahidov, R. "Application of machine learning techniques for supply chain demand forecasting", *European Journal of Operational Research*, Volume 184, Issue 3 (2008): 1140–1154.
2. Kraus, M., Feuerriegel, S., & Oztekin, A. "Deep learning in business analytics and operations research: Models, applications and managerial implications", *European Journal of Operational Research*, Volume 281, Issue 3 (2020): 628–641.
3. Boddu, R. S. K., Santoki, A. A., Khurana, S., Koli, P. V., Rai, R., & Agrawal, A. "An analysis to understand the role of machine learning, robotics and artificial intelligence in digital marketing", *Materials Today: Proceedings*, Volume 56, Part 4 (2022): 2288–2292.
4. Miklosik, A., & Evans, N. "Impact of big data and machine learning on digital transformation in marketing: A literature review," *IEEE Access*, volume 8 (2020): 101284–101292.
5. Sarker, I. H. "Machine learning: Algorithms, real-world applications and research directions", *SN Computer Science*, Volume 2 (2021): 160.
6. Saura, J. R. "Using data sciences in digital marketing: Framework, methods, and performance metrics", *Journal of Innovation & Knowledge*, Volume 6, Issue 2 (2021): 92–102.
7. Agarwal, R., & Dhar,V. "Editorial — Big data, data science, and Analytics: The opportunity and challenge for is research", *Information Systems Research*, Volume 25, Issue 3 (2014): 443–448.
8. Heikkila, J. "From supply to demand chain management: Efficiency and customer satisfaction", *Journal of Operations Management,* Volume 20, Issue 6 (2002): 747–767.
9. Lim, E.-P., Chen, H. & Chen, G. "Business intelligence and analytics: Research directions", *ACM Transactions on Management Information Systems*, Volume 3, Issue 4 (2013).
10. Lessmann, S., Baesens, B., Seow, H. V., & Thomas, L. C. "Benchmarking state-of-the-art classification algorithms for credit scoring: An update of research", *European Journal of Operational Research*, Volume 247, Issue 1 (2015): 124–136.
11. Li, W., Chen, H., & Nunamaker, J. F. "Identifying and profiling key sellers in the cyber carding community: Azsecure text mining system", *Journal of Management Information Systems*, Volume 33, Issue 4 (2017): 1059–1086.
12. Cox, A., Sanderson, J., & Watson, G. "Supply chains and power regimes: Toward an analytic framework for managing extended networks of buyer and supplier relationships", *Journal of Supply Chain Management*, Volume 37, Issue 2 (2001): 28–35.
13. Tan, K. C. "A framework of supply chain management literature", *European Journal of Purchasing & Supply Management* Volume 7, Issue 1 (2001): 39–48.
14. Bergen, K. J., Johnson, P. A., Maarten, V., & Beroza, G. C. "Machine learning for data-driven discovery in solid Earth geoscience", *Science* Volume 363 (2019).
15. Mitchell, T. *Machine Learning*, 1st ed., McGraw Hill (1997).
16. Zhu, X., & Goldberg, A. B. "Introduction to semi-supervised learning", *Synthesis Lectures on Artificial Intelligence and Machine Learning* Volume 3 (2009): 1–130.
17. Sarker, I. H., Abushark, Y. B., Alsolami, F., & Khan, A. "Intrudtree: A Machine Learning based cyber security intrusion detection model", *Symmetry* Volume 12, Issue 5 (2020): 754.
18. Ardabili, S. F., et al. "Covid-19 outbreak prediction with machine learning", *Algorithms* Volume13, Issue 10 (2020): 249.

19. De Amorim, R. C., "Constrained clustering with Minkowski weighted k-means", *IEEE 13th International Symposium on Computational Intelligence and Informatics*, (2012): 13–17.
20. Kamble, S. S., Gunasekaran, A., & Gawankar, S. A. "Achieving sustainable performance in a data-driven agriculture supply chain: a review for research and applications", *International Journal of Production Economics*, Volume 219 (2020):179–94.

19 Role of Fertamean Neutrosophic Sets for Decision Making Modelling in Machine Learning

R. Narmada Devi, Regan Murugesan, Nagadevi Bala Nagaram, Kala Raja Mohan, and Sathish Kumar Kumaravel

19.1 INTRODUCTION

Computers can learn on their own, without being taught, to carry out certain tasks, according to the core idea of machine learning. Given its capacity to automatically identify patterns in data and predict future outcomes, machine learning has become a technique that is increasingly in demand in recent years. This may be quite helpful for making judgments in a variety of areas, including financial trading and medical diagnosis.

In order to resolve ambiguous circumstances, fuzzy theory [8] is essential. After the fuzzy set was extended, a new set known as intuitionistic fuzzy sets [1] was developed. The most general idea, known as the Neutrosophic Sets (NS), was later developed by Smarandache [6]. There are other NS variations described in the literature, such as Pythagorean NS and Bipolar NS. Incomplete and ambiguous data in MCDM challenges have exposed the variability of approximative reasoning and pushed the decision-making process to new advancements and innovation. The Hausdorff distance between two NS's was determined by Broumi [2, 3]. Digital neutrosophic topological spaces were first suggested by Narmada Devi [4]. Antony Crispin Sweety is researching a brand-new notion of Fermatean Neutrosophic Sets (FNS) [5].

This chapter introduces and discusses the concept of the similarity measure known as tangent similarity measures for $\mathcal{FNS}$'s and also its characteristics. Using $\mathcal{FNS}$'s to represent uncertain MCDM issues using Python programming to analyse a real-world decision-making problem, we propose a decision-making method that may be applied to tackle such situations.

19.2 OPERATION OF FERMATEAN NEUTROSOPHIC SETS ($\mathcal{FNS}$)

Definition 2.1 A Fermatean neutrosophic set in short, $[\mathcal{FNS}]$ $A = \left\{ A_{\mathbb{T}}(\mathfrak{p}), A_{\mathbb{I}}(\mathfrak{p}), A_{\mathbb{F}}(\mathfrak{p}) : \mathfrak{p} \in X \right\}$ on $\mathbb{X}$ satisfies the following properties:

(i) $A_{\mathbb{T}}(\mathfrak{p}), A_{\mathbb{I}}(\mathfrak{p}), A_{\mathbb{F}}(\mathfrak{p}) \in [0,1]$
(ii) $0 \le A_{\mathbb{T}}(\mathfrak{p})^3 + A_{\mathbb{F}}(\mathfrak{p})^3 \le 1$
(iii) $0 \le A_{\mathbb{I}}(\mathfrak{p})^3 \le 1$
(iv) $0 \le A_{\mathbb{T}}(\mathfrak{p})^3 + A_{\mathbb{I}}(\mathfrak{p})^3 + A_{\mathbb{F}}(\mathfrak{p})^3 \le 2$, for all $\mathfrak{p} \in \mathbb{X}$.

DOI: 10.1201/9781003433309-19

The family of $\mathcal{FNS}$'s on X is denoted $\mathcal{FNS}(\mathbb{X})$.

Definition 2.2 Let $\mathcal{M}$ and $\mathcal{L}$ be any two $\mathcal{FNS}'s$ on $\mathbb{X}$. Then

(i) $\mathcal{M}^c = \left\{ \left\langle A_{\mathbb{F}}(\mathfrak{p}), 1 - A_{\mathbb{I}}(\mathfrak{p}), A_{\mathbb{T}}(\mathfrak{p}) \right\rangle : \mathfrak{p} \in X \right\}$ is the complement of $\mathcal{FN}$ set $\mathcal{M}$.

(ii) $\mathcal{M} \cap \mathcal{L} = \left\{ \left\langle \min\left(\mathcal{M}_{\mathbb{T}}(\mathfrak{p}), \mathcal{L}_{\mathbb{T}}(\mathfrak{p})\right), \min\left(\mathcal{M}_{\mathbb{F}}(\mathfrak{p}), \mathcal{L}_{\mathbb{F}}(\mathfrak{p})\right), \max\left(\mathcal{M}_{\mathbb{F}}(\mathfrak{p}), \mathcal{L}_{\mathbb{F}}(\mathfrak{p})\right) \right\rangle \right\}$.

(iii) $\mathcal{M} \cup \mathcal{L} = \left\{ \left\langle \max\left(\mathcal{M}_{\mathbb{T}}(\mathfrak{p}), \mathcal{L}_{\mathbb{T}}(\mathfrak{p})\right), \max\left(\mathcal{M}_{\mathbb{F}}(\mathfrak{p}), \mathcal{L}_{\mathbb{F}}(\mathfrak{p})\right), \min\left(\mathcal{M}_{\mathbb{F}}(\mathfrak{p}), \mathcal{L}_{\mathbb{F}}(\mathfrak{p})\right) \right\rangle \right\}$.

(iv) $0_{\mathcal{FN}} = \left\{ \left\langle 0, 0, 1 \right\rangle : \mathfrak{p} \in \mathbb{X} \right\}$ is a Null $\mathcal{FN}$ set on $\mathbb{X}$

(v) $1_{\mathcal{FN}} = \left\{ \left\langle 1, 1, 0 \right\rangle : \mathfrak{p} \in \mathbb{X} \right\}$ is a whole $\mathcal{FN}$ Set on $\mathbb{X}$

Definition 2.3 The similarity measure between $\mathcal{G}$ and $\mathcal{H}$ is defined as $\varpi(\mathcal{G}, \mathcal{H}) = \varpi_{\mathbb{T}}(\mathcal{G}, \mathcal{H}), \varpi_{\mathbb{I}}(\mathcal{G}, \mathcal{H}), \varpi_{\mathbb{F}}(\mathcal{G}, \mathcal{H})$ where $\varpi : \mathcal{FNS}(\mathbb{X}) \times \mathcal{FNS}(\mathbb{X}) \to [0,1]$ such that ϖ satisfies the following conditions:

(i) $\varpi(\mathcal{G}, \mathcal{H}) = \varpi(\mathcal{B} \ \mathcal{A}), \varpi(\mathcal{G}, \mathcal{H}) = (1, 0, 0) = 1$
(ii) if $\mathcal{A} = \mathcal{B}$ for all $\mathcal{G}, \mathcal{H} \in \mathcal{FNS}(\mathbb{X})$, $\varpi_{\mathbb{T}}(\mathcal{G}, \mathcal{H}) \geq 0$, $\varpi_{\mathbb{I}}(\mathcal{G}, \mathcal{H}) \geq 0$ and $\varpi_{\mathbb{F}}(\mathcal{G}, \mathcal{H}) \geq 0$
(iii) If $\mathcal{G} \subseteq \mathcal{H} \subseteq \mathcal{D}$ for all $\mathcal{G}, \mathcal{H}, \mathcal{D} \in \mathcal{FNS}(\mathbb{X})$, then $\varpi(\mathcal{G}, \mathcal{H}) \geq \varpi(\mathcal{G}, \mathcal{D})$ and $\varpi(\mathcal{H}, \mathcal{D}) \geq \varpi(\mathcal{G}, \mathcal{D})$.

where $\varpi_{\mathbb{T}}(\mathcal{G}, \mathcal{H})$, $\varpi_{\mathbb{I}}(\mathcal{G}, \mathcal{H})$ and $\varpi_{\mathbb{F}}(\mathcal{G}, \mathcal{H})$ are represented as the grade of $\mathbb{T}$-similarity, $\mathbb{I}$-similarity and $\mathbb{F}$-similarity between $\mathcal{G}$ and $\mathcal{H}$ respectively.

Definition 2.4 A tangent similarity measures between $\mathcal{FNS}'s$ $\mathcal{M}$ and $\mathcal{L}$ on $\mathbb{X}$ is defined as follows:

$$TS(\mathcal{M}, \mathcal{L}) = \frac{1}{n} \left[\sum_{i=1}^{n} 1 - \tan\left[\left(\frac{\pi}{12}\right) \left[\left| T_{\mathcal{M}}(\mathfrak{p}_i) - T_{\mathcal{L}}(\mathfrak{p}_i) \right| + \left| I_{\mathcal{M}}(\mathfrak{p}_i) - I_{\mathcal{L}}(\mathfrak{p}_i) \right| + \left| F_{\mathcal{M}}(\mathfrak{p}_i) - F_{\mathcal{L}}(\mathfrak{p}_i) \right| \right] \right] \right]$$

Proposition: 2.1 Let $\mathcal{M}$ and $\mathcal{L}$ be any two $\mathcal{FNS}'s$ on $\mathbb{X}$. Then

(i) $0 \leq TS(\mathcal{M}, \mathcal{L}) \leq 1$

(ii) $TS(\mathcal{M}, \mathcal{L}) = 1$ if and only if $\mathcal{M} = \mathcal{L}$;

(iii) $TS(\mathcal{M}, \mathcal{L}) = TS(\mathcal{L}, \mathcal{M})$

(iv) If $\mathcal{Q}$ is a FNS in X and $\mathcal{M} \subseteq \mathcal{Q} \subseteq \mathcal{L}$, then $TS(\mathcal{M}, \mathcal{Q}) \leq TS(\mathcal{M}, \mathcal{L})$ and $TS(\mathcal{M}, \mathcal{Q}) \leq TS(\mathcal{L}, \mathcal{Q})$.

Proof:

(i) All the three functions $\mathbb{T}$ and $\mathbb{F}$ of $\mathcal{FNS}'s$ and the value of the tangent function lie between 0 and 1 respectively. Hence $0 \leq TS(\mathcal{M}, \mathcal{L}) \leq 1$.

(ii)   For any two *FNSs* $\mathcal{M}$ and $\mathcal{L}$, if
$\mathcal{M} = \mathcal{L} \Rightarrow \mathbb{T}_{\mathcal{M}}(\mathfrak{x}) = \mathbb{T}_{\mathcal{L}}(\mathfrak{x}), \mathbb{I}_{\mathcal{M}}(\mathfrak{p}) = \mathbb{I}_{\mathcal{L}}(\mathfrak{x})$ and $\mathbb{F}_{\mathcal{M}}(\mathfrak{p}) = \mathbb{F}_{\mathcal{L}}(\mathfrak{p})$.
Therefore, $TS(\mathcal{M}, \mathcal{L}) = 1$.
Conversely, if $TS(\mathcal{M}, \mathcal{L}) = 1$.
$\Rightarrow \left| \mathbb{T}_{\mathcal{M}}(\mathfrak{p}_i) - \mathbb{T}_{\mathcal{L}}(\mathfrak{p}_i) \right| = \left| \mathbb{I}_{\mathcal{M}}(\mathfrak{p}_i) - \mathbb{I}_{\mathcal{L}}(\mathfrak{p}_i) \right| = \left| \mathbb{F}_{\mathcal{M}}(\mathfrak{p}_i) - \mathbb{F}_{\mathcal{L}}(\mathfrak{p}_i) \right| = 0$. Therefore, $\mathcal{M} = \mathcal{L}$.

(iii)   The proof is obvious.

(iv)   Since $\mathbb{T}_{\mathcal{M}}(\mathfrak{p}) \leq \mathbb{T}_{\mathcal{L}}(\mathfrak{p}) \leq \mathbb{T}_{\mathcal{Q}}(\mathfrak{p}), \mathbb{I}_{\mathcal{M}}(\mathfrak{p}) \leq \mathbb{I}_{\mathcal{L}}(\mathfrak{p}) \leq \mathbb{I}_{\mathcal{Q}}(\mathfrak{p})$ and

$$\mathbb{F}_{\mathcal{M}}(\mathfrak{p}) \geq \mathbb{F}_{\mathcal{L}}(\mathfrak{p}) \geq \mathbb{F}_{\mathcal{Q}}(\mathfrak{p}).$$

$$\Rightarrow \left| \mathbb{T}_{\mathcal{M}}(\mathfrak{p}) - \mathbb{T}_{\mathcal{L}}(\mathfrak{p}) \right| \leq \left| \mathbb{T}_{\mathcal{M}}(\mathfrak{p}) - \mathbb{T}_{\mathcal{Q}}(\mathfrak{p}) \right|,$$

$$\left| \mathbb{I}_{\mathcal{M}}(\mathfrak{p}) - \mathbb{I}_{\mathcal{L}}(\mathfrak{p}) \right| \leq \left| \mathbb{I}_{\mathcal{M}}(\mathfrak{p}) - \mathbb{I}_{\mathcal{Q}}(\mathfrak{p}) \right| \text{ and }$$

$$\left| \mathbb{F}_{\mathcal{M}}(\mathfrak{p}) - \mathbb{F}_{\mathcal{L}}(\mathfrak{p}) \right| \geq \left| \mathbb{F}_{\mathcal{M}}(\mathfrak{p}) - \mathbb{F}_{\mathcal{Q}}(\mathfrak{p}) \right|$$

$$\left| \mathbb{T}(\mathfrak{p}) - \mathbb{T}_{\mathcal{Q}}(\mathfrak{p}) \right| \leq \left| \mathbb{T}_{\mathcal{M}}(\mathfrak{p}) - \mathbb{T}_{\mathcal{Q}}(\mathfrak{p}) \right|, \left| \mathbb{I}_{\mathcal{L}}(\mathfrak{p}) - \mathbb{I}_{\mathcal{Q}}(\mathfrak{p}) \right| \leq \left| \mathbb{I}_{\mathcal{M}}(\mathfrak{p}) - \mathbb{I}_{\mathcal{Q}}(\mathfrak{p}) \right| \text{ and }$$

$$\left| \mathbb{F}_{\mathcal{L}}(\mathfrak{p}) - \mathbb{F}_{\mathcal{Q}}(\mathfrak{p}) \right| \geq \left| \mathbb{F}_{\mathcal{M}}(\mathfrak{p}) - \mathbb{F}_{\mathcal{Q}}(\mathfrak{p}) \right|$$

Hence, $TS(\mathcal{M}, \mathcal{Q}) \leq TS(\mathcal{M}, \mathcal{L})$ and $TS(\mathcal{M}, \mathcal{Q}) \leq TS(\mathcal{L}, \mathcal{Q})$.

19.3 $\mathcal{FN}$ DECISION MAKING ALGORITHM

Let $= \left\{ \mathring{A}_1, \mathring{A}_2, \dots, \mathring{A}_m \right\}$ be the collection of persons $\mathbb{C} = \left\{ \mathbb{C}_1, \mathbb{C}_2, \dots, \mathbb{C}_n \right\}$ be the set of parameters or criteria of each of the persons and $\mathcal{B} = \left\{ \mathcal{B}_1, \mathcal{B}_2, \dots, \mathcal{B}_l \right\}$ be the family of alternatives of persons.

Step-1 Estimate the connection between persons and attributes ($\mathfrak{R}_1$)
The connection between person $\mathring{A}_m$ (j = 1, 2, ..., m) and the attribute $\mathbb{C}_j$ (k = 1, 2, ..., n) can be given in Table 19.1 in terms of $\mathcal{FN}$ sets.

Step-2: Find the association between attributes and alternatives ($\mathfrak{R}_2$)
The association between attributes and alternatives $(\mathfrak{R}_2)$ can be represented in the form of decision Table 19.2 based on the $\mathcal{FN}$ sets.

Step-3 Calculation of the tangent similarity measures
By using the code in Python, calculate the tangent similarity measure between the two relations $\mathfrak{R}_1$ and $\mathfrak{R}_2$ based on $\mathcal{FN}$ sets using the formulas

TABLE 19.1
Connection between pupils and attributes

$\mathfrak{R}_1$	$\mathbb{C}_1$	$\mathbb{C}_2$	$\dots \mathbb{C}_n$
$\mathring{A}_1$	$\mathbb{T}_{11}, \mathbb{I}_{11}, \mathbb{F}_{11}$	$\mathbb{T}_{12}, \mathbb{I}_{12}, \mathbb{F}_{12}$	$\dots \mathbb{T}_{1n}, \mathbb{I}_{1n}, \mathbb{F}_{1n}$
$\mathring{A}_2$	$\mathbb{T}_{21}, \mathbb{I}_{21}, \mathbb{F}_{21}$	$\mathbb{T}_{22}, \mathbb{I}_{22}, \mathbb{F}_{22}$	$\dots \mathbb{T}_{2n}, \mathbb{I}_{2n}, \mathbb{F}_{2n}$
$\dots$	$\dots$	$\dots$	$\dots \dots$
$\mathring{A}_m$	$\mathbb{T}_{m1}, \mathbb{I}_{m1}, \mathbb{F}_{m1}$	$\mathbb{T}_{m2}, \mathbb{I}_{m2}, \mathbb{F}_{m2}$	$\dots \mathbb{T}_{mn}, \mathbb{I}_{mn}, \mathbb{F}_{mn}$

TABLE 19.2

Association between attributes and alternatives

$\mathfrak{R}_2$	$\mathcal{B}_1$	$\mathcal{B}_2$	... $\mathcal{B}_l$
$\mathbb{C}_1$	$\mathbb{T}_{11}, \mathbb{I}_{11}, \mathbb{F}_{11}$	$\mathbb{T}_{12}, \mathbb{I}_{12}, \mathbb{F}_{12}$	... $\mathbb{T}_{1l}, \mathbb{I}_{1l}, \mathbb{F}_{1l}$
$\mathbb{C}_2$	$\mathbb{T}_{21}, \mathbb{I}_{21}, \mathbb{F}_{21}$	$\mathbb{T}_{22}, \mathbb{I}_{22}, \mathbb{F}_{22}$	... $\mathbb{T}_{2l}, \mathbb{I}_{2l}, \mathbb{F}_{2l}$
...	...	...	......
$\mathbb{C}_n$	$\mathbb{T}_{n1}, \mathbb{I}_{n1}, \mathbb{F}_{n1}$	$\mathbb{T}_{n2}, \mathbb{I}_{n2}, \mathbb{F}_{n2}$	... $\mathbb{T}_{nl}, \mathbb{I}_{nl}, \mathbb{F}_{nl}$

TABLE 19.3

$\mathcal{TS}$ Measures

$\mathfrak{R}_1 \times \mathfrak{R}_2$	$\mathcal{B}_1$	$\mathcal{B}_2$	... $\mathcal{B}_l$
$Å_1$	TSm_{11}	TSm_{12}	... TSm_{1l}
$Å_2$	TSm_{21}	TSm_{22}	... TSm_{2l}
...	...	...	......
$Å_m$	TSm_{m1}	TSm_{m2}	... TSm_{ml}

$$TS(M,L) = \frac{1}{n}\left[\sum_{i=1}^{n} 1 - tan\left[\left(\frac{\pi}{12}\right)\left[\left|T_M(u_i) - T_L(u_i)\right| + \left|I_M(u_i) - I_L(u_i)\right| + \left|F_M(u_i) - F_L(u_i)\right|\right]\right]\right]$$

and interpret the score values in Table 19.3.

Step-4: Ordering the alternatives

Arrange the score values in descending order for the Table 19.3. The decision maker gives the preference order based the ordering of alternatives with respect to each pupils.

Step-5 Decision

Largest score value reflects the best alternative among the others.

19.4   APPLICATION OF TANGENT SIMILARITY MEASURE OF $\mathcal{FN}$ SETS

The pupils should be taken into consideration while choosing for an education program in higher secondary after passing their secondary exams. After Class XII, the students pursue the courses of their choosing and concentrate on improving their job chances. The majority of students make decisions at this critical moment that they subsequently regret because they are overly confused. The decision of which direction to go is frequently challenging for students.

Let $Å = \left\{ Å_1 = Raj\ Å_2 = Naruva\ Å_3 = Devi \right\}$ be set of all three pupils, $\mathcal{B} = \left\{\mathcal{B}_1, \mathcal{B}_2, \mathcal{B}_3, \mathcal{B}_4\right\}$ be a collection of instructional streams where $\mathcal{B}_1$ is a course in science, $\mathcal{B}_2$ is a course in the humanities or arts, $\mathcal{B}_3$ is a business programme and $\mathcal{B}_4$ is a type of career training respectively. Let $\mathbb{C} = \left\{\mathbb{C}_1, \mathbb{C}_2, \mathbb{C}_3, \mathbb{C}_4, \mathbb{C}_5\right\}$ be a collection of qualities where $\mathbb{C}_1$ is basic scientific and mathematical understanding, $\mathbb{C}_2$ is depth of language, $\mathbb{C}_3$ is successful performance in secondary exams, $\mathbb{C}_4$ is concentration and $\mathbb{C}_5$ is working diligently, respectively. Our approach helps to assess the pupils and determine which educational track is best for them based on $\mathcal{FN}$ sets and given the following Table 19.1 and Table 19.2. Using the steps below, the decision-making process is demonstrated.

TABLE 19.4
Connection between students and attributes

$\mathfrak{R}_1$	$\mathbb{C}_1$	$\mathbb{C}_2$	$\mathbb{C}_3$	$\mathbb{C}_4$	$\mathbb{C}_5$
$\AA_1 = $ *Raj*	(0.8, 0.2, 0.2)	(0.7, 0.2, 0.1)	(0.6, 0.3, 0.3)	(0.8, 0.3, 0.2)	(0.5, 0.4, 0.2)
$\AA_2 = $ *Naruva*	(0.7, 0.1, 0.3)	(0.6, 0.3, 0.2)	(0.7, 0.2, 0.1)	(0.7, 0.4, 0.1)	(0.8, 0.0, 0.2)
$\AA_3 = $ *Devi*	(0.5, 0.3, 0.3)	(0.6, 0.3, 0.2)	(0.7, 0.3, 0.1)	(0.8, 0.0, 0.2)	(0.8, 0.3, 0.1)

TABLE 19.5
Association between attributes and educational streams

$\mathfrak{R}_2$	$\mathcal{B}_1$	$\mathcal{B}_2$	$\mathcal{B}_3$	$\mathcal{B}_4$
$\mathbb{C}_1$	(0.8, 0.3, 0.1)	(0.9, 0.1, 0.2)	(0.7, 0.3, 0.2)	(0.8, 0.2, 0.2)
$\mathbb{C}_2$	(0.5, 0.4, 0.1)	(0.6, 0.3, 0.2)	(0.7, 0.4, 0.0)	(0.9, 0.1, 0.1)
$\mathbb{C}_3$	(0.6, 0.4, 0.1)	(0.7, 0.3, 0.3)	(0.7, 0.3, 0.2)	(0.9, 0.1, 0.2)
$\mathbb{C}_4$	(0.6, 0.2, 0.2)	(0.7, 0.2, 0.2)	(0.8, 0.2, 0.1)	(0.5, 0.4, 0.2)
$\mathbb{C}_5$	(0.7, 0.3, 0.2)	(0.6, 0.3, 0.2)	(0.7, 0.4, 0.1)	(0.6, 0.4, 0.1)

TABLE 19.6
$\mathcal{TS}$ Measures among students and educational streams

$\mathcal{TS}$	$\mathcal{B}_1$	$\mathcal{B}_2$	$\mathcal{B}_3$	$\mathcal{B}_4$
$\AA_1 = $ *Raj*	0.92127	**0.94757**	0.93707	0.92107
$\AA_2 = $ *Naruva*	**0.92644**	0.90014	0.90533	0.87874
$\AA_3 = $ *Devi*	0.92653	0.92096	**0.93177**	0.86284

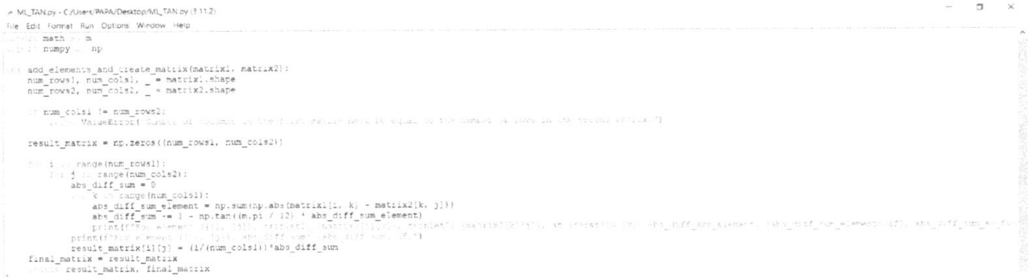

FIGURE 19.1 Coding in Python for Decision Table.

Step-1: The connection between pupils' qualities and their own in the form of $\mathcal{FN}$ sets is presented in the Table 19.4.

Step-2: The association between a student's characteristics and academic pathways in the form of $\mathcal{FN}$ sets is presented in the Table 19.5.

Step-3: Determine the tangent similarity measures between the Table 19.4 and Table 19.5 which are calculated using Python code as in Figure 19.1. The obtained score values are presented in Table 19.6.

```
ML_Tanline.py - C:/Users/PAPA/Desktop/ML_Tanline.py (3.11.2)

File  Edit  Format  Run  Options  Window  Help
# Library Import(numpy and matplotlib)
import matplotlib.pyplot as plot
import pandas as pd

# Make a data definition
_data = {'B1': [0.92127, 0.92644, 0.92653],
         'B2': [0.94757, 0.90014, 0.92096],
         'B3':[0.93707,0.90533, 0.93177],
         'B4':[0.92107,0.87874,0.86284]}
_df = pd.DataFrame(_data,columns=['B1', 'B2','B3', 'B4'], index = ['A1', 'A2', 'A3'])

# Multiple bar chart
_df.plot.bar()

# Display the plot
plot.show()
```

FIGURE 19.2 Coding in Python for Multiple Bar Chart Graph.

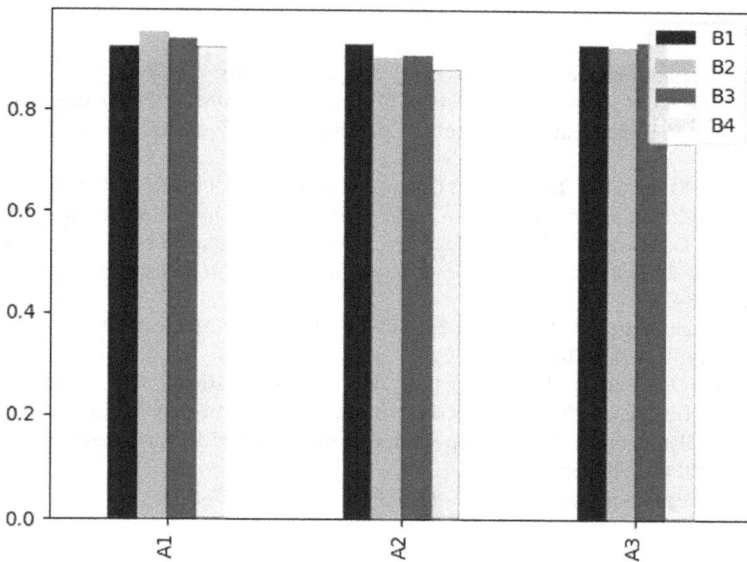

FIGURE 19.3 Multiple Bar Graph in Python.

Step-4: The highest TS measure values for each student with their respective educational stream are established and based on that decisions are taken and interpreted using Python coding in Figure 19.2 and Figure 19.3.

19.5 CONCLUSION

Based on TS measure values from the Table 19.7 pupils can choose their educational route in an appropriate manner. Therefore student *Raj* should select an arts route $\mathcal{B}_2$, student *Naruva*s should select a science route $\mathcal{B}_1$ and student *Devi* should select the science route $\mathcal{B}_3$.

The two fundamental foundations of the data mining process: similarity measure and machine leaching in uncertainty information, particularly $\mathcal{FN}$ environments are presented in this chapter. This reasoning might play a significant part in data mining strategies. The tangent similarity measure

TABLE 19.7
Highest Ranking Score Values

Highest Ranking Score Values	$\mathcal{TS}$
$\mathring{A}_1 = Raj$	0.94757 for $\mathcal{B}_2$
$\mathring{A}_2 = Naruva$	0.92644 for $\mathcal{B}_1$
$\mathring{A}_3 = Devi$	0.93177 for $\mathcal{B}_3$

is defined for attribute values for each of the alternatives using Python programming and this method approach for the data mining process contains large amounts of data with indeterminacy information. Future developments in machine learning will allow for this to be applied to the interval $\mathcal{FN}$ environment.

REFERENCES

(1) Atanassov, K. Intuitionistic Fuzzy Sets, *Fuzzy Sets and Systems*, 20(1986), 87–96.
(2) Jansi, R., Mohana, K., and Smarandache, F. Correlation Measure for Pythagorean Neutrosophic Sets with T and F as Dependent Neutrosophic Components, *Neutrosophic Sets and System*, 30 (2019), 202–212.
(3) Jansi, R., and Mohana, K. Tangent and Cotangent Similarity Measures of Pythagorean Neutrosophic Sets with T and F are Dependent Neutrosophic Components, *Infokara Research*, 8 (2019), 657–671.
(4) Rathinam, N. D., Rasappan, S., and Obaid, A. J. Novel on Digital Neutrosophic Topological Spaces, *Proceedings of 2nd International Conference on Mathematical Modeling and Computational Science*, 29 (2022), 213–219.
(5) Sanapati, T., and Yager, R. Fermatean Fuzzy Sets, *Journal of Ambient Intelligence and Humanized Computing*, 11 (2019), 663–674.
(6) Smarandache, F. (editor), Proceedings of the First International Conference on Neutrosophy, Neutrosophic Logic, *Set, Probability and Statistics*, University of New Mexico, Gallup (2002).
(7) Yager, R. R. Pythagorean membership grades in multicriteria decision making. *IEEE Trans Fuzzy System*, 22 (2014), 958–965.
(8) Zadeh, L. A. Fuzzy Sets, *Informatics and Control*, 8 (1965), 338–353.

20 Performance Evaluation of Machine Learning Algorithms in the Field of Security-Malware Detection

Aswathy K. Cherian, E. Poovammal, and M. Vaidhehi

20.1 INTRODUCTION

The digital age has brought about enormous progress and new possibilities, which have altered our everyday lives and modes of interaction. However, with the development of technological advances comes the risk of cyber threats and attacks, posing serious challenges to people, businesses, and governments. Malware presents itself as a persistent and widespread concern amongst the multiple risks in the cyber security [1] scene. Malware, an acronym for "malicious software," encompasses a broad category of malicious programs that aim to exploit security holes in computers and networks. Computers can be infiltrated by malicious software in many ways, such as through phishing emails, compromised websites, and tainted software downloads and installations. Malware can perform a variety of destructive behaviors while inside a system, including data theft, ransom demands, system interruption, and unauthorized access to sensitive information. Malware detection is critical in security and cannot be understated. As the volume and sophistication of cyber threats increase, effective malware detection [2] becomes critical to maintaining data and system confidentiality, integrity, and availability. Successful malware assaults can have devastating effects, resulting in financial losses, reputational damage, and potential legal responsibilities. Malware evolution presents cyber security professionals with an ever-increasing problem. Simple and static malware signatures were typically sufficient in the past to identify known threats [3]. However, fraudsters have mastered the art of obfuscating and altering malware code, making traditional signature-based detection systems ineffective against new and emerging malware strains. As cyber threats have become more complex, they can elude traditional security measures, necessitating the use of innovative and adaptive malware detection technologies. Considering modern malware is always evolving, it is essential to constantly monitor and analyze system behaviors, making machine learning an attractive solution. Examining and contrasting several machine-learning approaches to malware detection is the main thrust of this study. Malware may now be detected automatically and dynamically thanks to machine learning's use of statistical analysis and pattern recognition. By analyzing massive datasets comprising both harmful and benign files, machine learning algorithms can learn to distinguish between the two [4-6]. Members of the same malware family can be compared using either static or dynamic learning techniques to identify shared behavioral characteristics [7]. Unlike inert analysis, which assesses the contents of files without running them, dynamic analysis considers the behaviour of potentially harmful files by monitoring data flows, logging function calls, and embedding monitoring code into dynamic binaries [8]. This static and behavioral data can be used by machine learning algorithms to characterize the changing structure of modern malware, allowing them to detect more advanced malware attacks than can be detected by signature-based methods. When it comes to identifying brand new malware, machine learning-based solutions are far superior to signature-based approaches. Deep learning techniques that can do feature engineering on their own can be utilised to improve feature collection and representation [9]. Different malware analysis techniques are shown in Figure 20.1.

The purpose of this research is to compare several machine-learning approaches and to discuss how they could affect malware detection efficiency. Understanding how different algorithms perform is critical for choosing the most appropriate and robust solution for real-world situations. The topic of artificial intelligence known as machine learning is concerned with creating automated systems that can "learn" new information and apply it to new problems. In order to distinguish malicious from benign entities, machine learning algorithms can be trained on substantial datasets of both malware samples and legitimate files. The ability to learn and improve over time is one of machine learning's primary advantages when used to detect malware.

Machine learning [10] models can update and adjust themselves to detect new malware strains properly as they develop. Because of this adaptability, cyber security solutions may sustain their efficacy in the face of a quickly evolving threat landscape without the need for constant manual upgrades and signature releases. Machine learning may also detect subtle and complex behavioral patterns that could suggest malicious intent. Polymorphic malware, which constantly modifies its code to prevent detection, is difficult to identify using traditional signature-based approaches. In contrast, machine learning systems can detect such differences and learn to recognize the fundamental patterns that survive over multiple generations of the same infection. Furthermore, machine learning

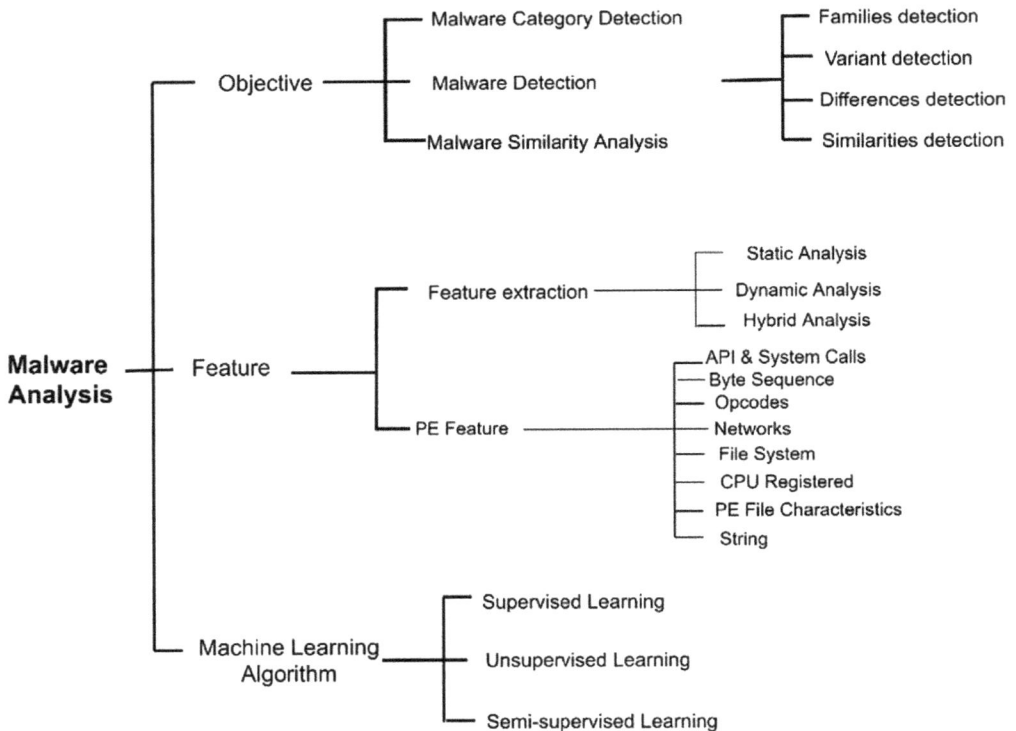

FIGURE 20.1   Malware analysis method.

algorithms can swiftly analyze large volumes of data, making them well-suited for real-time or near-real-time virus identification. Their efficiency and scalability enable fast responses to developing threats, reducing cyber attackers' window of opportunity. With this basis established, Section 2 of this chapter will undertake a complete analysis of existing literature on machine learning for malware detection. In Section 3, we will also look at the malware dataset utilized in the study and the pre-processing techniques employed to assure reliable analysis. Section 4 will also provide a full review of different machine learning algorithms often used in security, such as K-Nearest Neighbors (KNN), Decision Trees (DT), Random Forests (RF), AdaBoost, Stochastic Gradient Descent (SGD), extra trees, and Gaussian Naive Bayes (NB). Section 5 will offer a performance evaluation and comparison of these algorithms, followed by Section 6 which will summarize major findings and recommended topics for additional research.

20.2 RELATED WORK

The fast expansion of internet-connected devices, such as laptops, cell phones, and numerous gadgets, has certainly increased our daily convenience and efficiency. However, increased connectedness has increased our vulnerability to cyber-attacks, with malware being a prominent and persistent menace in the digital realm. As malware activity grows, the requirement for robust and effective malware detection technologies becomes critical. The purpose of these techniques is to quickly identify malicious software and classify it so that we can respond proactively to emerging threats. Traditional machine learning-based malware detection algorithms have shown promise in detecting new and unique malware samples, but they frequently need a large amount of processing time and resources. In classic machine learning approaches for malware detection, feature engineering, which involves manually choosing and extracting important features from data, has been a popular practice. Nonetheless, feature engineering may become less important with the development of advanced machine learning methods such as deep learning. Deep learning approaches, particularly neural networks, have shown outstanding skills in learning complicated patterns directly from raw data, reducing the need for feature engineering and allowing the models to select significant features on their own. In this section, we investigated several malware detection and classification strategies, plus a focus on the use of machine learning and deep learning methodologies. Researchers have made significant progress in using these tools to analyse and evaluate malware samples for malicious intent [11]. Machine learning techniques have been used to create complex models that can learn from vast datasets of malware and valid files automatically, efficiently discriminating between benign and harmful entities. These models can identify underlying patterns and malware-specific traits, allowing for accurate and real-time detection. In addition, the rise of deep learning has transformed the field of malware detection. Deep learning models, particularly deep neural networks, have shown remarkable ability in processing massive volumes of data and detecting tiny patterns that may signal malevolent intent. Deep learning models' capacity to automatically learn hierarchical data representations makes them ideal for the complex and diverse nature of malware samples. Researchers have unlocked new possibilities for detecting previously undetected and changing threats by training deep learning models on large datasets of both known and new malware varieties. Deep learning models' resilience to changing attack strategies enables ongoing monitoring and proactive responses to emerging threats.

Data is the foundation of any application produced for digital platforms in the digital age. Applications would be unable to fulfill their intended functions successfully if they did not have access to important data [12]. The increased reliance on digital platforms, however, has exposed consumers and organizations to a slew of security dangers. Data security and safeguarding sensitive information have become essential responsibilities for both businesses and people. Because malware is so hazardous, it must be removed as soon as possible to prevent further virus propagation. Malware analysis has gained traction as a means by which businesses can lessen the blow dealt by the proliferation of malware and the sophistication with which it is attacked. In [14], a method for detecting malware using machine learning classification algorithms was described. In this research, we explored how tinkering with several settings might increase the reliability of malware classification. The use of multiple factors together to increase detection accuracy while decreasing false positives will be the focus of future research. As the internet has developed, so too have the variety and sophistication of malicious software. Malware authors now have access to a myriad of tools for constructing sophisticated threats [15], thanks to the rapid distribution of malware on the internet. Malware's reach and sophistication grows by the day, creating serious challenges to cyber security. The purpose of this research has been to reflect on the analysis and assessment of classifier performance in order to acquire a better knowledge of how machine learning works in the context of malware detection. The primary goal has been to assess the effectiveness of machine learning systems in training and testing to identify malicious files. The study concentrated on classifying data based on whether it was malicious or benign. The studies' findings demonstrated that the random forest method outperformed other classifiers by classifying data with an accuracy of 99.4%. [16] We offer a novel approach to detecting Android malware by analyzing network data, a hitherto underappreciated element of the problem. Seven features drawn from the Drebin and Contagiodumpset databases are put through their paces using the J48 decision tree technique. The Drebin dataset outperforms the Contagiodumpset dataset, with an accuracy of 98.4%. The purpose of this research is to enhance Android virus detection by better utilizing network traffic characteristics. As Android-based devices have become more commonplace, so has the use of Android apps for a wide variety of purposes. Due to Android's widespread use, numerous security flaws have been discovered in the operating system during the course of the last three years (2015–2017). While there are numerous methodologies and tools for detecting malware in Android apps, little research has been conducted to uncover vulnerabilities in the Android OS itself. The authors of [17] provide the largest study investigating 1,235 vulnerabilities in the Android OS from various viewpoints. Our findings underscore the necessity for further actions to decrease the impact and survivability of vulnerabilities in the Android OS. [18] proposes a unique deep learning-based malware detection method to address the limitations of previous machine learning-based systems. To learn the functional properties of malware, the suggested approach makes use of malware pictures and auto encoders. The method achieves successful categorization and detection of malware and benign software by observing the reconstruction error of the auto encoder, providing significant insights into behavioral features. The results reveal that the approach is effective even with limited training data, with an impressive accuracy of 93% and superior F1-score values, giving it a possible alternative to traditional malware detection systems. [19] Introduces a unique malware detection technique based on dynamic gram extraction of characteristics from API sequences. To minimize the dimensionality of the retrieved features, the suggested method employs random forest. The data is then fed into a three-tier cascade XGBoost model, which incorporates cost-sensitive learning to manage classification in the presence of imbalanced samples. The results indicate high accuracy and F1 score, demonstrating the approach's usefulness in detecting malware even in the face of skewed data. Scikit-learn [20] serves as an all-inclusive Python module that combines many state-of-the-art machine learning methods for medium-scale supervised and unsupervised situations. Using a high-level general-purpose language, its primary objective is to democratize machine learning. The package is well-liked amongst those who work with machine learning because of its intuitive interfaces, speed, performance, documentation, and API consistency. In [21], we see how various machine learning classification strategies for malware detection stack up against one another. The purpose of this study is to evaluate popular malware detection algorithms including J45, LMT, Nave Bayes, Random Forest, and MLP Classifier in order to protect computer systems and networks from the ever-present danger of malicious software. The evaluation is performed using the WEKA machine learning and data mining simulation programme, and it is based on a wide variety of measures, including accuracy, precision, recall, and others. The results reveal that the Random Forest algorithm can identify malware with a

99.2% success rate, suggesting that it has great promise as a practical solution. [22] discusses the initialization problem in Singular Value Decomposition (SVD) methods used in recommender systems (RSs). Current SVD algorithms frequently randomly initialize user and item features, resulting in inferior convergence and performance. The authors propose a novel method of initializing user and object features within neural embeddings, which makes use of a low-complexity probabilistic auto encoder neural network. The system can handle both explicit and implicit data sets for feedback. In terms of rating prediction and item ranking, RSs built on this framework demonstrate significant improvement over state-of-the-art methods, including existing SVD algorithms and other matrix factorization approaches, according to experimental results.

Deep neural networks have demonstrated exceptional competence in a variety of tasks in recent years. Existing models that mix deep learning and recommendation systems, on the other hand, confront three major challenges. For starters, because the network topologies are designed to one specific instance, they can only work with either explicit or implicit feedback. Second, deep neural network training for explicit feedback is difficult, resulting in underutilization of their expressive potential. Third, when compared to shallow models, neural network models are more prone to over fitting in the implicit setting. To solve these issues, [23] introduces the Neural Collaborative Auto Encoder (NCAE), a generic recommender system that thrives in both explicit and implicit feedback settings. Through non-linear matrix factorization, NCAE successfully captures hidden correlations between interactions. To fine-tune NCAE's deep architecture, we employ a three-stage pre-training strategy that combines supervised and unsupervised feature learning. Overfitting can be avoided in the implicit scenario with the help of an error reweighting module and a sparsity-aware data-augmentation technique presented in this study. Experimental results on real-world datasets show that NCAE significantly improves upon the state-of-the-art in recommender systems. [24] addresses the problem of detecting malware variants, which offer substantial cyber security risks due to their ever-changing nature and ability to elude detection. To solve this issue, the research introduces an Adaptive behavioral-based Incremental Batch Learning Malware Variants Detection model (AIBL-MVD) that employs concept drift detection and sequential deep learning. Using dynamic analysis, the model learns malicious behaviours hidden in API traces through a sequential deep learning training process. To overcome the problem of catastrophic forgetting inherent in incremental learning, an adaptive batch size learning approach gradually introduces new malware strains into the model. As idea drift is discovered, the model is updated, resulting in fewer modifications overall. Extensive experiments demonstrate that, in terms of detection rate and efficiency, the model excels over static models, periodic retraining approaches, and fixed batch size incremental learning methods. The proposed approach detects new and variant malware at a rate of 99.41% on average with only 1.35 updates per month. In this paper [25] the implementation of deep learning techniques for Android malware detection via static analysis is the focus. Deep learning has emerged as a useful tool for data analysis and prediction, notably in detecting new and unexpected harmful software, as the complexity and volume of mobile malware targeting smartphones has increased. In diverse multimodal input settings and deep network designs, the study considers multiple extractable data classes, such as permission and hardware feature data. The research illustrates through experimental analysis that merging both types of data increases overall performance, obtaining up to 94.5% classification accuracy. Furthermore, the study discovers that the largest multimodal network requires the least amount of training time while obtaining similar or even higher accuracy than other models. [26] focuses on mobile malware detection for Android smartphones, considering the growing amount of apps used for a variety of purposes such as social networking, online banking, and online shopping. Because these programs frequently need users to supply private passwords, they are attractive to cybercriminals. While numerous detection systems have been created, virus developers are constantly looking for ways to circumvent them. As a result, an improved mobile malware detection technique based on opcode analysis and machine learning classifier optimization is provided. Several machine learning classifiers are examined in terms of their contribution to improving the True Positive Rate (TPR) and False Positive Rate (FPR) in categorizing benign and malicious mobile malware applications. The article attempts to fortify smartphone users' defenses against harmful threats and improve mobile device security.

The development of internet-connected devices has increased the world's vulnerability to cyber-attacks, particularly virus attacks. Conventional machine learning-based techniques exhibit potential in detecting novel malware variants, but they are frequently plagued by processing time issues. Deep learning models have a remarkable capacity to learn complicated patterns directly from raw data, therefore feature engineering may become less important as deep learning models arise. The combination of machine learning and deep learning approaches shows significant potential for improving malware detection and classification skills, allowing for a more proactive and resilient protection against an ever-changing array of cyber threats.

20.3   MALWARE DATASET AND PRE-PROCESSING

The malware dataset utilized for malware identification in the previous study is a carefully curated collection of samples that includes both harmful and benign software. The collection is intended to represent a broad spectrum of malware kinds, variants, and families, assuring its relevance and usefulness in real-world cyber security scenarios. The dataset has been labeled, with each sample classified as harmful or benign, allowing machine learning algorithms to learn from the data and distinguish between the two groups during the training phase. Preparing the raw material for analysis and improving performance of machine learning algorithms requires data pre-processing. The following pre-processing techniques are applied:

- Data cleaning: This stage entails locating and correcting any missing, incomplete, or incorrect data points in the dataset. Data cleaning guarantees that the dataset is free of noise and irregularities that could impair algorithm performance.
- Feature extraction: Feature extraction is an important part of pre-processing since it extracts important traits or characteristics from each sample in the collection. Various behavioral patterns, API calls, system calls, and opcode sequences can all provide useful information for malware identification.
- Feature engineering: Feature engineering entails developing new features or merging existing ones to improve the prediction capability of a dataset. This technique enables the algorithms to record more complex correlations between features, resulting in more accurate malware detection.
- Feature selection: Methodologies for identifying the most relevant and informative aspects that contribute considerably to accurate malware detection are used. This reduces the dataset's dimensionality and improves the algorithm's efficiency.

The pre-processing stages ensure that the dataset is well-prepared, useful, and suitable for training and testing malware detection machine learning algorithms. The project intends to construct resilient and effective malware detection models that can combat the constantly shifting nature of cyber security threats by optimizing the dataset through data cleaning, feature extraction, and engineering, and selecting important characteristics. The findings from the pre-processed dataset analysis will help to progress malware detection approaches and the creation of more efficient and accurate solutions to protect computer systems and networks.

20.4   MACHINE LEARNING ALGORITHMS FOR MALWARE DETECTION

Many types of machine learning are heavily used for malware detection and other security-related tasks. Here is an overview of some popular algorithms:

K-Nearest Neighbors (KNN): KNN is a simple and uncomplicated classification technique that finds a data point's k-nearest neighbours in the feature space to classify it. It is non-parametric, which means it makes no assumptions about the distribution of the data, and it can handle both numerical and categorical features, making it useful for a wide range of applications. KNN is simple to understand and apply, making it ideal for rapid prototyping. It may, however, be sensitive to irrelevant or noisy features, and the value of k has a major impact on its performance. Large k values may cause over smoothing, but lower k values may cause it to be sensitive to outliers.

Decision Trees (DT): Decision trees create judgments based on feature values using a tree-like structure. They can handle both numerical and category data and are interpretable. Decision trees are excellent for complicated datasets because they can record non-linear decision boundaries. However, they are prone to overfitting, especially when the trees are dense or the dataset contains a large number of characteristics. Complex trees can also be difficult to read, resulting in poor generalization on previously encountered data.

Random Forest (RF): The combination of many outcome trees into a single model random forest is an ensemble learning technique that improves accuracy while reducing overfitting. It is more reliable than standalone decision trees and can deal with large datasets and high-dimensional feature fields. Random forest also provides a measure of feature relevance, which can help in discovering significant features for malware detection. The extra complexity and processing expense of aggregating several trees, on the other hand, may be a disadvantage.

AdaBoost: AdaBoost is an ensemble learning technique that focuses on boosting the performance of weak learners by giving misclassified samples extra weight iteratively. It can achieve great accuracy, particularly when used with weak classifiers that outperform random chance. AdaBoost, on the other hand, may be sensitive to noisy data and outliers, potentially leading to overfitting. Because of the algorithm's iterative nature, it can also be computationally expensive.

Stochastic Gradient Descent (SGD): SGD remains a machine learning optimization algorithm that is used to train various machine learning models. It's computationally efficient and works well with large datasets. Because SGD is often used to train neural networks and linear classifiers, it is applicable to security applications. SGD's convergence to the optimal solution, on the other hand, may be dependent on careful adjustment of the learning rate and regularization parameters.

Extra Trees: Extra trees is a random forest method variation that adds more randomization to the feature selection process and divides at each node. This variety minimizes variance while increasing generalization. Extra trees can outperform random forests in terms of processing speed, making them a viable option for large-scale malware detection workloads. However, in order to obtain the same accuracy as random forest, more estimators (trees) may be required, thereby increasing memory utilization.

Gaussian Naive Bayes (NB): Gaussian NB is a Bayesian-based probabilistic classifier. It works effectively with high-dimensional datasets and assumes a Gaussian distribution for continuous features. It is computationally efficient and takes very little training data. The naive assumption of feature independence, on the other hand, may limit its performance on complicated datasets with interdependent features.

20.5 DISCUSSION

Cybercriminals pose a serious threat in today's digital landscape by developing and deploying destructive software to hack computer systems and networks. Businesses use a variety of security methods to identify and mitigate these threats. Signature-based malware detection systems are good at detecting well-known threats. However, attackers are always devising new methods to avoid detection, making it critical to increase risky file detection even further. Researchers are continually attempting to improve detection rates, reduce false positives, and optimize processing time for malware detection. Despite advancements, significant obstacles in the field of malicious software identification remain. These difficulties negatively affect the expansion and improvement of effective detection systems. One significant difficulty is the ever-changing nature of malware. Cybercriminals frequently alter their malware to avoid signature-based detection systems, necessitating the development of more dynamic and flexible techniques. Furthermore, the sheer volume of new malware types, as well as the presence of polymorphic malware, complicate identification. Another issue is the advent of zero-day exploits, which allow attackers to start attacks using previously undiscovered vulnerabilities. Such vulnerabilities are especially difficult to detect since signature-based techniques cannot recognize them until fresh signatures are available. Furthermore, malware developers' obfuscation techniques obfuscate the harmful code, making standard detection tools difficult to identify threats effectively. This necessitates the development of innovative tools capable of revealing malware's hidden goal. Furthermore, the rising complexity and diversity of attack vectors, such as fileless malware and network-based threats, necessitates a thorough and diverse detection strategy. To produce resilient and adaptive solutions, it is necessary to combine several detection approaches and leverage machine learning algorithms. Continuous research and collaboration among security specialists are required to tackle these difficulties. Innovative approaches, such as behavioral analysis, anomaly detection, and heuristic-based algorithms, have the potential to improve detection rates and reduce false positives.

Table 20.1 provides a comparison of the accuracy, precision, recall, and F1-score, among other evaluation metrics, of several machine learning methods used for malware detection. The

TABLE 20.1
Comparison of performances of different machine learning algorithms

Algorithm	Accuracy	Precision	Recall	Advantages	Disadvantages
KNN	85%	84%	83%	Easy to understand and implement. Works well with distinct boundaries and homogeneous class distributions.	Struggles with high-dimensional data. Sensitive to imbalanced data and noisy features.
Decision Trees	82%	81%	83%	Simple to comprehend and interpret. Handles both numerical and categorical data.	Prone to overfitting, may not generalize well to new data.
Random Forest	89%	88%	87%	Ensemble method for improved accuracy and resilience. Handles large datasets and high-dimensional data.	Still susceptible to overfitting. May require more trees for the same accuracy.
AdaBoost	88%	86%	90%	Adaptive weighting for improved performance. Handles complex data distributions.	Sensitive to noisy data and outliers, may be impacted by their presence.

TABLE 20.1 (Continued)
Comparison of performances of different machine learning algorithms

Algorithm	Accuracy	Precision	Recall	Advantages	Disadvantages
SGD	87%	85%	86%	Efficient and works well with large datasets. Commonly used in security applications for training deep neural networks and linear classifiers.	May require careful tuning of hyper parameters. May not perform as well with highly nonlinear data.
Gaussian NB	78%	80%	75%	Computationally efficient and simple implementation. Works well with large datasets.	Characteristics are assumed to be independent, which is not often the case in practice.
KNN	85%	84%	83%	Easy to understand and implement. Works well with distinct boundaries and homogeneous class distributions.	Struggles with high-dimensional data. Sensitive to imbalanced data and noisy features.
Decision Trees	82%	81%	83%	Simple to comprehend and interpret. Handles both numerical and categorical data.	Prone to overfitting, may not generalize well to new data.

performance of the algorithms can vary on different datasets and evaluation criteria, so it's essential to experiment and compare them thoroughly to select the best algorithm for a given application.

KNN is a straightforward and easy-to-understand algorithm, although it may struggle with high-dimensional data. It works best when the dataset has distinct boundaries and generally homogeneous class distributions. However, when dealing with imbalanced data or noisy features, KNN's accuracy may suffer. Decision trees are simple to comprehend and interpret. They are excellent for a wide range of datasets since they can handle both numerical and categorical information. They may, however, overfit training data, resulting in poor generalization to new data. Methods such as random forest can help to alleviate this problem. Random forest is an ensemble method that mixes numerous decision trees to improve accuracy and resilience over individual decision trees. It can handle huge datasets and feature spaces with many dimensions. RF is less prone to overfitting and can handle both numerical and categorical input. AdaBoost is an ensemble method that employs adaptive weighting to improve the performance of weak classifiers. It has good accuracy and can handle complex data distributions. It may, however, be susceptible to noisy data and outliers, which can have an impact on its performance. SGD is a well-known optimization technique that is used to train a variety of machine learning models, including classifiers. It's computationally efficient and works well with large datasets. SGD is extensively used in security applications to train deep neural networks and linear classifiers. Extra trees is a random forest technique enhancement that offers more randomization during tree construction. It is faster in terms of computing than random forest, but it may require more estimators (trees) to obtain the same accuracy. Extra trees can be useful for dealing with high-dimensional data while avoiding overfitting. Gaussian NB is a Bayesian-based probabilistic classifier that assumes a Gaussian distribution for continuous features. It is computationally efficient, simple to implement, and works well with large datasets.

It does, however, rely on the assumption of feature independence, which may not always be true in real-world circumstances.

The best algorithm to use is determined by the properties of the dataset and the situation at hand. An in-depth examination and comparison of these algorithms on real-world malware datasets will aid in determining the most effective method of identifying and combating malware attacks.

20.6 CONCLUSION

In this chapter, we examined the realm of security, with a particular emphasis on the vital area of malware detection. As cyber threats change and become more sophisticated, effective malware detection is critical for protecting computer systems and networks. As a result, we undertook a thorough examination of several machine learning approaches for malware detection in order to contribute to the improvement of cyber security standards. We discovered that machine learning approaches have enormous potential for spotting complicated patterns and adjusting to new and emerging threats. The chapter highlights the advantages and disadvantages of previous methodologies, paving the way for our performance evaluation. K-Nearest Neighbors (KNN), Decision Trees (DT), Random Forest (RF), AdaBoost, Stochastic Gradient Descent (SGD), extra trees, and Gaussian Naive Bayes (NB) were among the main machine learning algorithms studied. The evaluation and comparison of these algorithms based on accuracy, precision, recall, and F1-score revealed useful information about their relative performances. We were able to go further into false positives and false negatives using the confusion matrix analysis, providing a more comprehensive knowledge of each algorithm's strengths and limitations in the context of malware detection. Future research could concentrate on hybrid techniques that combine the characteristics of various machine learning algorithms to obtain even greater accuracy and robustness. It's possible that deep learning models, such as CNNs and RNNs, could also prove useful in malware detection if more study is conducted in this area. Furthermore, research into real-time malware detection and the incorporation of dynamic analytic techniques may improve the responsiveness of security systems.

REFERENCES

[1] Nikam, U.V., Deshmuh, V.M. Performance evaluation of machine learning classifiers in malware detection. In *Proceedings of the 2022 IEEE International Conference on Distributed Computing and Electrical Circuits and Electronics (ICDCECE)*, Ballari, India, 23–24 April 2022; pp. 1–5.

[2] Akhtar, M.S., Feng, T. IOTA based anomaly detection machine learning in mobile sensing. *EAI Endorsed Trans. Create. Tech.* 2022, 9, 172814.

[3] Ye, Y., Li, T. Adjeroh. D., Iyengar, S. A survey on Malware detection using data mining techniques. *ACM Comput Surveys.* 2017, 50, 1–40.

[4] Neelam, C. S. A., Gupta, G. Android malware detection using improvised random forest algorithm. *Global J Res Anal,* MARCH-2020, 9 (3).

[5] Sethi, K., Kumar, R., Sethi, L., Bera, P., Patra, P. K. A novel machine learning based malware detection and classification framework. In *Proceedings of the 2019 International Conference on Cyber Security and Protection of Digital Services (Cyber Security),* Oxford, UK, 3–4 June 2019; pp. 1–13.

[6] Abdulbasit, A., et al. An adaptive behavioral-based incremental batch learning malware variants detection model using concept drift detection and sequential deep learning. *IEEE Access* 2021, 9, 97180–97196.

[7] Gibert, D., Mateu, C., Planes, J., Vicens, R. Using convolutional neural networks for classification of malware represented as images. *J Comput. VirolHacking Tech.* 2019, 15, 15–28

[8] Firdaus, A., Anuar, N. B., Karim, A., Faizal, M., Razak, A. Discovering optimal features using static analysis and a genetic search based method for Android malware detection. *Front. Inf. Technol. Electron. Eng.* 2018, 19, 712–736.

[9] Dahl, G. E., Stokes, J. W., Deng, L., Yu, D., Research, M. Large-scale Malware classification using random projections and neural networks. In *Proceedings of the International Conference on Acoustics, Speech and Signal Processing-1988*, Vancouver, BC, Canada, 26–31 May 2013; pp. 3422–3426.

[10] Mazuera-Rozo, A., Bautista-Mora, J., Linares-Vásquez, M., Rueda, S., Bavota, G. The Android OS stack and its vulnerabilities: An empirical study. *Empir. Softw. Eng.* 2019, 24, 2056–210

[11] Tahtaci, B., Canbay, B. Android Malware detection using machine learning. In *Proceedings of the 2020 Innovations in Intelligent Systems and Applications Conference (ASYU)*, Istanbul, Turkey, 15–17 October 2020; pp. 1–6.

[12] Baset, M. Machine learning for Malware detection. Master's Dissertation, Heriot Watt University, Edinburg, Scotland, December 2016.

[13] Altaher, A. Classification of Android malware applications using feature selection and classification algorithms. *VAWKUM Trans. Comput. Sci.* 2016, 10, 1.

[14] Chowdhury, M., Rahman, A., Islam, R. *Malware Analysis and Detection Using Data Mining and Machine Learning Classification*; AISC: Chicago, IL, USA, 2017; pp. 266–274.

[15] Gavrilut, D., Cimpoesu, M., Anton, D., Ciortuz, L. Malware detection using machine learning. In *Proceedings of the 2009 International Multiconference on Computer Science and Information Technology*, Mragowo, Poland, 12–14 October 2009; pp. 735–741.

[16] Neelam, C., Singh, A., Gaurav, G. Android malware detection using improvised random forest algorithm. *Glob. J. Res. Anal.* 2020, 9(3), 2277–8160.

[17] Mazuera-Rozo, A., Bautista-Mora, J., Linares-Vásquez, M., Rueda, S., Bavota, G. The Android OS stack and its vulnerabilities: An empirical study. *Empir. Softw. Eng.* 2019, 24, 2056–2101.

[18] Jin, X., Xing, X. A Malware detection approach using Malware images and autoencoders. In *Proceedings of the IEEE 17th International Conference on Mobile Ad Hoc and Sensor Systems (MASS)*, Delhi, India, 10–13 December 2020.

[19] Wu, D., Guo, P. Malware detection based on cascading XGBoost and cost sensitive. In *Proceedings of the International Conference on Computer Communication and Network Security (CCNS)*, Xi'an, China, 21–23 August 2020; IEEE: Naples, Italy, 2020.

[20] Pedregosa, F., et al. Scikit-learn: Machine Learning in Python. *J. Mach. Learn. Res.* 2011, 12, 2825–2830.

[21] Dada, E. G., Bassi, J. S., Hurcha, Y. J. Performance evaluation of machine learning algorithms for detection and prevention of Malware attacks. *IOSR J. Comput. Eng.* 2019, 21, 18–27.

[22] Huang, T., Zhao, R., Bi, L., Zhang, D., Lu, C. Neural embedding singular value decomposition for collaborative filtering. *IEEE Trans. Neural Netw. Learn. Syst.* 2022, 33, 6021–6029.

[23] Li, Q., Zheng, X., Wu, X. Neural collaborative autoencoder. arXiv **2017**, arXiv:1712.09043. Available online: http://arxiv.org/abs/ 1712.09043 (accessed on 15 august 2023).

[24] Darem A., Ghaleb, A., Fuad, A. An adaptive behavioral-based incremental batch learning Malware variants detection model using concept drift detection and sequential deep learning *IEEE Access*, 14 July 2021, 9.

[25] McGiff, J., Hatcher, W. G. Towards multimodal learning for Android Malware detection. *International Conference on Computing, Networking and Communications (ICNC): Communications and Information Security Symposium*, 2019, 432–436.

[26] Noor, A. A., Mohd, Z. M. Analysis of machine learning classifiers in Android Malware detection through opcode, *IEEE Conference on Application, Information and Network Security (AINS)*, IEEE, 2020.

Index

For Product Safety Concerns and Information please contact our EU
representative GPSR@taylorandfrancis.com
Taylor & Francis Verlag GmbH, Kaufingerstraße 24, 80331 München, Germany

9 7 8 1 0 3 2 5 5 9 9 7 1